Advanced and Multivariate Statistical Methods

Practical Application and Interpretation

Second Edition

Craig A. Mertler
Bowling Green State University

Rachel A. Vannatta
Bowling Green State University

 Pyrczak Publishing
P.O. Box 39731 • Los Angeles, CA 90039

Cover design by Robert Kibler and Larry Nichols.

Editorial assistance provided by Sharon Young, Brenda Koplin, Cheryl Alcorn, Monica Lopez, Randall R. Bruce, and Elaine Parks.

Printed in the United States of America by McNaughton and Gunn, Inc.

ISBN 1-884585-41-8

CONTENTS

Detailed Chapter Contents

PREFACE

Nearly all graduate programs in the social and behavioral sciences (i.e., education, psychology, sociology, and so on) offer an advanced or multivariate statistics course covering concepts beyond those addressed in an introductory-level statistics course. This textbook addressees these advanced statistical concepts.

Purpose of this Textbook and Intended Audience

This textbook provides conceptual and practical information regarding multivariate statistical techniques to students who do not *necessarily* need the technical and/or mathematical expertise of these methods. Instead, these students are required to understand the basic concepts and practical applications of these advanced methods in order to interpret the results of research studies that have utilized such methods as well as apply these analytical methods to their own data analyses as part of their thesis or dissertation research studies. This text is not intended for use with students who are majoring in statistical or research methodologies, although it could certainly be used as a reference by these students. In contrast, this text is appropriate for students taking a multivariate statistics course as part of a graduate degree program where the course is viewed as a research tool and not as a content or "core" requirement. Examples of degree programs for which this textbook might be appropriate include—but would definitely not be limited to—psychology, education, sociology, criminal justice, social work, mass communication, and nursing.

This textbook has three main purposes. The first is to facilitate conceptual understanding of multivariate statistical methods, and to do so by limiting the technical nature of the discussion of those concepts and, instead, focusing on their practical applications. The multivariate statistical methods covered in this text are:

- factorial analysis of variance (ANOVA),
- analysis of covariance (ANCOVA),
- multivariate analysis of variance (MANOVA),
- multivariate analysis of covariance (MANCOVA),
- multiple regression,
- path analysis,
- factor analysis,
- discriminant analysis, and
- logistic regression.

The second purpose is to provide students with the skills necessary to interpret research articles that have employed multivariate statistical techniques. A critical component of their graduate research projects is a review of the research literature related to their topic. It is crucial for students to be able to understand not only what multivariate statistical techniques were used in a particular research study, but also to appropriately interpret the results of that study for the purposes of synthesizing the existing research as background for their study. The acquisition of these skills will benefit students not only during the time that they are conducting their graduate research studies, but also long after that time as they review current research as part of their professional career activities.

The third purpose of this text is to prepare graduate students to apply multivariate statistical methods to the analysis of quantitative data of their own or their institutions, such that they are able to complete the following for each particular technique:

- understand the limitations of the technique,
- fulfill the basic assumptions of the technique,
- conduct the appropriate steps (including the selection of various options available through the use of computer analysis software),
- interpret the results, and
- write the results in appropriate research reporting format.

Special Features of this Textbook

There are currently no textbooks available that take a more practical, truly *applied* approach to multivariate statistics. Most texts tend to be too technical and mathematical in their orientation. While this approach is highly appropriate for students who are majoring in statistical and research methodologies, it is not nearly as appropriate for students representing other majors in the social and behavioral sciences. We see a "mismatch" between what professors are trying to teach in applied advanced statistics courses and the material (content) presented in current textbooks. Oftentimes, professors are required to "water down" the content coverage of the statistical techniques when using one of the existing texts. An alternative is for professors to utilize an introductory statistics text and supplement the content coverage with lectures and examples. This approach ultimately harms students because they oftentimes do not have access to these supplementary resources in order to process the information at their own pace outside of class.

Our main reason for writing this book was that we were unable to find an appropriate text to use in teaching our courses in Advanced Statistical Methods. Our students have only had an introductory statistics course and find many of the texts previously available to be far too technical. To our knowledge, no text was previously available that combined advanced/multivariate statistical methods with a practical approach to conducting and interpreting such tests.

Approach to Content

To facilitate student understanding of the practical applications and interpretation of advanced/multivariate statistical methods, each chapter of the text focuses on a specific statistical test and includes sections on:
- Practical View
    - Purpose
    - Example Research Questions
- Assumptions and Limitations
    - Methods of Testing Assumptions
- The Process and Logic
    - The Logic Behind the Test
    - Interpretation of Results
    - Writing Up Results
- Sample Study
    - Problem
    - Method
    - Output and Interpretation of Results
    - Presentation of Results
- SPSS "How To"
- Summary

The first section, *Practical View*, describes the purpose of the test in very conceptual terms and provides example research questions and scenarios. The section *Assumptions and Limitations* presents the assumptions that must be fulfilled in order to conduct the specific test, suggested methods for testing those assumptions, and limitations of the test. *Process and Logic* explains the theory behind the test in very practical terms, the statistics that are generated from the test, how to determine significance, and how the results are written. The *Sample Study* provides a demonstration of the process of the test, how it was conducted using SPSS, the SPSS output generated, how to interpret the output, and how to summarize the results. A unique feature of our text is the highlighting of key test statistics and their implications for the test within the SPSS output. This has been accomplished by highlighting relevant test statistics and describing them with dialogue boxes within example outputs. The sample studies are based on analysis of SPSS data sets, which can be downloaded from this SPSS Web site:

(http://www.spss.com/tech/DataSets.html#spss).

Data sets are also described in Appendix A, which includes specification of variables and their measurement scales. The section on *SPSS How To* demonstrates the steps used in conducting the test by means of SPSS. Within each chapter, pertinent SPSS menu screens are pictured to display the sequence of steps involved in the analysis, as well as the options available in conducting the test. Finally, the *Summary* provides a step-by-step checklist for conducting the specific statistical procedure.

Pedagogical Features

All chapters, with the exception of Chapters 1 and 2, include SPSS example output as well as menu screens to demonstrate various program options. Tables that display sample study results have been included. Figures, wherever appropriate, have been used to graphically represent the process and logic of the statistical test. Since the data sets are accessible via the SPSS Web site, students can conduct the sample analyses themselves as well as practice the procedures using other data sets. Assignments that utilize these data sets have also been included.

It is our hope that all material included in this text is accurate. We take full responsibility for any errors or omissions related to its content.

ACKNOWLEDGMENTS

We would like to sincerely thank Fred Pyrczak for believing in us and in this project. Without his assistance and timely responsiveness, this text would not exist. We would also like to express our gratitude to the entire staff at Pyrczak Publishing. Appreciation is also extended to Sumangali Krishnan-Banerjee for her assistance with editorial activities and the development of several appendices. We would certainly be remiss if we were not to acknowledge the editorial feedback—both grammatical and substantive in nature—provided by our doctoral students in the College of Education and Human Development at BGSU.

Craig A. Mertler
Rachel A. Vannatta

DEDICATIONS

For
Kate, Addison, my mother, Barb,
and the memory of my father, Chuck
~ *C.A.M.*

For
Paul
~ *R.A.V.*

CHAPTER 1

INTRODUCTION TO MULTIVARIATE STATISTICS

For many years, multivariate statistical techniques have "simplified" the analysis of complex sets of data. As a collective group, these techniques enable researchers, evaluators, policy analysts, and others to analyze data sets that include numerous independent variables (IVs) as well as dependent variables (DVs). In other words, they allow researchers to analyze data sets where the subjects have been "described" by several demographic variables and also have been measured on a variety of outcome variables. For example, a researcher may want to compare the effectiveness of four alternative approaches to reading instruction on measures of reading comprehension, word recognition, and vocabulary, while controlling for initial reading ability. The most appropriate method of analyzing this data is to examine the relationships and potential interactions between all variables simultaneously. Relying on univariate statistical procedures would prevent proper examination of these data. Due to the increasingly complex nature of research questions in the social sciences, and to the advent—and continued refinement—of computer analysis programs (e.g., SPSS, SAS, BMDP, and SYSTAT), the results of multivariate analyses are appearing more and more frequently in academic journals.

The purpose of this book is to provide the reader with an overview of multivariate statistical techniques by examining for each technique its purpose, the logic behind the test, practical applications of the technique, and interpretations of results. The authors' major goal is to prepare students to apply and interpret the results of various multivariate statistical analysis techniques. It is not our intent to inundate the student with mathematical formulae, but rather to provide an extremely practical approach to the use and interpretation of multivariate statistics.

SECTION 1.1 MULTIVARIATE STATISTICS: SOME BACKGROUND

Multivariate statistical techniques are used in a variety of fields, including research in the social sciences (i.e., education, psychology, and sociology), hard sciences, and medical fields. Their use has become more commonplace due largely to the increasingly complex nature of research designs and related research questions. It is oftentimes unrealistic to examine the effects of an isolated treatment condition on a single outcome measure, especially in the social sciences where the subjects of research studies are nearly always human beings.

As we all know, human beings are complex entities complete with knowledge, beliefs, feelings, opinions, attitudes, etc. Studying human subjects by examining a single independent variable (IV) and a single dependent variable (DV) is truly impractical since these variables do not co-exist in *isolation* as part of the human mind or set of behaviors. These two variables may affect or be affected by several other variables. In order to be able to draw conclusions and offer accurate explanations of the phenomenon of interest, the researcher should be willing to examine many variables simultaneously.

Stevens (1992) offers three reasons for using multiple outcome measures (i.e., DVs) in research studies, specifically those involving examinations of the effects of varying treatments (e.g., teaching methods, counseling techniques, etc.). These reasons are:

1.   Any treatment will usually affect subjects in more than one way. Examining only one criterion measure is too limiting. To fully understand the effects of a treatment condition, the researcher must look at various ways that subjects may respond to the conditions.

2.   By incorporating multiple outcome measures, the researcher is able to obtain a more complete and detailed description of the phenomenon under investigation.

3.   Treatments can be expensive to implement, but the cost of obtaining measures on several dependent variables (within the same study) is often quite small and allows the researcher to maximize the information gain.

It should be noted that a study appropriate for *multivariate* statistical analysis is typically defined as one with several dependent variables (as opposed to *univariate* studies, which have only one dependent variable). However, the authors have included several techniques in this book that would typically be classified as "advanced" univariate techniques (e.g., multiple regression, factorial analysis of variance, analysis of covariance, etc.). The reason for their inclusion here is because they are ordinarily not included in an introductory course in statistical analysis but are nonetheless important techniques for students to understand.

Research Designs

The basic distinction between experimental and nonexperimental research designs is whether the levels of the independent variable(s) have been manipulated by the researcher. In a true experiment, the researcher has control over the levels of the IVs; that is, the researcher decides to which conditions subjects will be exposed. For example, if a researcher is conducting an experiment to investigate the effectiveness of three different counseling techniques, she would randomly assign each subject to one of the three conditions. In essence, she has *controlled* which subjects receive which treatment condition.

In nonexperimental research (e.g., descriptive, correlational, survey, or causal-comparative designs), the researcher has no control over the levels of the IVs. The researcher can define the IV, but cannot assign subjects to the various levels of it; the subjects enter the study already "belonging" to one of the levels. For example, suppose a researcher wanted to determine the extent to which groups differed on some outcome measure. A simple scenario might involve an examination of the extent to which boys and girls differed with respect to their scores on a statewide proficiency test. The independent variable, "gender" in this case, cannot be manipulated by the researcher; all subjects enter the study already categorized into one of the two levels of the IV. However, notice that in both experimental and nonexperimental research designs, the levels of the independent variable have defined the groups that will ultimately be compared on the outcome DV.

Another important distinction between these two types of research designs is the ability of the researcher to draw conclusions with respect to causality. In an experimental research study, if the researcher finds a statistically significant difference between two or more of the groups representing different treatment conditions, he can have some confidence in attributing causality to the IV. Manipulating the levels of the IV by randomly assigning subjects to those levels permits the researcher to draw causal inferences from the results of his study. However, since there is no manipulation or random assignment in a nonexperimental research study, the researcher is able to conclude that the IV and DV are related to each other, but causal inference is limited.

The choice of statistical analysis technique is extraneous to the choice of an experimental or nonexperimental design. The various multivariate statistical techniques described in this book are appropriate for situations involving experimental as well as nonexperimental designs. The computer analysis programs will "run" and the statistics will "work" in either case. However, the decision of the researcher to attribute causality from the IV(s) to the DV(s) is ultimately dependent upon the initial decision of whether the study will be experimental or nonexperimental.

The Nature of Variables

The authors have been using the terms "independent" and "dependent" variables throughout the beginning of this chapter, so a review of these terms—and others related to the nature of variables—is undoubtedly in order. Variables can be classified in many ways. The most elementary classification scheme dichotomizes variables into either independent or dependent variables. Independent variables (IVs) consist of the varying treatment conditions (e.g., a new medication versus a standard medication) to which subjects are exposed, or differing characteristics that the subjects bring into the study with them (e.g., school location, defined as urban, suburban, and rural). In an experimental situation, the IVs may also be referred to as "predictor" or "causal" variables because they have the potential of causing differing scores on the DV, which is sometimes referred to as the "criterion" or "outcome" variable. The reader should also be aware that a specific variable is neither inherently an IV or a DV; an IV in one study might be a DV in another study, and vice versa. "Univariate" statistics refers to analyses where there is one or more IVs and only one DV. "Factorial" analyses are appropriate in situations when there are two or more IVs and one DV. "Bivariate" statistics refers to analyses that involve two variables where neither is identified as an IV or a DV. Finally, "multivariate" statistics refers to situations where there is more than one DV and there may be one or more IVs.

Another way to classify variables refers to the level of measurement represented by the variable. Variables may be quantitative, categorical, or dichotomous. "Quantitative" variables are measured on a scale that has a smooth transition across all possible values. The numerical value represents the amount of the variable possessed by the subject. Examples of quantitative variables include age, income, and temperature. Quantitative variables are also referred to as "continuous" or "interval" variables.

"Categorical" variables consist of separate, indivisible categories. There are no values between neighboring categories of a categorical variable. Categorical variables are often used to classify subjects. Examples of categorical variables would include gender (male or female), type of school (urban, suburban, or rural), and categories of religious affiliation. Categorical variables may also be referred to as "nominal," "discrete," or "qualitative." A specific type of categorical variable is one that is "dichotomous." A dichotomous variable is one that has only two possible levels or categories. For example, "gender" is a categorical variable that is also dichotomous. Oftentimes, for purposes of addressing specific research questions, quantitative or categorical variables may be dichotomized. For example, age is a quantitative variable, but one could recode the values so that it would be "transformed" into a dichotomous variable. Age could be dichotomized into two categories, say "less than 35 years of age" and "35 years of age and older." Oftentimes, a transformation of data such as this allows the researcher to be more flexible in terms of the analysis techniques she can use.

When conducting a multivariate analysis, researchers sometimes have a tendency to want to include too many variables. Prior consideration to the analysis is crucial in determining on which variables to collect data and to include. The best recommendation is to obtain the solution with the fewest number of variables (Tabachnick & Fidell, 1996). This is known as a "parsimonious" solution. Arguments for the inclusion of variables should be based on the feasibility (i.e., cost and availability) of collecting data on them, and the nature of the theoretical relationships among the variables being considered.

Data Appropriate for Multivariate Analyses

Obviously, the data for multivariate analyses must be numerical. Continuous variables consist of the scores themselves on specific variables. The values for discrete variables consist of the codes assigned by the researcher. For example, for the variable school location, urban schools might be assigned a "1," suburban schools would be assigned a "2," and rural schools would be assigned a "3."

There are many forms in which data can be submitted for analysis using multivariate techniques. The majority of the time, a data matrix will be analyzed. A *data matrix* is an organization of raw scores or data, where the rows represent subjects, or cases, and the columns represent variables. Another possible format in which data may appear for analysis is a correlation matrix. Readers who have completed an introductory course in statistics are probably somewhat familiar with this type of matrix. A *correlation matrix* is a square, symmetrical matrix where each row and each column represents a different variable and the intersecting cells contain the correlation coefficient between two variables. A third option is a *variance-covariance matrix*, which is also a square, symmetrical matrix where the elements on the main diagonal (i.e., the intersection of a variable with itself) represent the variance of each variable and the elements on the off-diagonals represent the covariances between variables. Finally, a *sum-of-squares and cross-products matrix* is the precursor to the variance-covariance matrix. Specifically, it is a matrix consisting of deviation values that have not yet been averaged.

The mathematical calculations involved in multivariate statistical analyses may be performed on any of the previously mentioned matrices. However, the calculations are rather complex and involve a set of skills known as matrix algebra. *Matrix algebra* is somewhat different from scalar algebra—i.e., addition, subtraction, multiplication, and division of a single number—with which the reader is undoubtedly more familiar. Matrix algebra is an extension of scalar algebra where mathematical operations are performed on an ordered array of numerical values. Since, as stated earlier in this chapter, it is not the intent of the authors to deluge the reader with intricate, and oftentimes convoluted, mathematical calculations, matrix algebra will not be discussed further in this text. If the reader is interested in learning more about matrix algebra and its applications in multivariate statistical analyses, several excellent resources include Johnson & Wichern (1998), Tabachnick & Fidell (1996), Stevens (1992), and Tatsuoka (1988).

The reader should be aware that in multivariate statistics, as in univariate statistics, the quality of the data is crucial. Fortunately, advanced computer analysis programs make the computations easy. However, there is a downside to this wonderful feature: The programs will provide output to the requested analysis, including beautifully formatted graphs and tables, regardless of the quality of the data on which the analyses were performed. For example, assume a researcher has data that has not been reliably collected, contains data entry errors, and includes "strange" values that will surely influence the results. In this case, the old adage—"garbage in, garbage out"—would hold true. However, by simply examining the output, the researcher usually would be unable to discern that the results were of poor quality. Prior to analysis, the researcher must take measures to ensure that the data are of the highest possible quality (techniques will be discussed in Chapter 3). Only by doing so can one have assurance in the quality of the results and confidence in the subsequent conclusions drawn.

Standard and Sequential Analyses

The benefits of—and the disadvantages associated with—multivariate statistics are often direct results of the relationships among the variables in a given data set. A lack of relationship among variables typically enables the researcher to interpret the results of an analysis with more clarity. For this reason, orthogonality is an important concept in the application of multivariate statistical analyses. *Orthogonality* is perfect nonassociation between variables. If we know the value for an individual on a given variable, and if that variable has an orthogonal relationship with a second variable, knowing the value of the first variable provides no information in determining the value of the second variable. In other words, the correlation between the two variables is equal to zero.

Orthogonality is often a desirable quality for multivariate statistical analyses. For example, assume we are interested in examining the nature of the relationships among a set of IVs and a single DV. If all pairs of IVs in the set are orthogonal, then each IV would add a distinctively unique component to the prediction of the DV. As a simple example, assume that we are investigating the effects that two IVs (years of education and motivation) have on a single DV (income). If years of education and motivation are orthogonal, then each would contribute separately, and in additive fashion, to the prediction of income. For example, if 25% of the variability in income can be predicted by years of education and 40% can be predicted by motivation, then 65% of the variability in income can be predicted from years of education and motivation taken together. This relationship can easily be shown through the use of a Venn Diagram (see Figure 1.1).

Figure 1.1 Venn Diagram for Income, Years of Education, and Motivation.

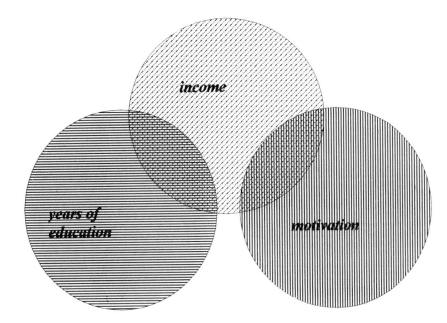

Figure 1.2 Venn Diagram for a Standard Analysis of the Relationship Between Income, Years of Education, and Motivation.

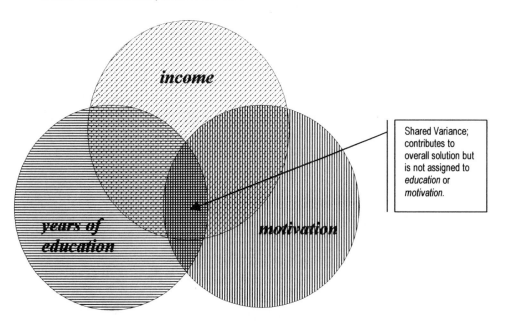

Having a data set with orthogonal variables is the ideal situation. However, most variables with which social science researchers work are correlated to some degree; that is, they are nonorthogonal. When variables are correlated, they have overlapping, or shared, variance. Returning to our previous example, if years of education and motivation were correlated, the contribution of years of education to income would still be 25%, and the contribution of motivation would remain at 40%. However, their combined contribution—which could no longer be determined by means of an additive procedure—would be

Figure 1.3 Venn Diagram for a Sequential Analysis of the Relationship
Between Income, Years of Education, and Motivation.

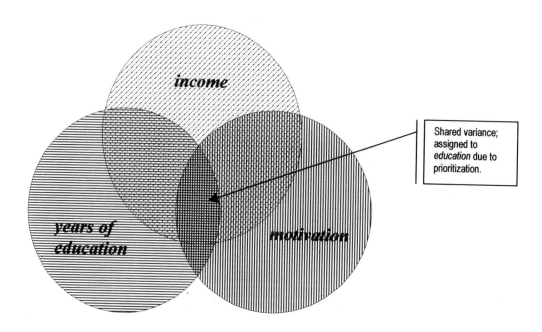

less than 65%. This is due to the fact that the two IVs share some amount of variance. There are two basic strategies for handling this situation in order to determine the contribution of individual IVs to a DV.

Using a *standard analysis* approach, the overlapping portion of variance is included in the overall summary statistics of the relationship of the set of IVs to the DV, but that portion is not assigned to either of the IVs as part of their individual contribution. The overlapping variance is completely disregarded when evaluating the contribution of each IV, taken separately, to the overall solution. Figure 1.2 is a Venn Diagram of a situation where years of education and motivation are nonorthogonal and they share some variance. For this shared portion, we are investigating the effects that two IVs (years of education and motivation) have on a single DV (income). Using a standard analysis approach, the shared variance is included in the total variability explained by the set of IVs, but is not assigned to either years of education or motivation when examining their *individual* contributions.

An alternative approach, *sequential analysis*, requires the researcher to prioritize the entry of IVs into the equation or solution. The first variables entered into the equation will be assigned both its unique variance and any additional variance that it shares with any lower-priority variables; the second variable would be assigned its unique variance and any overlapping variance with lower-priority variables; and so on. Figure 1.3 is a Venn Diagram showing this type of approach where years of education has been assigned the highest priority and therefore is "credited" both with its unique variance and that which it shares with motivation. Notice that, in this situation, the total amount of variance remains the same; however, years of education now has a stronger relationship to income than it did in the standard analysis, yet the contribution of motivation remains the same.

Difficulties in Interpreting Results

The need to understand the nature of relationships among numerous variables measured simultaneously makes multivariate analysis an inherently difficult subject (Johnson & Wichern, 1998). One of the major difficulties in using multivariate statistical analyses is that it is sometimes nearly impossible to get a firm statistical answer to your research questions (Tabachnick & Fidell, 1996). This is due largely to the increased complexity of the techniques. Oftentimes, results may be ambiguous. Two or more statistical indices resulting from one "computer run" may contradict each other. The researcher must then play a very important role in determining the most appropriate way to interpret the results of the analysis. There exists the potential for the introduction of some subjectivity into this process. But, rest assured, we believe that the benefits of being able to examine complex relationships among a large set or sets of variables make multivariate procedures well worth the time and effort required to master them.

SECTION 1.2   REVIEW OF DESCRIPTIVE AND INFERENTIAL STATISTICS

The purpose of the remainder of this chapter is to provide the reader with a brief review of descriptive and inferential statistics. While it certainly is not our intention to provide thorough coverage of these topics, the discussions should serve as a good refresher of material already mastered by the reader prior to beginning a course in multivariate statistics.

Descriptive Statistics

The first step in nearly any data analysis situation is to describe or summarize the data collected on a set of subjects that constitute the sample of interest. In some studies, such as simple survey research, the entire analysis may involve only descriptive statistics. However, most studies begin with a summarization of the data using descriptive techniques and then move on to more advanced techniques in order to address more complex research questions. There are four main types of descriptive statistics: measures of central tendency, variability, relative position, and relationship.

Measures of Central Tendency. Measures of central tendency permit the researcher to describe a set of data with a single, numerical value. This value represents the average, or typical, value. The three most commonly used measures of central tendency are the mode, the median, and the mean. The *mode* is the most frequently occurring score in a distribution. There is no calculation involved in obtaining the mode; one simply examines the distribution of scores and determines which score was obtained by the subjects most often. The mode does have limited use and is most appropriately used for variables measured at a nominal level.

The *median* is the score in the distribution that divides the upper 50% of scores from the lower 50%. Like the mode, the median is also of limited use since it does not take into consideration all values in the distribution. The values of extreme scores, both positive and negative, are completely ignored. The median is most appropriate for ordinal measures.

The most frequently used measure is the *mean*, which is simply the arithmetic average of a set of scores. It is the preferred measure of central tendency, since it takes into account the actual values of all scores in a distribution. If there are extreme scores in the distribution, the mean can be unduly influenced (e.g., an extreme high score will increase the value of the mean, thus making it less representative of the distribution). In this case, the median may be the more appropriate measure. However, when data are measured on an interval or ratio scale, the mean remains the favored method of describing central tendency.

Measures of Variability. Oftentimes, a measure of central tendency is not enough to adequately describe a distribution of scores. A researcher may also want to know the degree to which the scores are spread around the mean, or other measure of central tendency. This amount of spread is indicated by one of three measures of variability. The most basic measure of variability is the range. The *range* is simply the difference between the highest score and the lowest score in the distribution. The range is not a good indicator of the amount of spread in scores since it is based solely on the largest and smallest values. It is typically used only as a rough estimate of the variability in a set of scores.

When a distribution of scores contains some extreme values, an alternative to the range is the quartile deviation. The *quartile deviation* is defined as one-half of the difference between the

3rd quartile (i.e., the 75th percentile) and the 1st quartile (i.e., the 25th percentile). The resulting value is actually the amount of spread in the scores that are located within a range defined by the median ±12.5% of the cases. A limitation of the quartile deviation is that it does not take into consideration all values in the distribution.

The *standard deviation* is an appropriate measure of variability when variables are measured on an interval or ratio scale. The standard deviation is defined as a special type of average distance of scores away from the mean. It is the most stable measure of variability since it takes into account every score in the distribution. It is obtained by first subtracting the mean from each score, squaring the resulting differences, summing the squared differences, and finally finding the average of that summed value. This value is called the *variance*, and one must simply find the square root of the variance in order to obtain the standard deviation. A large standard deviation indicates that the scores in the distribution are spread out away from the mean, and a small standard deviation indicates that the scores are clustered closer together about the mean. The mean and standard deviation taken together do a fairly good job of describing a set of scores.

Measures of Relative Position.   Measures of relative position indicate where a specific score is located in relation to the rest of the scores in the distribution. Interpretation of these measures allows a researcher to describe how a given individual performed when compared to all others measured on the same variable(s). The two most common measures of relative position are percentile ranks and standard scores.

Many of us have seen our performances on standardized tests reported as percentile ranks. A *percentile rank* indicates the percent of scores that fall at or below a given score. If a raw score of 75 points corresponds to a percentile rank of 88, then 88% of the scores in the distribution were equal to or less than 75 points. Percentile ranks are most appropriate for ordinal measures, although they are often used for interval measures as well.

There are several types of standard scores that can be used to report or describe relative position. A *standard score* is derived from the manipulation of a raw score that expresses how far away from the mean a given score is located, usually reported in standard deviation units. Since the calculation of a standard score involves some algebraic manipulation, the use of standard scores is appropriate when data are measured at an interval or ratio level. Two of the most common types of standard scores are z-scores and T-scores. A *z-score* indicates the distance away from the mean a score is in terms of standard deviation units and is calculated by subtracting the mean from the raw score and then dividing the value by the standard deviation. If a raw score is equal to the mean, it would have a z-score equal to 0; if a raw score was two standard deviations greater than the mean, it would have a z-score equal to +2.00. If a raw score was one standard deviation below the mean, it would have a z-score equal to −1.00. Recall that the sign is an important component of a reported z-score since it serves as a "quick" indicator of whether the score is located above or below the mean.

A *T-score* is simply a z-score expressed on a different scale. In order to convert a z-score to a T-score, simply multiply the z-score by 10 and add 50. For example, if we had a distribution with a mean of 65 and a standard deviation of 5, an individual who obtained a raw score of 75 would have a z-score equal to +2.00 and a T-score equal to 70. The reader should be aware that all three measures used in this example (i.e., the raw score, the z-score, and the T-score) indicate a score that is equivalent to two standard deviations above the mean.

Measures of Relationship.   Measures of relationship indicate the degree to which two quantifiable variables are related to each other. These measures do not describe—or even *imply*—a causal relationship. They only verify that a relationship exists. Degree of relationship between two variables is expressed as a correlation coefficient ranging from −1.00 to +1.00. If the two variables in question are not related, a coefficient at or near zero will be obtained; if they are highly related, a coefficient near +1.00 or −1.00 will be obtained. Although there are many different types of correlation coefficients, depending on the scale of measurement being used, two commonly used measures of relationship are the *Spearman rho* and the *Pearson r*.

If data for one or both of the variables are expressed as ranks (i.e., ordinal data) instead of scores, the Spearman rho is the appropriate measure of correlation. The interpretation is the same as previously discussed, with values ranging from –1.00 to +1.00. If a group of subjects produced identical ranks on the two variables of interest, the correlation coefficient would be equal to +1.00, indicating a perfect relationship.

If data for both variables represent interval or ratio measures, the Pearson r is the appropriate measure of correlation. Like the mean and standard deviation, the Pearson r takes into account the value of every score in both distributions. The Pearson r assumes that the relationship under investigation is a linear one; if in reality it is not, then the Pearson r will not yield a valid measure of the relationship.

Inferential Statistics

Recall that inferential statistics deal with collecting and analyzing information from samples in order to draw conclusions, or inferences, about the larger population. The adequacy, or representativeness, of the sample is a crucial factor in the validity of the inferences drawn as the result of the analyses; the more representative the sample, the more generalizable the results will be to the population from which the sample was selected. Assume we are interested in determining whether or not two groups differ from each other on some outcome variable. If we take appropriate measures to ensure that we have a representative sample (i.e., use a random sampling technique), and we find a difference between the group means at the end of our study, the ultimate question in which we are interested is whether a similar difference exists in the population from which the samples were selected. It is possible that no real difference exists in the population, and the one that we found between our samples was due simply to chance. Perhaps if we had used two different samples, we would not have discovered a difference. However, if we do find a difference between our samples and we conclude that the difference is large enough to infer that a real difference exists in the population (i.e., the difference was *statistically significant*), then what we really want to know is "how likely is it that our inference is incorrect?" This idea of "how likely is it" is the central concept in inferential statistics. In other words, if we inferred that a true difference exists in the population, how many times out of 100 would we be wrong? Another way of looking at this concept is to think of selecting 100 random samples, testing each of them, and then determining for how many our inference would be wrong.

There are several key underlying concepts to the application of inferential statistics. One of those is the concept of standard error. Any given sample will, in all likelihood, not perfectly represent the population. In fact, if we selected several random samples from the same population, each sample mean probably would be different from the other sample means and probably none of them would be equal to the population mean. This expected, chance variation among sample means is known as *sampling error*. Sampling error is inevitable and cannot be eliminated. Even though sampling errors are random, they behave in a very orderly fashion. If enough samples are selected and means are calculated for each sample, all samples will not have the same mean, but those means will be normally distributed around the population mean; this is called the *distribution of sample means*. A mean of this distribution of sample means can be calculated and will provide a good estimate of the population mean. Furthermore, as with any distribution of scores, not only the mean but a measure of variability can also be obtained. The standard deviation of the sample means is usually referred to as the *standard error*. The standard error of the mean tells us by how much we would expect our sample means to differ if we used other samples from the same population. This value, then, indicates how well our sample represents the population from which it was selected; obviously, the smaller the standard error, the better. With a smaller standard error, we can have more confidence in the inferences that we draw about the population based on sample data. In reality, we certainly would not have the time or resources to select countless random samples, nor do we need to. Only the sample size and the sample standard deviation are required in order to calculate a good estimate of the standard error.

The main goal of inferential statistics is to draw inferences about populations based on sample data, and the concept of standard error is central to this goal. In order to draw these inferences with confidence, a researcher must ensure that a sample is representative of the population. In *hypothesis testing*, we are testing predictions we have made regarding our sample. For example, suppose the difference between two means was being examined. The *null hypothesis* (H_o) explains the chance occurrence that we have just discussed and predicts that the only differences that exist are chance differences that represent only random sampling error. In other words, the null hypothesis states that there is no true difference to be found in the population. In contrast, the *research or alternative hypothesis* (H_1) states that one method is expected to be better than the other or, in other words, that the two group means are not equal and therefore represent a true difference in the population. In inferential statistics, it is the null hypothesis that we are testing, since it is easier to disprove the null than to prove the alternative.

Null hypotheses are tested through the application of specific statistical criteria known as *significance tests*. Significance tests are the procedures used by the researcher to determine if the difference between sample means is substantial enough to rule out sampling error as an explanation for the difference. A test of significance is made at a predetermined *probability level* (i.e., the probability that the null hypothesis is correct), which obviously allows the researcher to pass judgment on the null hypothesis. For example, if the difference between two sample means is not large enough to convince us that a real difference exists in the population, the statistical decision would be to "fail to reject the null hypothesis." In other words, we are not rejecting the null hypothesis, which stated that there was no real difference, other than a difference due to chance, between the two population means. On the other hand, if the difference between sample means was substantially large (i.e., large enough to surpass the statistical criteria), we would "reject the null hypothesis" and conclude that a real difference, beyond chance, exists in the population.[1] There are a number of tests of significance that can be used to test hypotheses including, but not limited to, the *t* test, analysis of variance, the chi square test, and tests of correlation.

Based on the results of these tests of significance, the researcher must decide whether to reject or fail to reject the null hypothesis. The researcher can never know with 100% certainty whether the statistical decision was correct, only that he or she was *probably* correct. There are four possibilities with respect to statistical decisions—two reflect correct decisions and two reflect erroneous conclusions:

1. The null hypothesis is actually true (i.e., there is no difference), and the researcher concludes that it is true (*fail to reject H_o*) – correct decision.
2. The null hypothesis is actually false (i.e., a real difference exists), and the researcher concludes that it is false (*reject H_o*) – correct decision.
3. The null hypothesis is actually true, and the researcher concludes that it is false (*reject H_o*) – incorrect decision.
4. The null hypothesis is actually false, and the researcher concludes that it is true (*fail to reject H_o*) – incorrect decision.

If it is concluded that a null hypothesis is false when it is actually true (number 3 above), a *Type I Error* has been committed by the researcher; if a null hypothesis is actually false when it is concluded to be true (number 4 above), a *Type II Error* has been made.

When a researcher makes a decision regarding the status of a null hypothesis, she or he does so with a pre-established (*a priori*) probability of being incorrect. This probability level is referred to as the *level of significance* or *alpha (α) level*. This value determines how large the difference between means must be in order to be able to be declared significantly different, thus resulting in a decision to reject the null hypothesis. The most common probability levels used in behavioral science settings are $\alpha = .05$ or $\alpha = .01$. The selected significance level (α) determines the probability of committing a Type I Error; in other words, the risk of being wrong that is assumed by a researcher. The probability of committing a

---

[1] For testing the difference(s) between mean(s), the size of the difference(s) is a major consideration. Sample size and the amount of variability within the samples also play a major role in significance testing.

Type II Error is symbolized by β (beta) but is not arbitrarily set as is alpha. To determine the value for β, a complex series of calculations is required. Many beginning researchers assume that it is best to set the alpha level as small as possible (thereby reducing the risk of a Type I Error to almost zero). However, the probability levels of committing Type I and Type II Errors have a complimentary relationship. If one reduces the probability of committing a Type I Error, the probability of committing a Type II Error increases (Harris, 1998). These factors must be weighed and levels established prior to the implementation of a research study.

The *power* of a statistical test is the probability of rejecting H_0 when H_0 is, in fact, false; in other words, making a correct decision (number 2 above). Power is appropriately named since this is exactly what the researcher hopes to accomplish during hypothesis testing. Therefore, it is desirable for a test to have high power (Agresti & Finlay, 1997). Power is determined in the following manner:

$$\text{Power} = 1 - \beta$$

Power, as with α, is established arbitrarily and should be set at a high level since the researcher is hoping to reject a null hypothesis that is not true and wants to have a high probability of doing so (Brewer, 1978).

Another factor related to hypothesis testing is *effect size*. Effect size (often denoted as "ES" or partial η^2) is defined as the size of the treatment effect the researcher wishes to detect with respect to a given level of power. In an experimental study, ES is equal to the difference between the population means of the experimental and control groups divided by the population standard deviation for the control group. In other words, it is a measure of the amount of difference between the two groups reported in standard deviation units (it is a standardized or transformed score and is, therefore, metric-free). Effect sizes can also be calculated for correlation coefficients or for mean differences resulting from nonexperimental studies (Harris, 1998). Effect size, like α and power, is set *a priori* by the researcher, but also involves strong consideration of what the researcher hopes to find in the study as well as what constitutes important and trivial differences.

A more powerful statistical test will be able to detect a smaller effect size. Cohen (1988) established a rule of thumb for evaluating effect sizes: An ES of .2 is considered small, one of .5 is considered medium, and one of .8 is considered large. A researcher would want to design a study and statistical analysis procedures that would be powerful enough to detect the smallest effect size that would be of interest and nontrivial (Harris, 1998).

Sample size (n) is a final factor whose value must be considered when conducting a research study and must be done prior to data collection. The required sample size for a study is a function of alpha, power, and effect size. Since sample size has several relationships with these three factors, values for the factors must be set prior to the selection of a sample. For example, for a fixed α-level, the probability of a Type II Error decreases when the sample size increases. Additionally, for a fixed α-level, power increases as sample size increases. If sample size is held constant and α is lowered (e.g., in an attempt to reduce the probability of committing a Type I Error), power will also decrease. This fact provides partial justification for not setting α near zero; the power of a test would be too low and the researcher may be much less likely to reject a null hypothesis that is really false (Agresti & Finlay, 1997). However, the solution to this dilemma is not to obtain the largest sample possible; a huge sample size might produce such a powerful test that even the slightest, trivial difference could be found to be statistically significant (Harris, 1998). In summation, as n increases, ES, α, and β will decrease, causing power to increase (Brewer, 1978). The reader is reminded, however, that obtaining the largest possible sample size need not be the case because the most appropriate sample involves a balanced combination of α, ES, and power. Tables have been developed for a diverse range of values for n, α, ES, and power and provide optimum sample sizes (Cohen, 1969).

SECTION 1.3 ORGANIZATION OF THE BOOK

The remainder of this textbook is organized in the following manner. Chapter 2 presents a guide to various multivariate techniques, in addition to reviewing several univariate techniques. Included for each is an overview of the technique and descriptions of research situations appropriate for its use. Chapter 3 addresses the assumptions associated with multivariate statistical techniques and also discusses methods for determining if any of those assumptions have been violated by the data and, if so, how to deal with those violations. The concepts and procedures discussed in Chapter 3 are requisite to conducting any of the statistical analyses in subsequent chapters.

Chapters 4 through 11 present specific multivariate statistical procedures. Included in each chapter (i.e., for each technique) are a practical description of the technique, examples of research questions, as well as assumptions and limitations, the logic behind the technique, and how to interpret and present the results. A sample research study, from problem statement through analyses and presentation of results, is also included in each chapter. Finally, a step-by-step guide to conducting the analysis procedure using SPSS is presented.

CHAPTER 2

A GUIDE TO MULTIVARIATE TECHNIQUES

One of the most difficult tasks for students conducting quantitative research is identifying the appropriate statistical technique to be utilized for a particular research question. Fortunately, if an accurate and appropriate research question has been generated, the process of determining the statistical technique is really quite simple. The primary factor that determines the statistical test to be used is the variable—more specifically, the type or scale of variables (categorical or quantitative) and the number of independent and dependent variables, both of which influence the nature of the research question being posed. To facilitate this identification process, we provide two decision-making tools so that the reader may select one that is most comfortable. The Table of Statistical Tests begins with the identification of the numbers and scales of independent and dependent variables. In contrast, the Decision-Making Tree is organized around the four different types of research questions: degree of relationship among variables, significance of group differences, prediction of group membership, and structure. This chapter presents these decision-making tools and provides an overview of the statistical techniques addressed in this text as well as basic univariate tests, all of which will be organized by the four types of research questions. Please note that there are additional multivariate techniques that are not addressed in this book.

SECTION 2.1 DEGREE OF RELATIONSHIP AMONG VARIABLES

When investigating the relationship among two or more quantitative variables, correlation and/or regression is the appropriate test. Three statistical tests are presented that address this type of research question.

Bivariate Correlation and Regression

Bivariate correlation and regression evaluate the degree of relationship between two quantitative variables. The Pearson correlation coefficient (r), the most commonly used bivariate correlation technique, measures the association between two quantitative variables without distinction between the independent and dependent variables (e.g., What is the relationship between SAT scores and freshman college GPA?). In contrast, bivariate regression utilizes the relationship between the independent and dependent variables to predict the score of the dependent variable from the independent variable (e.g., To what degree do SAT scores [IV] predict freshman college GPA [DV]?). An overview of bivariate correlation and regression is provided in Chapter 7.

When to use bivariate correlation/regression?

      1   IV (quantitative)   ————————▶   relationship/prediction
      1   DV (quantitative)

Multiple Regression

Multiple regression identifies the best combination of predictors (IVs) of the dependent variable. Consequently, it is used when there are several independent quantitative variables and one dependent quantitative variable (e.g., Which combination of risk-taking behaviors [amount of alcohol use, drug use, sexual activity, and violence—IVs] best predicts the amount of suicide behavior [DV] among adolescents?). To produce the best combination of predictors of the dependent variable, a sequential multiple regression selects independent variables, one at a time, by their ability to account for the most variance in the dependent variable. As a variable is selected and entered into the group of predictors, the relationship between the group of predictors and the dependent variables is reassessed. When no more variables are left that explain a significant amount of variance in the dependent variable, then the regression model is complete. Multiple regression is discussed in Chapter 7.

When to use multiple regression?

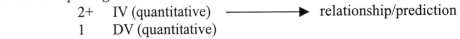

2+   IV (quantitative) ⟶ relationship/prediction
1   DV (quantitative)

Path Analysis

Path analysis utilizes multiple applications of multiple regression to estimate causal relations, both direct and indirect, among several variables and to test the acceptability of the causal model hypothesized by the researcher (e.g., What are the direct and indirect effects of reading ability, family income, and parents' education [IVs] on students' GPA [DV]?). Before any data analysis is conducted, the researcher must first hypothesize the causal model, which is usually based upon theory and previous research. This model is then graphically represented in a path diagram. Path coefficients are calculated to estimate the strength of the relationships in the hypothesized causal model. A further discussion of path analysis is presented in Chapter 8.

When to use path analysis?

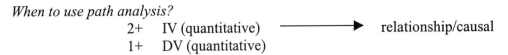

2+   IV (quantitative) ⟶ relationship/causal
1+   DV (quantitative)

SECTION 2.2 SIGNIFICANCE OF GROUP DIFFERENCES

A primary purpose of testing for group differences is to determine a causal relationship between the independent and dependent variables. Comparison groups are created by the categories identified in the IV(s). The number of categories in the IV, the number of IVs, and the number of DVs determine the appropriate test.

t Test

The most basic statistical test that measures group differences is the *t* test, which analyzes significant differences between two group means. Consequently, a *t* test is appropriate when the IV is defined as having two categories and the DV is quantitative (e.g., Do males and females [IV] have significantly different SAT scores [DV]?). Further explanation of *t* tests is provided in most introductory level statistical texts and therefore is not included in this text.

When to use a t test?

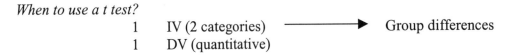

1   IV (2 categories) ⟶ Group differences
1   DV (quantitative)

One-Way Analysis of Variance

One-way analysis of variance (ANOVA) tests the significance of group differences between two or more means as it analyzes variation between and within each group. ANOVA is appropriate when the IV is defined as having two or more categories and the DV is quantitative, (e.g., Do adolescents from low, middle, and high socioeconomic status families [IV] have different scores on an AIDS knowledge test [DV]?). Since ANOVA only determines the significance of group differences and does not identify which groups are significantly different, post hoc tests are usually conducted in conjunction with ANOVA. An overview of ANOVA is provided in Chapter 4.

When to use a one-way ANOVA?

1	IV (2+ categories) ⟶	Group differences
1	DV (quantitative)	

One-Way Analysis of Covariance

One-way analysis of covariance (ANCOVA) is similar to ANOVA in that two or more groups are being compared on the mean of some DV, but ANCOVA additionally controls for a variable (covariate) that may influence the DV (e.g., Do preschoolers of low, middle, and high socioeconomic status [IV] have different literacy test scores [DV] after adjusting for family type [covariate]?) Many times the co-variate may be pretreatment differences in which groups are equated in terms of the covariate(s). In general, ANCOVA is appropriate when the IV is defined as having two or more categories, the DV is quantitative, and the effects of one or more covariates need to be removed. Further discussion of one-way ANCOVA is provided in Chapter 5.

When to use a one-way ANCOVA?

1	IV (2+ categories) ⟶	Group differences
1	DV (quantitative)	
1+	covariate	

One-Way Multivariate Analysis of Variance

Similar to ANOVA in that both techniques test for differences among two or more groups as defined by a single IV, one-way multivariate analysis of variance (MANOVA) is utilized to simultaneously study two or more related DVs while controlling for the correlations among the DVs (Vogt, 1993). If DVs are not correlated, then it is appropriate to conduct separate ANOVAs. Since groups are being compared on several DVs, a new DV is created from the set of DVs that maximizes group differences. After this linear combination of the original DVs is created, an ANOVA is then conducted to compare groups based on the new DV. A MANOVA example follows: Does ethnicity [IV] significantly affect reading achievement, math achievement, and overall achievement [DVs] among 6th grade students? Chapter 6 discusses one-way and factorial models of MANOVA and MANCOVA.

When to use a one-way MANOVA?

1	IV (2+ categories) ⟶	Group differences
2+	DVs (quantitative)	

One-Way Multivariate Analysis of Covariance

An extension of ANCOVA, multivariate analysis of covariance (MANCOVA) investigates group differences among several DVs while also controlling for covariate(s) that may influence the DVs (e.g., Does ethnicity [IV] significantly affect reading achievement, math achievement, and overall achievement [DVs] among 6th grade students after adjusting for family income [covariate]?).

When to use a one-way MANCOVA?

    1   IV (2+ categories)   ⟶   Group differences
    2+   DVs (quantitative)
    1+   covariate

Factorial Multivariate Analysis of Variance

Factorial multivariate analysis of variance (factorial MANOVA) extends MANOVA to research scenarios with two or more IVs that are categorical (e.g., Does ethnicity and learning preference [IVs] significantly affect reading achievement, math achievement, and overall achievement [DVs] among 6th grade students?). Since several independent variables are used, different combinations of DVs are created for each main effect and interaction of the IVs.

When to use a factorial MANOVA?

    2+   IVs (categorical)   ⟶   Group differences
    2+   DVs (quantitative)

Factorial Multivariate Analysis of Covariance

Factorial multivariate analysis of covariance (factorial MANCOVA) extends factorial MANOVA to research scenarios that require the adjustment of one or more covariates on the DVs (e.g., Does ethnicity and learning preference [IVs] significantly affect reading achievement, math achievement, and overall achievement [DVs] among 6th grade students after adjusting for family income [covariate]?).

When to use a factorial MANCOVA?

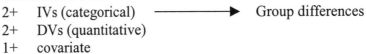

    2+   IVs (categorical)   ⟶   Group differences
    2+   DVs (quantitative)
    1+   covariate

SECTION 2.3   PREDICTION OF GROUP MEMBERSHIP

The primary purpose of predicting group membership is to identify specific IVs that best predict group membership as defined by the DV. Consequently, the following statistical techniques are appropriate when the DV is categorical.

Discriminant Analysis

Discriminant analysis is often seen as the reverse of MANOVA in that it seeks to identify which combination of quantitative IVs best predict group membership as defined by a single DV that has two or more categories (e.g., Which risk-taking behaviors [amount of alcohol use, drug use, sexual activity, violence—IVs] distinguish suicide attempters from nonattempters [DV]?). In contrast, MANOVA identifies group differences on a combination of quantitative DVs. Discriminant analysis seeks to interpret the pattern of differences among the predictors (IVs); consequently, the analysis will often produce several sets or combinations of IVs that predict group membership. Each IV set, referred to as a function, represents a mathematical attempt to maximize a linear combination of the IVs to discriminate among groups. Discriminant analysis is best used when groups are formed naturally based on some characteristic and not randomly. Chapter 10 discusses discriminant analysis in further detail.

When to use discriminant analysis?

2+   IVs (quantitative)
1   DV (2+ categories) ——————————→ Group prediction

Logistic Regression

Logistic regression is similar to discriminant analysis in that both identify a set of IVs that best predict group membership. Although SPSS provides both binary and multinomial logistic regression, our discussion will address only the binary logistic regression in which the DV is a dichotomous (having only two categories) variable. The IVs may be categorical and/or quantitative. Since the DV consists of only two categories, logistic regression estimates the odds probability of the DV occurring as the values of the IVs change. For example, a research question that would utilize logistic regression is: To what extent do certain risk-taking behaviors (amount of alcohol use, drug use, sexual activity, and the presence of violent behavior—IVs) increase the odds of a suicide attempt (DV) occurring? Logistic regression is discussed in Chapter 11.

When to use logistic regression?

2+   IVs (categorical/quantitative)
1   DV (2 categories) ——————————→ Group prediction

SECTION 2.4   STRUCTURE

When the researcher questions the underlying structure of an instrument or is interested in reducing the number of IVs, factor analysis and/or principal components are appropriate methods. Although factor analysis and principal components are different techniques, they are very similar and will be presented together under the heading of factor analysis. Both of these techniques will be discussed in Chapter 9.

Factor Analysis and Principal Components Analysis

Factor analysis allows the researcher to explore underlying structures of an instrument or data set and is often used to develop and test theory. Principal components is generally used to reduce the number of IVs, which is advantageous when conducting multivariate techniques in which the IVs are highly correlated. For example, principal components can reduce a 100-item instrument to ten factors that will then be utilized as IVs in subsequent data analysis. This IV reduction can also aid the researcher in exploring, developing, and testing theories based upon how the items are grouped. Consequently, factor analysis/principal components combine several related IVs into fewer, more basic underlying factors. Independent variables that share common variance are grouped together. Once factors are created, they are often adjusted (rotated) so that these factors are not highly related to one another and more accurately represent the combined IVs. Since research questions that utilize factor analysis/principal components typically only address IVs, this statistical technique is not included in the Table of Statistical Tests, which relies upon the identification of both IVs and DVs. An example research question that would utilize factor analysis/principal components is as follows: What underlying structure exists among the variables of male life expectancy, female life expectancy, birth rate, infant mortality rate, fertility rate among women, number of doctors, number of radios, number of telephones, number of hospital beds, and gross domestic product?

SECTION 2.5 THE TABLE OF STATISTICAL TESTS

The Table of Statistical Tests is presented in Figure 2.1. This tool organizes statistical methods by the number and type (categorical versus quantitative) of IVs and DVs. Steps for using this table are listed as follows:

1. Identify the variables in the research question.
2. Indicate which variables are the independent and dependent variables and covariates.
3. Determine the type (categorical or quantitative) of all variables. If a variable is categorical, determine the number of categories.
4. Use the table to identify:

    - the appropriate row for IVs;
    - the appropriate column for the DVs;
    - the row and column intersection indicates the statistical test to be used.

These steps are applied to the following research question: Does ethnicity significantly affect reading achievement, math achievement, and overall achievement among 6th grade students after adjusting for family income?

Step 1:   Underline the variables in the research question.

Does ethnicity significantly affect reading achievement, math achievement, and overall achievement among 6th grade students after adjusting for family income.

Step 2: Identify which variables are the independent and dependent variables and the covariates. It is helpful to examine the sentence order of variables since the first variables are usually the IVs. The verb of the research question can also help in the identification process.

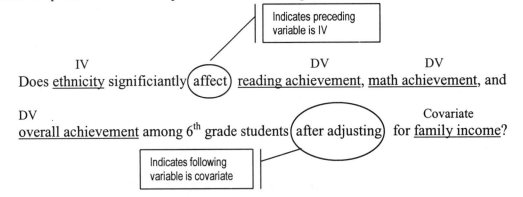

Step 3:   Determine the type (categorical or quantitative) of all variables. This is dependent on how you decide to operationalize your variables.

(IV) 3+ categories   (DV) quantitative   (DV) quantitative
Does ethnicity significantly affect reading achievement, math achievement, and

(DV) quantitative   (covariate) quantitative
overall achievement among 6th grade students after adjusting for family income?

Consequently, this research question includes the following: one IV (3+ categories), 3 DVs (all quantitative), and one covariate (quantitative).

Step 4: Use the table to:

- identify the appropriate row for IVs.
    Example: IV→categorical→one IV→2+ categories with one covariate
- identify the appropriate column for the DVs.
    Example: DV→quantitative→several DVs
- the row and column intersection indicates the statistical test to be used.
    Example: The intersection of the preceding row and column indicates that one-way
    MANCOVA should be conducted.

SECTION 2.6 THE DECISION-MAKING TREE FOR STATISTICAL TESTS

The Decision-Making Tree for Statistical Tests is presented in Figure 2.2. This tool organizes statistical methods by the purpose of the research question. Once the purpose has been identified, the process is then guided by the number and type of variables. Although the Decision-Making Tree begins with the purpose of the research question, we recommend first identifying the number and types of variables as this will guide the process of determining the purpose. The steps for using the Decision-Making Tree are listed as follows.

1. Identify the variables in the research question.
2. Indicate which variables are the independent and dependent variables and covariates.
3. Determine the type (categorical or quantitative) of all variables. If a variable is categorical, determine the number of categories.
4. Determine the purpose of the research question: degree of relationship, group differences, prediction of group membership, and structure. Here are a few helpful hints in using the variable information to determine the research question purpose.

    - When the IVs and DVs are all quantitative, the purpose is degree of relationship.
    - When the IVs are categorical and the DVs are quantitative, the purpose is group differences.
    - When the DVs are categorical, the purpose is predicting group membership.

5. Apply the information from the preceding steps to the Decision-Making Tree following the process of decisions—research question, number and type of DV, number and type of IV, and covariates—to the appropriate test.

These steps are applied to the following research question: Which combination of risk-taking behaviors [amount of alcohol use, drug use, sexual activity, and violence] best predict the amount of suicide behavior among adolescents?

Step 1: Identify the variables in the research question.
    Which combination of risk-taking behaviors [amount of alcohol use, drug use, sexual activity, and violence] best predict the amount of suicide behavior among adolescents?

Step 2: Indicate which variables are the independent and dependent variables and covariates.
                                                                    IV   IV
    Which combination of risk-taking behaviors [amount of alcohol use, drug use,

        IV   IV   DV
    sexual activity, and violence] best (predict) the amount of suicide behavior among

    adolescents?

> Indicates that previous variables are IVs

19

Step 3:   Determine the type (categorical or quantitative) of all variables.

                                                  (IV) quantitative   (IV) quantitative

Which combination of risk-taking behaviors [amount of alcohol use, drug use,

(IV) quantitative (IV) quantitative   (DV) quantitative

sexual activity, and violence] best predict the amount of suicide behavior among

adolescents?

Consequently, this research question includes the following: four IVs (all quantitative) and one DV (quantitative).

Step 4:   Determine the purpose of the research question: degree of relationship, group differences, prediction of group membership, and structure. Since all our variables are quantitative, the purpose of the research question is degree of relationship.

Step 5:   Apply the information from the preceding steps to the Decision-Making Tree: research question, number and type of DV, number and type of IV, and covariates. Continuing with the example, the decisions would be as follows:
degree of relationship→1 DV (quant.)→2+ IVs (quant.)→ multiple regression

SUMMARY

    Determining the appropriate statistical technique relies upon the identification of the type of variables (categorical or quantitative) and the number of IVs and DVs, all of which influence the nature of the research questions being posed. This chapter introduces the statistical tests to be presented in the upcoming chapters. The statistical methods are organized under four purposes of research questions: degree of relationship, significance of group differences, prediction of group membership, and structure. Statistical tests that analyze the degree of relationship include bivariate correlation and regression, multiple regression and path analysis. Research questions addressing degree of relationship have all quantitative variables. Methods that examine the significance of group differences are *t* test, one-way and factorial ANOVA, one-way and factorial ANCOVA, one-way and factorial MANOVA, and one-way and factorial MANCOVA. Research questions that address group differences have categorical IV(s). Statistical tests that predict group membership are logistic regression and discriminant analysis. Research questions that address prediction of group membership have a categorical DV. Statistical tests that address the purpose of structure are factor analysis and principal components; questions that address structure usually do not distinguish between independent and dependent variables.

    Two decision-making tools are provided to assist in identifying which statistical method to utilize—the Table of Statistical Tests and the Decision-Making Tree. The Table of Statistical Tests is organized by the type and number of IVs and DVs, while the Decision-Making Tree is organized by the purpose of the research question.

Figure 2.1 Table of Statistical Tests

INDEPENDENT VARIABLE(s)				DEPENDENT VARIABLE(s): Categorical — 2 categories	DEPENDENT VARIABLE(s): Categorical — 2+ categories	DEPENDENT VARIABLE(s): Quantitative — One DV	DEPENDENT VARIABLE(s): Quantitative — Several DVs
Categorical	One IV	2 categories				*t* Test	One-way MANOVA
Categorical	One IV	2+ categories				One-way ANOVA	One-way MANOVA
Categorical	One IV	With covariate				One-way ANCOVA	One-way MANCOVA
Categorical	Several IVs	No covariate		Logistic Regression		Factorial ANOVA	Factorial MANOVA
Categorical	Several IVs	With covariate				One-way ANCOVA	Factorial MANCOVA
Quantitative	One IV					Bivariate Correlation Bivariate Regression	
Quantitative	Several IVs			Discriminant Analysis Logistic Regression	Discriminant Analysis	Multiple Regression Path Analysis	Path Analysis

Figure 2.2 Decision-Making Tree

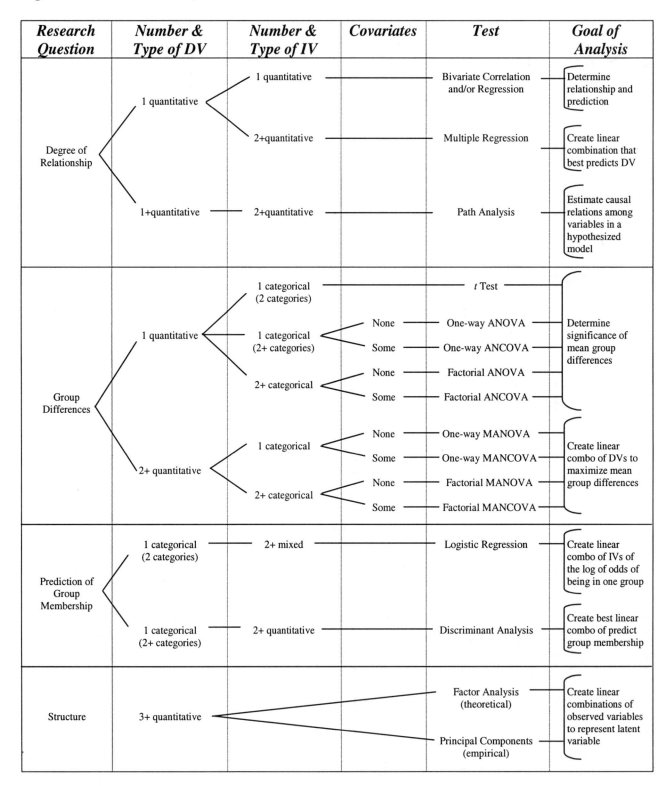

Research Question	Number & Type of DV	Number & Type of IV	Covariates	Test	Goal of Analysis
Degree of Relationship	1 quantitative	1 quantitative		Bivariate Correlation and/or Regression	Determine relationship and prediction
		2+quantitative		Multiple Regression	Create linear combination that best predicts DV
	1+quantitative	2+quantitative		Path Analysis	Estimate causal relations among variables in a hypothesized model
Group Differences	1 quantitative	1 categorical (2 categories)		*t* Test	Determine significance of mean group differences
		1 categorical (2+ categories)	None	One-way ANOVA	
			Some	One-way ANCOVA	
		2+ categorical	None	Factorial ANOVA	
			Some	Factorial ANCOVA	
	2+ quantitative	1 categorical	None	One-way MANOVA	Create linear combo of DVs to maximize mean group differences
			Some	One-way MANCOVA	
		2+ categorical	None	Factorial MANOVA	
			Some	Factorial MANCOVA	
Prediction of Group Membership	1 categorical (2 categories)	2+ mixed		Logistic Regression	Create linear combo of IVs of the log of odds of being in one group
	1 categorical (2+ categories)	2+ quantitative		Discriminant Analysis	Create best linear combo of predict group membership
Structure	3+ quantitative			Factor Analysis (theoretical)	Create linear combinations of observed variables to represent latent variable
				Principal Components (empirical)	

Exercises for Chapter 2

Directions: The research questions below are ones used as examples throughout this chapter. Identify the appropriate statistical test for each question by using both decision-making tools. Determine the tool with which you are most comfortable.

1. To what degree do SAT scores predict freshman college GPA?

2. Does ethnicity significantly affect reading achievement, math achievement, and overall achievement among 6th grade students?

3. What are the causal effects (direct and indirect) among number of school absences due to illness, reading ability, semester GPA, and total score on Iowa Test of Basic Skills among 8th grade students?

4. Do males and females have significantly different SAT scores?

5. What is the relationship between SAT scores and freshman college GPA?

6. Which risk-taking behaviors [amount of alcohol use, drug use, sexual activity, violence] distinguish nonsuicide attempters from suicide attempters?

7. Do preschoolers of low, middle, and high socioeconomic status have different literacy test scores after adjusting for family type?

8. Does ethnicity significantly affect reading achievement, math achievement, and overall achievement among 6th grade students after adjusting for family income?

9. Which combination of risk-taking behaviors [amount of alcohol use, drug use, sexual activity, and violence] best predict the amount of suicide behavior among adolescents?

10. Do preschoolers of low, middle, and high socioeconomic status have different literacy test scores?

11. Does ethnicity and learning preference significantly affect reading achievement, math achievement, and overall achievement among 6th grade students?

12. To what extent do certain risk-taking behaviors (amount of alcohol use, drug use, sexual activity, and the presence of violent behavior) increase the odds of a suicide attempt occurring?

13. Do ethnicity and learning preference significantly affect reading achievement, math achievement, and overall achievement among 6th grade students after adjusting for family income?

14. What underlying structure exists among the following variables: amount of alcohol use, drug use, sexual activity, school misconduct, cumulative GPA, reading ability, and family income?

CHAPTER 3

PRE-ANALYSIS DATA SCREENING

In this chapter, we discuss several issues related to the quality of data that a researcher wishes to subject to a multivariate analysis. These issues must be carefully considered and addressed *prior* to the actual statistical analysis—they are essentially an analysis *within* the analysis! Only after these quality assurance issues have been examined can the researcher be confident that the main analysis will be an honest one, which will ultimately result in valid conclusions being drawn from the data.

SECTION 3.1 WHY SCREEN DATA?

There are four main purposes for screening data prior to conducting a multivariate analysis. The first of these deals with the accuracy of the data that have been collected. Obviously, the results of any statistical analysis are only as good as the data that were analyzed. If inaccurate data are used, the computer program will run the analysis (in all likelihood), and the researcher will obtain her output. However, the researcher will not be able to discern the extent to which the results are valid simply by examining the output—the results will appear to be legitimate (e.g., values for test statistics will appear, accompanied by significance values, etc.). The researcher will then proceed to interpret the results and draw conclusions; however, unknown to her, they are erroneous conclusions because they have been based on the analysis of inaccurate data.

With a small data file, simply printing the entire data set and proofreading it against the actual data is probably an easy and efficient method of determining the accuracy of data. This can be accomplished by using the **SPSS List** procedure. However, if the data set is rather large, this process would be overwhelming. In this case, examination of the data using frequency distributions and descriptive statistics would be a more realistic method. Both frequency distributions and descriptive statistics can be obtained by using the **SPSS Frequencies** procedure. For quantitative variables, a researcher might examine the range of values to be sure that no cases have values outside the range of possible values. Assessment of the means and standard deviations (i.e., are they plausible?) would also be beneficial. For categorical variables, the researcher would also want to make sure that all cases have values that correspond to the coded values for the possible categories.

The second purpose deals with missing data and attempts to assess the effect of and ways to deal with incomplete data. Missing data occur when measurement equipment fails, subjects do not complete all trials or respond to all items, or errors occur during data entry. The amount of missing data is less crucial than the pattern of missing data (Tabachnick & Fidell, 1996). Missing values that are randomly scattered throughout a data set sometimes are not serious because their pattern is random. Nonrandom missing data, on the other hand, create problems with respect to the generalizability of the results. Since these missing values are nonrandom, there is likely some underlying reason as to their occurrence. Unfortunately, there are no firm guidelines for determining how much missing data is too much for a given sample size. Those decisions still rest largely on the shoulders of the researcher. Methods for dealing with missing data are discussed in Section 3.2.

The third purpose deals with assessing the effects of extreme values (i.e., *outliers*) on the analysis. Outliers are cases with such extreme values on one variable or on a combination of variables that they distort the resultant statistics. Outliers often create critical problems in multivariate data analyses.

There are several causes for a case to be defined as an extreme value, some of which are far more serious than others. These various causes and methods for addressing each will be discussed in Section 3.3.

Finally, all multivariate statistical procedures are based on assumptions, to some degree. The fourth purpose of screening data is to assess the adequacy of fit between the data and the assumptions of a specific procedure. Some multivariate procedures have "unique" assumptions (which will be discussed in those specific chapters) upon which they are based, but nearly all techniques include three basic assumptions: normality, linearity, and homoscedasticity. These assumptions will be defined and methods for assessing the adequacy of the data with respect to each will be discussed in Sections 3.4, 3.5, and 3.6, respectively. Techniques for implementing these methods using SPSS will be described in Sections 3.7 and 3.8.

SECTION 3.2 MISSING DATA

Many researchers tend to assume that any missing data that occur within their data sets is random in nature. This may or may not be the case; if it is not the case, serious problems can arise when trying to generalize to the larger population from which the sample was obtained. The best thing to do when a data set includes missing data is to examine it. Using data that are available, a researcher should conduct tests to see if patterns exist in the missing data. To do so, one could create a dichotomous dummy variable, coded so that one group includes cases with values on a given variable and the other group contains cases with missing values on that variable. For example, if respondents on an attitudinal survey are asked to provide their income and many do not supply that information (for reasons unknown to us at this time), those who provided an income level would be coded "0" and those who did not would be coded "1." Then the researcher could run a simple independent samples t-test to determine if there are significant mean differences in attitude between the two groups. If significant differences do exist, there is an indication that those who did not provide income information possess different attitudes than those who did report their income. In other words, there exists a pattern in the missing responses.

If a researcher decides that the missing data are important and need to be addressed, there are several alternative methods to handle these data. (For a discussion on additional techniques to use when there are missing data, the reader is advised to refer to Tabachnick and Fidell, 1996.) The first of these alternatives involves deleting the cases or variables that have created the problems. Any case that has a missing value is simply dropped from the data file. If only a few cases have missing values, this is a good alternative. Another option involves a situation where the missing values may be concentrated to only a few variables. In this case, an entire variable may be dropped from the data set, provided it is not central to the main research questions and subsequent analysis. However, if missing values are scattered throughout the data and are abundant, deletion of cases and/or variables may result in a substantial loss of data, either in the form of subjects or measures. Sample size may begin to decrease rather rapidly and, if the main analysis involves group comparisons, some groups may approach dangerously low sample sizes inappropriate for some multivariate analyses.

A second alternative to handling missing data is to estimate the missing values and then use these values during the main analysis. There are three main methods of estimating missing values. The first of these is for the researcher to use *prior knowledge*, or a well-educated guess, for a replacement value. This method should be used only when a researcher has been working in the specific research area for quite some time and is very familiar with the variables and the population being studied.

Another method of estimating missing values involves the calculation of the means, using available data, for variables with missing values. Those mean values are then used to replace the missing values prior to the main analysis. When no other information is available to the researcher, the mean is the best estimate for the value on a given variable. This is somewhat of a conservative procedure since the overall mean does not change by inserting the mean value for a case, and no guessing on the part of the researcher is required. However, the variance is reduced somewhat since the "real" value probably would not have been precisely equal to the mean. This is usually not a serious problem unless there are numerous missing values. In this situation, a possible concession is to insert a group mean, as opposed to the overall mean, for a missing value. This procedure is more appropriate for situations involving group comparison analyses.

Finally, a third alternative to handling missing data also estimates the missing value, but does so using a *regression* approach. Regression is discussed extensively in Chapter 7. In regression, several IVs are used to develop an equation that can be used to predict the value on a DV. For missing data, the variable with missing values becomes the DV. Cases with complete data are used to develop this prediction equation. The equation is then used to predict missing values on the DV for incomplete cases. An advantage to this procedure is that it is more objective than a researcher's guess and factors in more information than simply inserting the overall mean. One disadvantage of regression is that the predicted scores are better than they actually would be; since the predicted values are based on other variables in the data set, they are more consistent with those scores than a real score would be. Another disadvantage to regression is that the IVs must be good predictors of the DV in order for the estimated values to be accurate; otherwise, this amounts to simply inserting the overall mean in place of the missing value (Tabachnick & Fidell, 1996).

If any of the above methods is used to estimate missing values, a researcher should consider repeating the analysis using only complete cases (i.e., conduct the main analysis with the missing values and repeat the analysis with no missing values). If the results are similar, one can be confident in the results. However, if they are different, an examination of the reasons for the differences should be conducted. The researcher should then determine which of the two represents the "real world" more accurately, or consider reporting both sets of results.

SECTION 3.3   OUTLIERS

Cases with unusual or extreme values at one or both ends of a sample distribution are known as *outliers*. There are three fundamental causes for outliers: (1) data entry errors were made by the researcher, (2) the subject is not a member of the population for which the sample is intended, or (3) the subject is simply different from the remainder of the sample (Tabachnick & Fidell, 1996).

The problem with outliers is that they can distort the results of a statistical test. This is due largely to the fact that many statistical procedures rely on squared deviations from the mean (Aron & Aron, 1997). If an observation is located far from the rest of the distribution (and, therefore, far from the mean), the value of its deviation would be large. Imagine by how much a deviation increases when squared! Generally speaking, statistical tests are quite sensitive to outliers. An outlier can exert a great deal of influence on the results of a statistical test. A single outlier, if extreme enough, can cause the results of a statistical test to be significant when, in fact, it would not have been if it had been based on all values other than the outlier. The complementary situation can also occur: An outlier can cause a result to be insignificant when, without the outlier, it would have been significant. Similarly, outliers can seriously affect the values of correlation coefficients. As researchers, it is vital that the results of our statistical analyses represent the majority of the data and not be largely influenced by one, or a few, extreme observations. It is for this reason that it is crucial for researchers to be able to identify outliers and decide how to handle them (Stevens, 1992).

Outliers can exist in both univariate and multivariate situations, among dichotomous and continuous variables, and among IVs as well as DVs (Tabachnick & Fidell, 1996). Univariate outliers are cases with extreme values on one variable; multivariate outliers are cases with unusual combinations of scores on two or more variables. With data sets consisting of a small number of variables, detection of univariate outliers can be relatively simple. This can be accomplished by visually inspecting the data, either by examining a frequency distribution or by obtaining a histogram and looking for unusual values. One would simply look for values that appear far from the others in the data set. In Figure 3.1, Case #3 would clearly be identified as an outlier since it is located far from the rest of the observations.

Figure 3.1 Sample Data Set (a) and Corresponding Histogram (b), Indicating One Outlier.

(a)

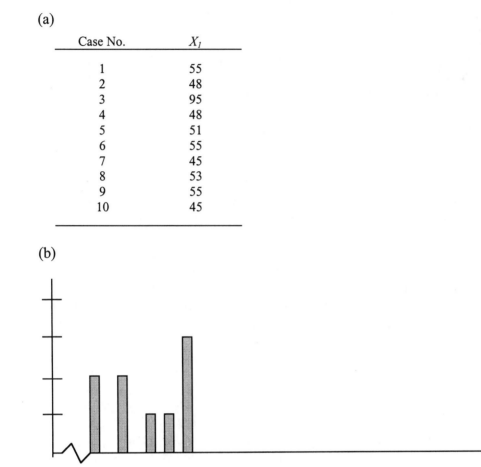

Case No.	X_1
1	55
2	48
3	95
4	48
5	51
6	55
7	45
8	53
9	55
10	45

(b)

45	50	55	60	65	70	75	80	85	90	95

Univariate outliers can also be detected through statistical methods by standardizing all raw scores in the distribution. This is most easily accomplished by transforming the data to z-scores. If a normal distribution is assumed, approximately 99% of the scores will lie within three standard deviations of the mean. Therefore, any z value greater than +3.00 or less than -3.00 indicates an unlikely value and the case should be considered an outlier. However, with large sample sizes (e.g., $n > 100$), it is likely that a few subjects could have z-scores in excess of ±3.00. In this situation, the researcher might want to consider extending the rule to $z > +4.00$ and $z < -4.00$ (Stevens, 1992). For small sample sizes (e.g., $n \leq 10$), any data point with a z value greater than 2.50 should be considered as a possible outlier.

Univariate outliers can also be detected by means of graphical methods (Tabachnick & Fidell, 1996). Box plots literally "box in" cases that are located near the median value; extreme values are located far away from the box. Figure 3.2 presents a sample box plot.

As shown in Figure 3.2, the box portion of the plot extends from the 25th to the 75th percentiles, with the dark line representing the median value for the distribution. The lines above and below the box include all values within 1.5 box lengths. Cases with values between 1.5 and 3 box lengths from the upper or lower edges of the box are outliers and are designated by a small circle (o). The specific case number is also listed next to the symbol. Although not depicted in the figure, cases with values greater than 3 box lengths from the edges are also identified and are designated with asterisks (*) and the specific case number.

Figure 3.2 Sample Box Plot Indicating One Outlier.

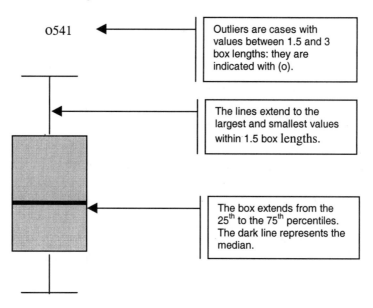

Multivariate outliers, as previously stated, consist of unusual combinations of scores on two or more variables. Individual z-scores may not indicate that the case is a univariate outlier (i.e., for each variable, the value is within the expected range), but the combination of variables clearly separates the particular case from the rest of the distribution. Multivariate outliers are more subtle and, therefore, more difficult to identify, especially by using any of the previously mentioned techniques. Fortunately, a statistical procedure (known as *Mahalanobis distance*) exists which can be used to identify outliers of any type (Stevens, 1996). Mahalanobis distance is defined as the distance of a case from the centroid of the remaining cases where the centroid is the point created by the means of all the variables (Tabachnick & Fidell, 1996).

For multivariate outliers, Mahalanobis distance is evaluated as a chi-square (χ^2) statistic with degrees of freedom equal to the number of variables in the analysis (Tabachnick & Fidell, 1996). The accepted criterion for outliers is a value for Mahalanobis distance which is significant at $p < .001$, determined by comparing the obtained value for Mahalanobis distance to the chi-square critical value.

Once outliers have been identified, it is necessary to investigate them further. First, the researcher must determine whether the outlier was due to an error in data entry. In this situation, of course, the value would be corrected and the data reanalyzed. However, if the researcher determines that the extreme value was correctly entered and that it may be due to an instrumentation error or that the subject is simply different from the rest of the sample, then it is appropriate to drop the case from the analysis. If it cannot be determined that either of these situations resulted in the extreme value, one should not drop the case from the analysis, but rather should consider reporting two analyses (one with the outlying case included and the other after the case has been deleted) (Stevens, 1996). Remember that outliers should not be viewed as being "bad" because they often represent interesting cases. Care must be taken so that the outlying case is not automatically dropped from the analysis. The case and its value(s) on the variable(s) may be perfectly legitimate.

If the researcher decides that a case with unusual values is legitimate and should remain in the sample, steps may be taken to reduce the relative influence of those cases. Variables may be transformed (i.e., the scales may be changed so that the distribution appears more normal), thus reducing the impact of extreme values. Data transformations are discussed in greater detail in the next section. For a more thorough discussion of variable transformations, the reader is advised to refer to Johnson and Wichern (1998), Stevens (1992), and Tabachnick & Fidell (1996).

SECTION 3.4   NORMALITY

As previously mentioned, there are three general assumptions involved in multivariate statistical testing: normality, linearity, and homoscedasticity.   There are consequences of applying statistical analyses—particularly inferential testing—to data that do not conform to these assumptions.   If one or more assumptions are violated, the results of the analysis may be biased (Kennedy & Bush, 1985).   It is critical then to assess the extent to which the sample data meet the assumptions.   The issue at hand is one of test robustness.   *Robustness* refers to the relative insensitivity of a statistical test to violations of the underlying inferential assumptions. In other words, it is the degree to which a statistical test is still appropriate to apply when some of its assumptions are not met:

> If in the presence of marked departures from model assumptions, little or no discrepancy between nominal and actual levels of significance occurs, then the statistical test is said to be robust with respect to that particular violation (Kennedy & Bush, 1985, p. 144).

The first of these assumptions is that of a normal sample distribution.   Prior to examining multivariate normality, one should first assess univariate normality.   Univariate normality refers to the extent to which all observations in the sample for a given variable are distributed normally.   There are several ways, both graphical and statistical, to assess univariate normality.   A simple graphical method involves the examination of the histogram for each variable.   Although somewhat oversimplified, this does give an indication as to whether or not normality might be violated.   One of the most popular graphical methods is the *normal probability plot*.   In a normal probability plot, also known as a *normal Q-Q plot*, the observations are arranged in increasing order of magnitude and plotted against the expected normal distribution values (Stevens, 1996).   The plot shows the variable's observed values along the x-axis and the corresponding predicted values from a standard normal distribution along the y-axis (Norusis, 1998). If normality is defensible, the plot should resemble a straight line.

Among the statistical options for assessing univariate normality are the use of skewness and kurtosis coefficients.   As a reminder to the reader, *skewness* is a quantitative measure of the degree of symmetry of a distribution about the mean; *kurtosis* is a quantitative measure of degree of peakedness of a distribution.   A variable can have significant skewness, kurtosis, or both. When a distribution is normal, the values for skewness and kurtosis are both equal to zero.[1]   If a distribution has a positive skew (i.e., a skewness value > zero), there is a clustering of cases to the left and the right tail is extended with only a small number of cases.   In contrast, if a distribution has a negative skew (i.e., a skewness value < zero), there is a clustering of cases to the right and the left tail is extended with only a small number of cases.   Values for kurtosis that are positive indicate that the distribution is too peaked with long, thin tails (a condition known as *leptokurtosis*); kurtosis values that are negative indicate that the distribution is too flat, with many cases in the tails (a condition known as *platykurtosis*).   Significance tests for both skewness and kurtosis values should be evaluated at an alpha level of .01 or .001 for small to moderate sample sizes, using a table of critical values for skewness and kurtosis, respectively.   Larger samples may show significant skewness and/or kurtosis values, but often may not deviate enough from normal to make a meaningful difference in the analysis (Tabachnick & Fidell, 1996).

Another specific statistical test that is used to assess univariate normality is the *Kolmogorov-Smirnov statistic*, with Lilliefors significance level. The Kolmogorov-Smirnov statistic tests the null hypothesis that the population is normally distributed. A rejection of this null hypothesis based on the value of the Kolmogorov-Smirnov statistic and associated observed significance level serves as an indication that the variable is not normally distributed.

Multivariate normality refers to the extent to which all observations in the sample for all combinations of variables are distributed normally.   Similar to the univariate examination, there are several

---

[1] The mathematical equation for kurtosis gives a value of 3 when the distribution is normal, but statistical packages subtract 3 before printing so that the expected value is equal to zero.

ways, both graphical and statistical, to assess multivariate normality. It is difficult to completely describe multivariate normality but, suffice to say, "normality on each of the variables separately is a necessary but not sufficient condition for multivariate normality to hold" (Stevens, 1996, p. 245). Since univariate normality is a necessary condition for multivariate normality, it is recommended that all variables be assessed based on values for skewness and kurtosis, as previously described.

Other characteristics of multivariate normality include:

1. Each of the individual variables must be normally distributed;
2. Any linear combination of the variables must be normally distributed; and
3. All subsets of the set of variables (i.e., every pairwise combination) must have a multivariate normal distribution (this is known as *bivariate normality*).

Bivariate normality implies that the scatterplots for each pair of variables will be elliptical. An initial check for multivariate normality would consist of an examination of all bivariate scatterplots to check that they are approximately elliptical (Stevens, 1996). A specific graphical test for multivariate normality exists, but requires a special computer program be written, as it is not available in standard statistical software packages (Stevens, 1996).

If the researcher determines that the data have substantially deviated from normal, he or she can consider transforming the data. ***Data transformations*** involve the application of mathematical procedures to the data in order to make them appear "more normal." Once data have been transformed, provided all other assumptions have been met, the results of the statistical analyses will be more accurate. It should be noted that there is nothing unethical about transforming data; transformations are nothing more than a reexpression of the data in different units (Johnson & Wichern, 1998). The transformations are performed on every subject in the data set, so the order and relative position of observations is not affected (Aron & Aron, 1997).

A variety of data transformations exists, depending on the shape (e.g., extent of deviation from normal) of the original raw data. For example, if a distribution differs only moderately from normal, a square root transformation should be tried initially. If the deviation is more substantial, a log transformation is obtained. Finally, if a distribution differs severely, an inverse transformation is tried. The direction of the deviation must also be considered. The above transformations are appropriate for distributions with positive skewness. If the distribution has a negative skew, the appropriate strategy is to "reflect" the variable and then apply the transformation procedure listed above. Reflection involves finding the largest score in the distribution and adding one to it to form a constant that is larger than any score in the distribution. A new variable is then created by subtracting each score from the constant. In effect, this process converts a distribution with negative skewness to one with positive skewness. It should be noted that interpretation of the results of analyses of this variable must also be reversed (Tabachnick & Fidell, 1996). Transformations can be easily obtained in various statistical packages, including SPSS. The transformations discussed here, along with the SPSS language for the computation of new variables, are summarized in Figure 3.3.

Once variables have been transformed, it is important to reevaluate the normality assumption. Following the confirmation of a normal or near-normal distribution, the analysis may proceed typically, resulting in vastly improved results (Tabachnick & Fidell, 1996). Additionally, the researcher should be cognizant of the fact that any transformations performed on the data must be discussed in the methods section of any research report.

It should be understood that the topic of data transformation is much too broad to be adequately addressed here. Should one require further details and examples of these various transformations, it is recommended that the reader refer to Tabachnick and Fidell (1996).

Figure 3.3 Summary of Common Data Transformations to Produce Normal Distributions.

Original Shape	Transformation	SPSS Compute Language
Moderate positive skew	Square root	NEWX=SQRT(X)
Substantial positive skew	Logarithm	NEWX=LG10(X)
With value < 0	Logarithm	NEWX=LG10($X + C$) [a]
Severe positive skew	Inverse	NEWX=1/X
With value < 0	Inverse	NEWX=1/($X + C$) [a]
Moderate negative skew	Reflect & square root	NEWX=SQRT($K - X$) [b]
Substantial negative skew	Reflect & logarithm	NEWX=LG10($K - X$) [b]
Severe negative skew	Reflect & inverse	NEWX=1/($K - X$) [b]

[a] C = a constant added to each score in order to bring the smallest value to at least 1.

[b] K = a constant from which each score is subtracted so that the smallest score equals 1.

SECTION 3.5 LINEARITY

The second assumption, that of linearity, presupposes that there is a straight line relationship between two variables. These two variables can be individual raw data variables (e.g., "drug dosage" and "length of illness") or can be combinations of several raw data variables (i.e., *composite* or *subscale scores*, such as eight items additively combined to arrive at a score for "self-esteem"). The assumption of linearity is important in multivariate analyses due to the fact that many of the analysis techniques are based on linear combinations of variables. Furthermore, statistical measures of relationship such as Pearson's r capture only linear relationships between variables and ignore any substantial nonlinear relationships that may exist (Tabachnick & Fidell, 1996).

There are essentially two methods of assessing the extent to which the assumption of linearity is supported by data. In analyses that involve predicted variables (e.g., multiple regression as presented in Chapter 7), nonlinearity is determined through the examination of residuals plots. *Residuals* are defined as the portions of scores not accounted for by the multivariate analysis; they are also referred to as "prediction errors" since they serve as measures of the differences between obtained and predicted values on a given variable. If standardized residual values are plotted against the predicted values, nonlinearity will be indicated by a curved pattern to the points (Norusis, 1998). In other words, residuals will fall above the zero line for some predicted values and below the line for other predicted values (Tabachnick & Fidell, 1996). Therefore, a relationship that does not violate the linearity assumption would be indicated by the points clustering around the zero line. A nonlinear relationship is depicted in Figure 3.4 (a).

A second, and more crude, method of assessing linearity is accomplished by inspection of bivariate scatterplots. If both variables are normally distributed and linearly related, the shape of the scatterplot will be elliptical. If one of the variables is not normally distributed, the relationship will not be linear, and the scatterplot between the two variables will not be oval-shaped. Assessing linearity by means of bivariate scatterplots is an extremely subjective procedure, at best. The process can become even more cumbersome when data sets with numerous variables are being examined. In situations where nonlinearity between variables is apparent, the data can once again be transformed in order to enhance the linear relationship.

Figure 3.4 Sample Standardized Residuals Plots Showing a Strong Nonlinear Relationship (a) and a Linear Relationship (b).

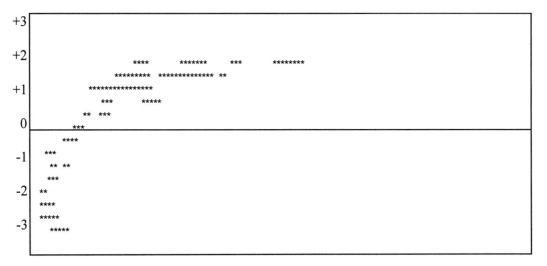

Predicted values

(a) nonlinear relationship

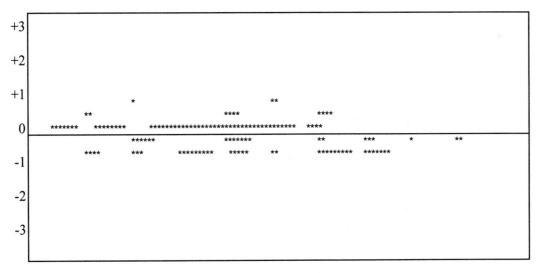

Predicted values

(b) linear relationship

SECTION 3.6 HOMOSCEDASTICITY

The third and final assumption is the assumption of homoscedasticity. *Homoscedasticity* is the assumption that the variability in scores for one continuous variable is roughly the same at all values of another continuous variable. This concept is analogous to the univariate assumption of homogeneity of variance (i.e., the variability in a continuous dependent variable is expected to be roughly consistent at all levels of the independent, or discrete grouping, variable). In the univariate case, homogeneity of variances is assessed statistically with *Levene's test*. This statistic provides a test of the hypothesis that the samples come from populations with the same variances. If the observed significance level for Levene's test is small (i.e., $p < .05$), one should reject the null hypothesis that the variances are equal. It should be noted that a violation of this assumption, based on a "reject" decision of Levene's test, is not fatal to the analysis. Furthermore, Levene's test provides a sound means for assessing univariate homogeneity since it is not affected by violations of the normality assumption (Kennedy & Bush, 1985).

Homoscedasticity is related to the assumption of normality because if the assumption of multivariate normality is met, the two variables must be homoscedastic (Tabachnick & Fidell, 1996). The failure of the relationship between two variables to be homoscedastic is caused either by the nonnormality of one of the variables or by the fact that one of the variables may have some sort of relationship to the transformation of the other variable (Tabachnick & Fidell, 1996). Errors in measurement, which are greater at some levels of the independent variable than at others, may also cause a lack of homoscedasticity.

Heteroscedasticity, or the violation of the assumption of homoscedasticity, can be assessed through the examination of bivariate scatterplots. Within the scatterplot, the collection of points between variables should be approximately the same width across all values with some bulging toward the middle. Although subjective in nature, homoscedasticity is best assessed through the examination of bivariate scatterplots. In multivariate situations, homoscedasticity can be assessed statistically by using *Box's M test for equality of variance-covariance matrices*. This test allows the researcher to evaluate the hypothesis that the covariance matrices are equal. If the observed significance level for Box's M test is small (i.e., $p < .05$), one should reject the null hypothesis that the covariance matrices are equal. It should be noted, however, that Box's M test is very sensitive to nonnormality; thus one may reject the assumption that covariance matrices are equal due to a lack of multivariate normality, not because the covariance matrices are different (Stevens, 1996). Therefore, it is recommended that the tenability of the multivariate normality assumption be assessed prior to examining the results of the Box's M test as a means of assessing possible violations of the assumption of homoscedasticity. Violations of this assumption can be corrected by transformation of variables; however, it should be noted that a violation of the assumption of homoscedasticity, similar to a violation of homogeneity, will not prove fatal to an analysis (Tabachnick & Fidell, 1996; Kennedy & Bush, 1985). The linear relationship will still be accounted for, although the results will be greatly improved if the heteroscedasticity is identified and corrected (Tabachnick & Fidell, 1996).

Because screening data prior to multivariate analysis requires univariate screening, we have provided two univariate examples and one multivariate example.

SECTION 3.7 USING SPSS TO EXAMINE DATA FOR UNIVARIATE ANALYSIS

The following univariate examples explain the steps for using SPSS to examine missing values, outliers, normality, linearity, and homoscedasticity for both grouped data and ungrouped data. Both examples utilize the data set *gssft.sav* from the SPSS Web site.

Univariate Example with Grouped Data

Suppose one is interested in investigating income (*rincom91*) differences between individuals who are either satisfied or not satisfied with their job (*satjob2*). Since this research question compares groups, screening procedures must also examine data for each group.

Before screening of data begins, we must first address a coding problem within the variable *rincom91*. This variable represents income levels ranging from 1-21; however, 22 represents "refusal to report" and 98 and 99 represent "not applicable." Since these values could be misinterpreted as income levels, they should be recoded as missing values. To do so, open the following menus:

Transform
        Recode
                Into Different Variable

Recode into Different Variables Dialogue Box (see Figure 3.5)

We recommend recoding *rincom91* into a different variable, since this provides a record of both the original and altered variables. (Since variables may be transformed numerous times, we will name our new variable *rincom_2*). Once in this dialogue box, indicate the new name for the variable, then click **Change**. Then click **Old and New Values** to specify the transformations.

Figure 3.5 Recode Dialogue Box.

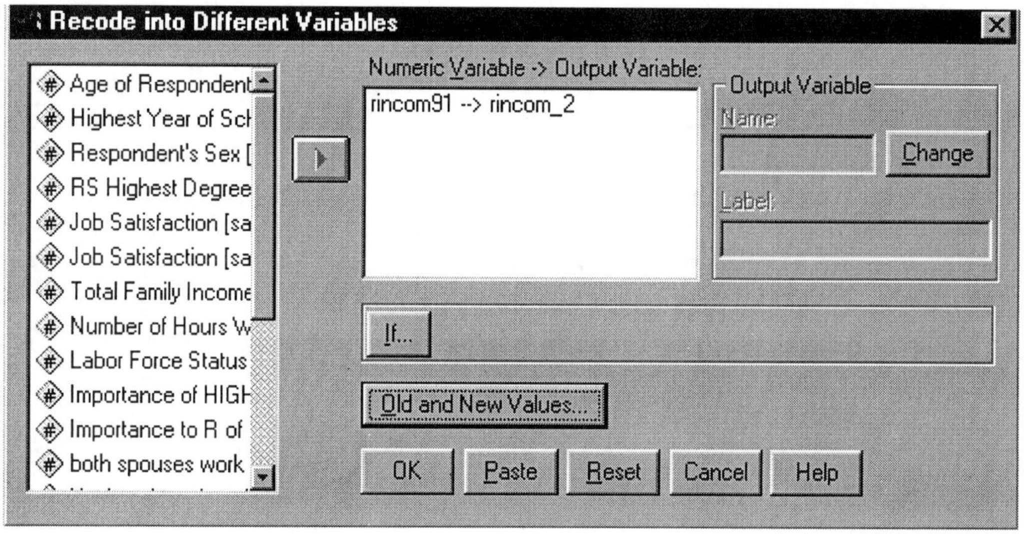

Recode into Different Variables: Old and New Values Dialogue Box (see Figure 3.6)

The only cases to be changed are those with values of 22, 98, and 99; all other values will remain the same. To indicate these transformations, click **Value** under Old Value. In the blank, type the value of 22. Under New Value, click **System Missing** then click **Add**. Continue this procedure for both the values of 98 and 99. Once this transformation has been made, be sure to indicate that all other values should be copied. (Specifically, click "All of the values," then click on "Copy old value," then click on "Add," and then click on "Continue" and "OK.") Now examination for missing data may begin.

Figure 3.6 Recode Old and New Values Dialogue Box.

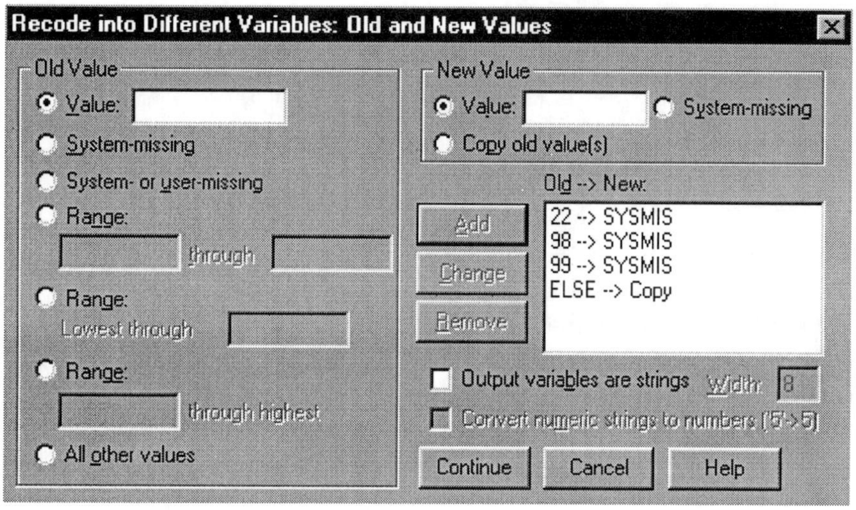

Missing Data

SPSS has several procedures within the analysis process for deleting cases or subjects that have missing values. For most analyses, the **Option** Dialogue Box typically displays the default of **Listwise** in which subjects with missing values are removed only if the missing values are critical to the variables being analyzed. **Pairwise** is another method of deleting subjects with missing values. This method removes subjects with missing values from any and all analyses, even if the missing values are not critical to the variables being studied. Consequently, most researchers utilize the **Listwise** method since it allows for the maximum number of subjects within each analysis.

Examination of missing data in categorical variables can be done by creating a frequency table using **Frequencies**. To determine the extent of missing values within the variable of *satjob2*, the following menus would be selected:

    Analyze
        Descriptive Statistics
            Frequencies

Quantitative variables with missing data can be examined by creating a table of **Descriptive Statistics**. To evaluate missing values in *rincom_2,* open the following menus:

    Analyze
        Descriptive Statistics
            Descriptives

For our example, the frequency and descriptive tables reveal zero missing values for *satjob2* and 37 missing values for *rincom_2*. Typically, if a categorical variable has less than 5% of cases missing, the **Listwise** default would be utilized to delete the cases during the analyses. If a categorical variable has 5-15% of cases with missing data, an additional level or category would be created within the variable so that missing data would be recoded with this new level. Since SPSS no longer detects the missing values and does not recognize the new category as providing meaningful information for the variable being analyzed, these cases would not be included in the analysis.

SPSS also provides a variety of options for handling missing values in quantitative data. In our example, data is missing for 37 cases in the variable *rincom_2*. Since less than 5% of the cases have missing values, the **listwise** default will be used to delete the missing cases. If 5% or more of the values were missing, the method of replacement would be utilized. The most common method is to replace the missing values with the mean score of available cases for that variable. The replacement procedure also allows for other types of replacement values (e.g., median of nearby points, mean of nearby points). Typically, replacing 15% or less of the subjects will have little effect on the outcome of the analysis. However, if a certain subject or variable has more that 15% missing data, you may want to consider dropping the subject or variable from the analysis. To replace missing values with an estimated value, select the following menus:

> **Transform**
>> **Replace Missing Values**

Replace Missing Values Dialogue box (see Figure 3.7)

Once in this dialogue box, identify the targeted variable and move it to the New Variable Box. Notice that a name for the new variable has been generated; this may be changed accordingly. Next, select the Method of replacement. Five options are available in which missing values are replaced by:

Series Mean—The mean of all available cases for the specific variable. This is the default.

Mean of Nearby Points—The mean of surrounding values. You can designate the number of surrounding values to use under **Span of Nearby Points**. The default span is two values.

Median of Nearby Points—The median of surrounding values, the number of which can be designated.

Linear Interpolation—The value midway between the surrounding two values.

Linear Trend at Point—A value consistent with a trend that has been established (e.g., values increasing from the first to the last case).

Once the method of replacement has been determined for the variable, you may also identify additional variables for replacement by using the **Change** button. This will allow you to identify another variable as well as another replacement method.

Figure 3.7 Replace Missing Values Dialogue Box.

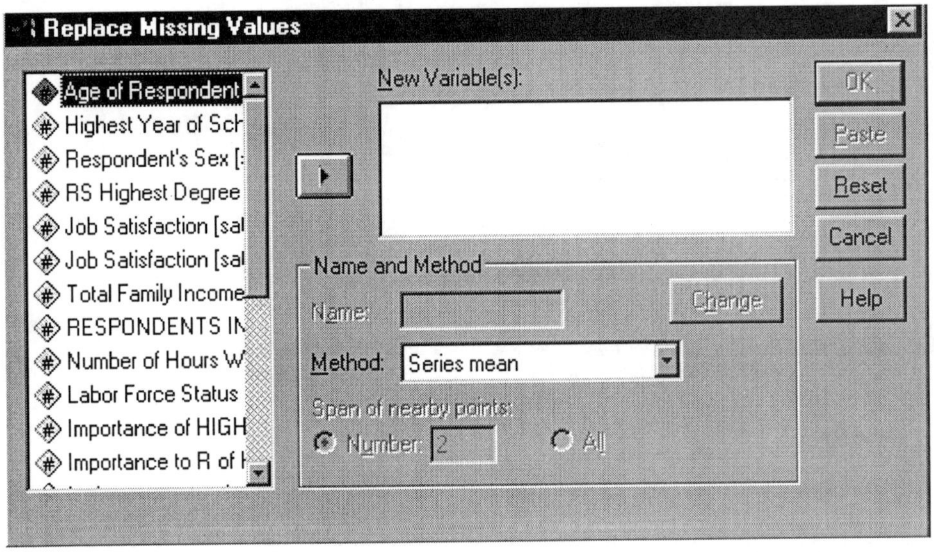

Missing values in quantitative variables can also be estimated by creating a regression equation in which the variable with missing data serves as the dependent variable. Since this method is fairly sophisticated, we will discuss how to use predicted values in Chapter 7 on Multiple Regression.

Finally, if publishing the results of analyses that have utilized replacement of missing values, one should present the procedure(s) for handling such data.

Outliers

Since univariate outliers are subjects or cases with extreme values for one variable, identification of such cases is fairly easy. The **Explore** menu under **Descriptive Statistics** offers several options for such examination. To identify outliers in the categorical variable of *satjob2*, **Frequencies** could be used to detect very uneven splits in categories, splits that typically produce outliers. Categorical variables with 90-10 splits between categories are usually deleted, since scores in the category with 10% of the cases influence the analysis more than those in the category with 90% of the cases. Because our example research question investigates group differences in income, both the IV (*satjob2*) and DV (*rincom_2*) can be examined for outliers using **Explore**. This procedure will allow us to identify outliers for income within each group. To do so, select the following menus:

> **Analyze**
> > **Descriptive Statistics**
> > > **Explore**

Explore Dialogue Box (see Figure 3.8)

Within this dialogue box, move the DVs into the Dependent List. Move IVs into the Factor List. After you have defined the variables, click the **Statistics** button.

Figure 3.8 Explore Dialogue Box.

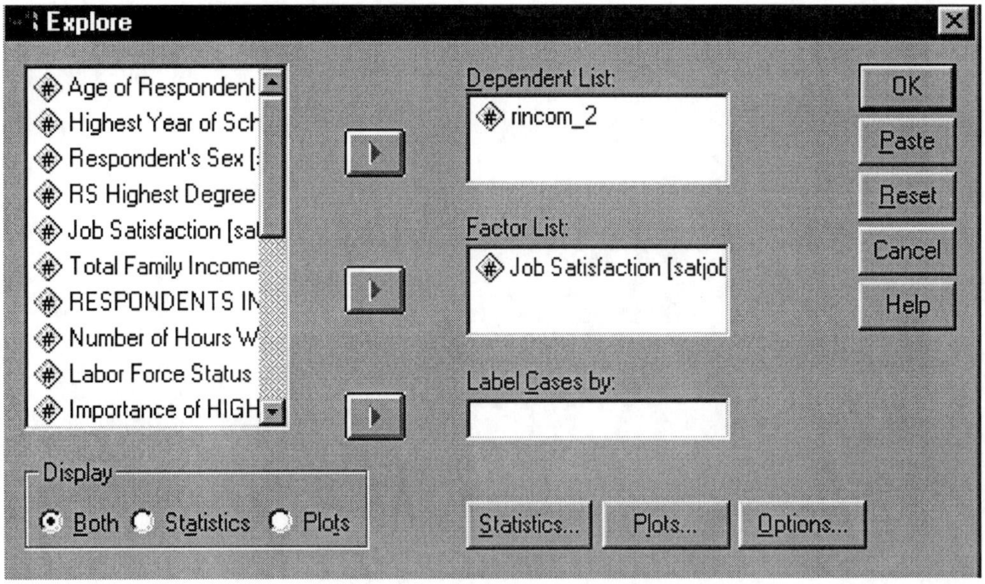

Explore Statistics Dialogue Box (see Figure 3.9)

This box provides the following options for examining outliers:

Descriptives—Calculates descriptive statistics for all subjects and identified categories in the data. This is selected by default.

M-Estimators—Assigns weights to cases depending upon their distance from the center.

Outliers—Identifies the five highest and five lowest cases for the DV by group.

Percentiles—Displays the 5th, 10th, 25th, 50th, 75th, 90th, and 95th percentiles for the DV by group.

For our example, we selected **Descriptives** and **Outliers**. Click **Continue,** then click **Plots.**

Figure 3.9 Explore Statistics Dialogue Box.

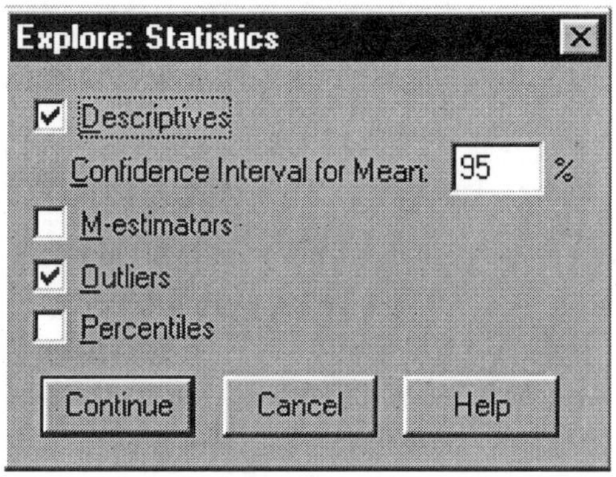

Explore Plots Dialogue Box (see Figure 3.10)

This box provides several options for creating graphic representations of the data. For our example, we will select **Boxplot** and **Stem-and-Leaf Plot.** Since it is best to examine normality after outliers have been addressed, other selections such as **Normality Plots with Tests** and **Histograms** will be conducted later.

Figure 3.10 Explore Plots Dialogue Box.

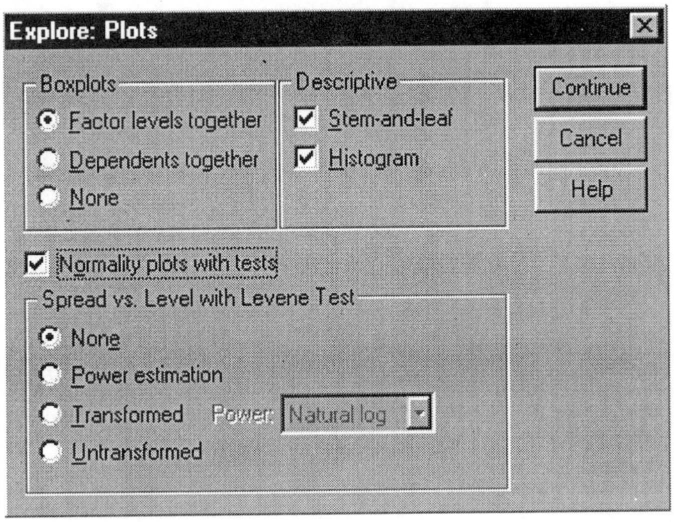

The output reveals some outlier problems within the example. The case summary shows category splits in that 44% of the sample is very satisfied while 56% is not satisfied. This split is not severe enough to delete this variable. The table generated on extreme values (see Figure 3.11) identifies the five highest and lowest scores for each group; keep in mind that these values are not necessarily outliers. The boxplot (see Figure 3.12) generated reveals that both groups have some outliers. The stem-and-leaf plots (see Figure 3.13) support this finding but provide more information regarding the number of outliers. The first plot indicates that 16 subjects who are very satisfied reported extreme income values of 3 or less. In contrast, the second plot displays 22 subjects who are not very satisfied reported extreme in-

come values of 3 or less. Since the number of outlying cases for both groups is fairly small, these outliers could either be deleted using the case numbers identified in the boxplot or be altered to a value that
is within the extreme tail in the accepted distribution. In this example, outliers will be altered by replacing them with a maximum/minimum value (depending on the direction of outliers) that falls within the
accepted distribution. To alter the outliers in *rincom_2*, the stem-and-leaf plot (see Figure 3.13) helps
one identify the specific outlying values to be altered and the accepted minimum value to be used as the
replacement value. Cases that have an income level of 3 or less will be replaced with the accepted value
of 4. To alter outliers, complete the following steps:

> **Transform**
>> **Recode**
>>> **Into Different Variable**

Figure 3.11 Extreme Values Table for Income (*rincom_2*) by Job Satisfaction (*satjob2*).

Extreme Values

Job Satisfaction				Case Number	Value
RINCOM_2	Very satisfied	Highest	1	69	21.00
			2	579	21.00
			3	388	21.00
			4	479	21.00
			5	62	.[a]
		Lowest	1	716	.00
			2	663	.00
			3	363	.00
			4	649	.00
			5	691	.[b]
	Not very satisfied	Highest	1	463	21.00
			2	89	21.00
			3	64	21.00
			4	502	21.00
			5	41	.[a]
		Lowest	1	670	.00
			2	419	.00
			3	206	.00
			4	507	.00
			5	184	.[b]

[a.] Only a partial list of cases with the value 21 are shown in the table of upper extremes.

[b.] Only a partial list of cases with the value 0 are shown in the table of lower extremes.

Figure 3.12 Boxplot for Income (*rincom_2*) by Job Satisfaction (*satjob2*).

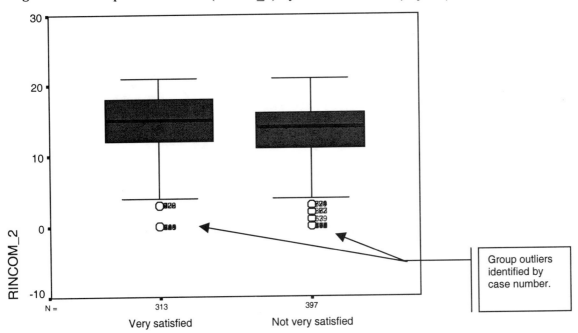

Recode into Different Variables Dialogue Box (see Figure 3.14)

Recoding *rincom_2* into a different variable will allow us to conduct our analysis with the original variable (*rincom91*), the first altered variable (*rincom_2*) and the second altered variable (*rincom_3*). This joint analysis helps determine if altering the outliers had an impact on the results. Once in this dialogue box, identify the variable(*rincom_2*) to be altered and move to the Input (*or* Numeric Variable)→Output Variable box. Indicate a new name for the variable (*rincom_3*), then click **Change**. Click **Old and New Values** to specify the transformations.

Recode Old and New Values Dialogue Box (see Figure 3.15)

The only cases to be changed are those with values 3 or less; all other values will remain the same. To indicate these transformations, under Old Value, click **Range: Lowest through X**. In the blank, indicate the cutoff value of 3. Under New Value, type in 4 and click **Add**. These commands have transformed the outliers (those 3 or less) to a value of 4. The next step is to indicate that all other values will stay the same. To do so, under Old Value click **Else**.[2] Under New Value, click **Copy old value(s)**, then click **Add**. Once cases have been altered, you can proceed with further data examination and analysis. But remember, when conducting the analyses, do so with both original and altered variables.

---

[2]In some versions of SPSS, "**Else**" is "**All other values**"

Figure 3.13 Stem-and-Leaf Plots for Income (*rincom_2*) by Job Satisfaction (*satjob2*).

```
RINCOM_2 Stem-and-Leaf Plot for
SATJOB2= Very satisfied

 Frequency    Stem &  Leaf

    16.00 Extremes    (=<3.0)
     1.00      4 .  0
     2.00      5 .  00
     2.00      6 .  00
     3.00      7 .  000
     7.00      8 .  0000000
     9.00      9 .  000000000
    11.00     10 .  00000000000
    22.00     11 .  0000000000000000000000
    13.00     12 .  0000000000000
    17.00     13 .  00000000000000000
    21.00     14 .  000000000000000000000
    36.00     15 .  000000000000000000000000000000000000
    41.00     16 .  00000000000000000000000000000000000000000
    25.00     17 .  0000000000000000000000000
    30.00     18 .  000000000000000000000000000000
    23.00     19 .  00000000000000000000000
     8.00     20 .  00000000
    26.00     21 .  00000000000000000000000000

 Stem width:      1.00
 Each leaf:       1 case(s)
```

16 subjects are outliers with values of 3 or less.

```
RINCOM_2 Stem-and-Leaf Plot for
SATJOB2= Not very satisfied

 Frequency    Stem &  Leaf

    22.00 Extremes    (=<3.0)
     8.00      4 .  00000000
     4.00      5 .  0000
     4.00      6 .  0000
     4.00      7 .  0000
     9.00      8 .  000000000
    24.00      9 .  000000000000000000000000
    24.00     10 .  000000000000000000000000
    33.00     11 .  000000000000000000000000000000000
    29.00     12 .  00000000000000000000000000000
    27.00     13 .  000000000000000000000000000
    35.00     14 .  00000000000000000000000000000000000
    41.00     15 .  00000000000000000000000000000000000000000
    36.00     16 .  000000000000000000000000000000000000
    24.00     17 .  000000000000000000000000
    27.00     18 .  000000000000000000000000000
    15.00     19 .  000000000000000
    14.00     20 .  00000000000000
    17.00     21 .  00000000000000000

 Stem width:      1.00
 Each leaf:       1 case(s)
```

22 subjects are outliers with values of 3 or less.

Figure 3.14 Recode into Different Variables Dialogue Box.

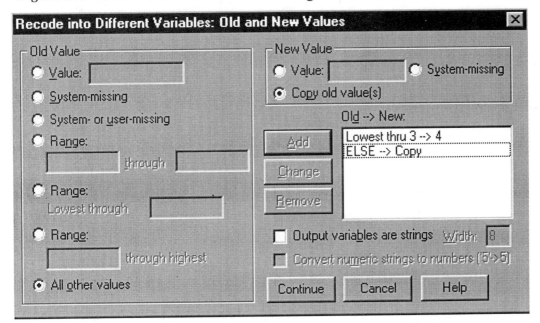

Figure 3.15 Recode Old and New Values Dialogue Box.

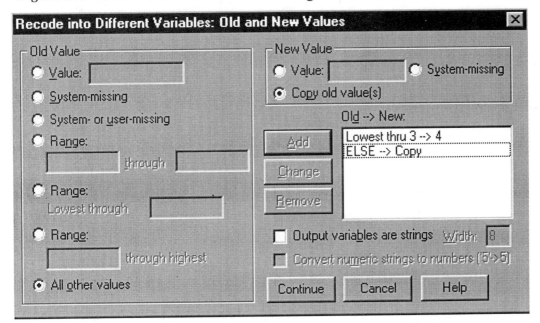

Normality, Linearity, and Homoscedasticity

The **Explore** procedure also provides several options for examining normality and is usually conducted after addressing outliers. To conduct this procedure using the DV (*rincom_3*) and IV (*sat-job2*), return to the previous directions for Explore. Within the **Explore Statistics** Dialogue Box, be sure to check **Descriptives**. In the **Explore Plots** Dialogue Box, be sure to check **Histograms** and **Normality plots with tests**. These are most helpful in examining normality. Descriptive statistics (see Figure 3.16) present skewness and kurtosis values, which also imply negative distributions. Typically, skewness and kurtosis values should lie between +1 and –1. Histograms (see Figure 3.17) display moderate, negatively skewed distributions for both groups. The normal

Q-Q plots support this finding as the observed values deviate somewhat from the straight line (see Figure 3.18). Tests of normality were also calculated. Specifically, the Kolmogorov-Smirnov test (see Figure 3.19) significantly rejects the hypothesis of normality of income for the populations of both groups. Thus, the variable of *rincom_3* must be transformed again. To decrease the moderate negative skewness, the transformation procedure will reflect and take the square root of the variable. Steps for such transformation follow:

Transform
>   **Compute**

Figure 3.16 Descriptive Statistics for Income (*rincom_2*) by Job Satisfaction (*satjob2*).

Descriptives

	Job Satisfaction			Statistic	Std. Error
RINCOM_3	Very satisfied	Mean		14.6166	.2475
		95% Confidence Interval for Mean	Lower Bound	14.1297	
			Upper Bound	15.1035	
		5% Trimmed Mean		14.8518	
		Median		15.0000	
		Variance		19.167	
		Std. Deviation		4.3780	
		Minimum		4.00	
		Maximum		21.00	
		Range		17.00	
		Interquartile Range		6.0000	
		Skewness		-.739	.138
		Kurtosis		.078	.275
	Not very satisfied	Mean		13.2972	.2238
		95% Confidence Interval for Mean	Lower Bound	12.8573	
			Upper Bound	13.7372	
		5% Trimmed Mean		13.3938	
		Median		14.0000	
		Variance		19.881	
		Std. Deviation		4.4588	
		Minimum		4.00	
		Maximum		21.00	
		Range		17.00	
		Interquartile Range		5.5000	
		Skewness		-.395	.122
		Kurtosis		-.404	.244

For a normal distribution, kurtosis and skewness values will be close to zero but can range between −1 and +1.

Figure 3.17 Histograms for Income (*rincom_3*) by Job Satisfaction (*satjob2*).

Histogram

For SATJOB2= Very satisfied

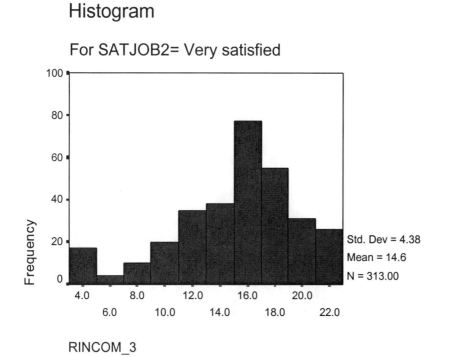

Std. Dev = 4.38
Mean = 14.6
N = 313.00

RINCOM_3

Histogram

For SATJOB2= Not very satisfied

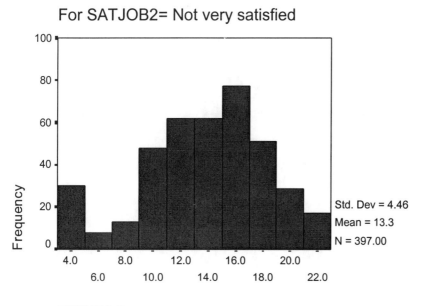

Std. Dev = 4.46
Mean = 13.3
N = 397.00

RINCOM_3

Figure 3.18 Normal Q-Q Plots for Income (*rincom_3*) by Job Satisfaction (*satjob2*).

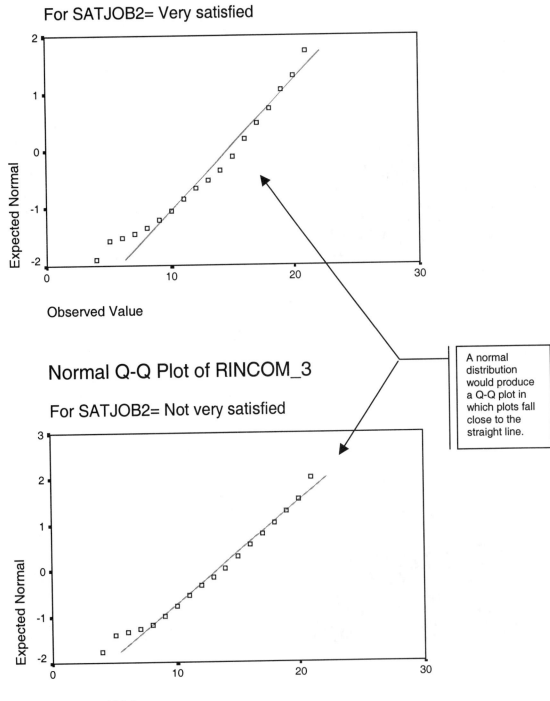

Figure 3.19 Tests for Normality for Income (*rincom-3*) by Job Satisfaction (*satjob2*).

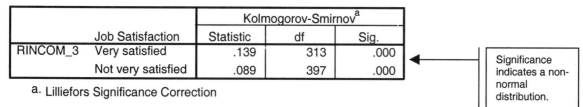

Tests of Normality

	Job Satisfaction	Kolmogorov-Smirnov[a]		
		Statistic	df	Sig.
RINCOM_3	Very satisfied	.139	313	.000
	Not very satisfied	.089	397	.000

a. Lilliefors Significance Correction

> Significance indicates a non-normal distribution.

Compute Variable Dialogue Box (see Figure 3.20)

Within the Compute Variable Dialogue Box, create a name for the new variable you are creating. For our example, we have used *rincom_4*. Then identify the appropriate function to be used in the transformation and move it to the Numeric Expression box. Since the example calls for reflect with square root, the equation to apply is: NewVar=SQRT(K-OldVar) in which K is the largest score of the OldVar plus one. For the *rincom_3*, the largest value is 21; thus K=22. Once the function has been inserted into the Numeric Expression box, the remaining parts of the transformation equation must be inserted creating the expression: *rincom_4*=SQRT(22-*rincom_3*).

Figure 3.20 Compute Variable Dialogue Box.

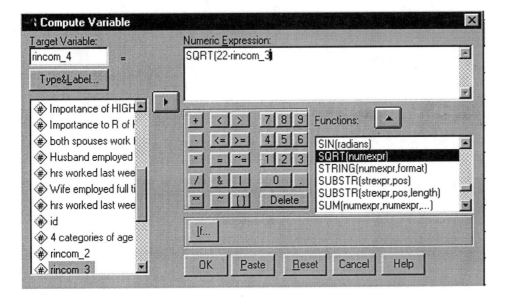

Once data has been transformed, examination of normality should be conducted again using the **Explore** procedure. Although the Kolmogorov-Smirnov test is still significant, the skewness and kurtosis values (see Figure 3.21) are much closer to zero. In addition, histograms (see Figure 3.22) and normal Q-Q plots (see Figure 3.23) reveal distributions for both groups that are much more normal. Consequently, we will assume the transformation was successful.

Since our research question involves comparing groups on a single quantitative variable, linearity cannot be examined. However, homoscedasticity, also known as *homogeneity of variance* when comparing groups, can be assessed by determining if the variability for the DV (*rincom_4*) is about the same within each category of the IV (*satjob2*). This can be completed when conducting the group comparison analyses (e.g., *t*-test, ANOVA). Within these statistical procedures, Levene's test for equal variances is automatically calculated. Figure 3.24 presents output from an independent *t*-test conducted with our example variables. The Levene's statistic is 0.139 with a *p* value of 0.709. Thus, the hypothesis for equal variances is not rejected, which indicates that variances are fairly equivalent between the groups.

Figure 3.21 Descriptive Statistics for Income (*rincom_4*) by Job Satisfaction (*satjob2*).

Descriptives

	Job Satisfaction			Statistic	Std. Error
RINCOM_4	Very satisfied	Mean		2.5877	4.694E-02
		95% Confidence Interval for Mean	Lower Bound	2.4953	
			Upper Bound	2.6800	
		5% Trimmed Mean		2.5839	
		Median		2.6458	
		Variance		.690	
		Std. Deviation		.8304	
		Minimum		1.00	
		Maximum		4.24	
		Range		3.24	
		Interquartile Range		1.1623	
		Skewness		-.013	.138
		Kurtosis		-.362	.275
	Not very satisfied	Mean		2.8384	4.039E-02
		95% Confidence Interval for Mean	Lower Bound	2.7590	
			Upper Bound	2.9178	
		5% Trimmed Mean		2.8593	
		Median		2.8284	
		Variance		.648	
		Std. Deviation		.8048	
		Minimum		1.00	
		Maximum		4.24	
		Range		3.24	
		Interquartile Range		.9409	
		Skewness		-.291	.122
		Kurtosis		-.265	.244

Skewness and kurtosis values are closer to zero, indicating a more normal distribution.

Figure 3.22 Histograms for Income (*rincom_4*) by Job Satisfaction (*satjob2*).

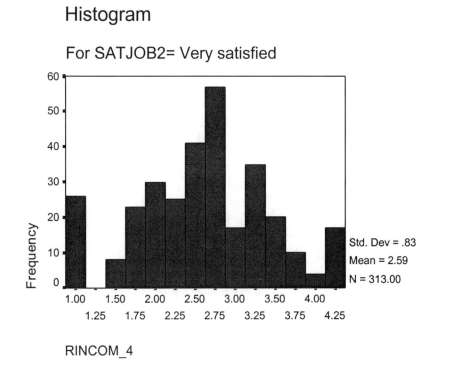

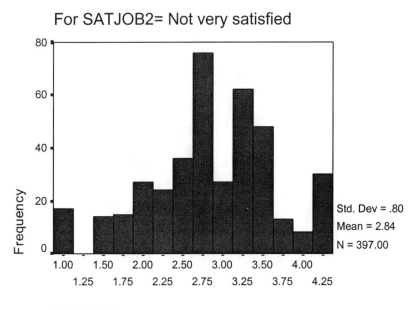

Figure 3.23 Normal Q-Q Plots for Income (*rincom_4*) by Job Satisfaction (*satjob2*).

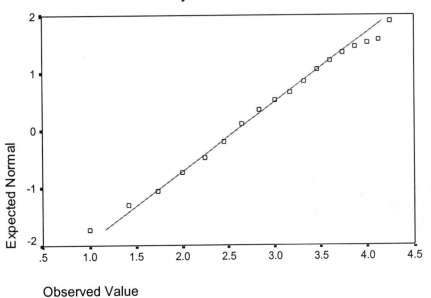

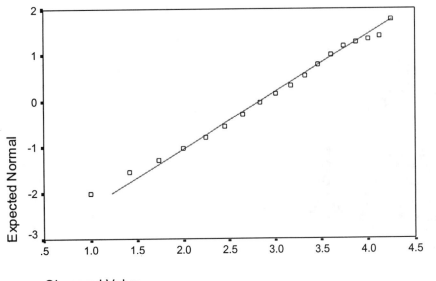

Figure 3.24 Levene's Test for Equality of Variances.

Independent Samples Test

		Levene's Test for Equality of Variances		t-test for Equality of Means						
									95% Confidence Interval of the Difference	
		F	Sig.	t	df	Sig. (2-tailed)	Mean Difference	Std. Error Difference	Lower	Upper
RINCOM_	Equal variance assumed	.139	.709	-4.065	708	.000	-.2508	6.169E-02	-.3719	-.1297
	Equal variance not assumed			-4.050	659.997	.000	-.2508	6.169E-02	-.3724	-.1292

Nonsignificant value indicates homogeneity of variance.

Univariate Example of Ungrouped Data

In this example, we seek to investigate the degree to which the variables of years of education (*educ*), age (*age*), and hours worked weekly (*hrs1*) predict income levels (*rincom_4*).

Missing Data and Outliers

Missing data is analyzed for each of the four variables using the methods described in the previous example. However, the identification of outliers requires a different method since several variables are in question. Although this is not a multivariate example, analysis for multivariate outliers will occur in order to examine the variables together with respect to outliers. The most common method is calculating Mahalanobis distance within the **Regression** procedure. To calculate Mahalanobis distance, complete the following steps to conduct the regression:

> **Analyze**
> > **Regression**
> > > **Linear**

Linear Regression Dialogue Box (see Figure 3.25)

Identify all four quantitative variables to be analyzed and move them to the Independent(s) Box. Utilize a case number or ID variable for the Dependent, since this procedure is not influenced by the DV. Next, click the **Save** box. All other procedures regarding regression will be discussed further in Chapter 7.

Figure 3.25 Linear Regression Dialogue Box.

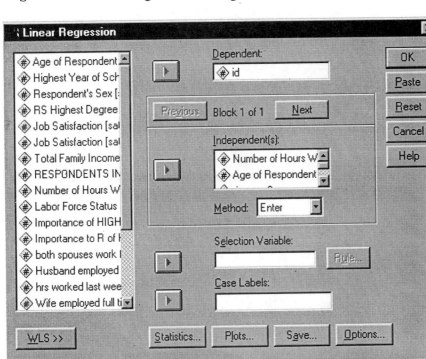

Figure 3.26 Linear Regression Save Dialogue Box.

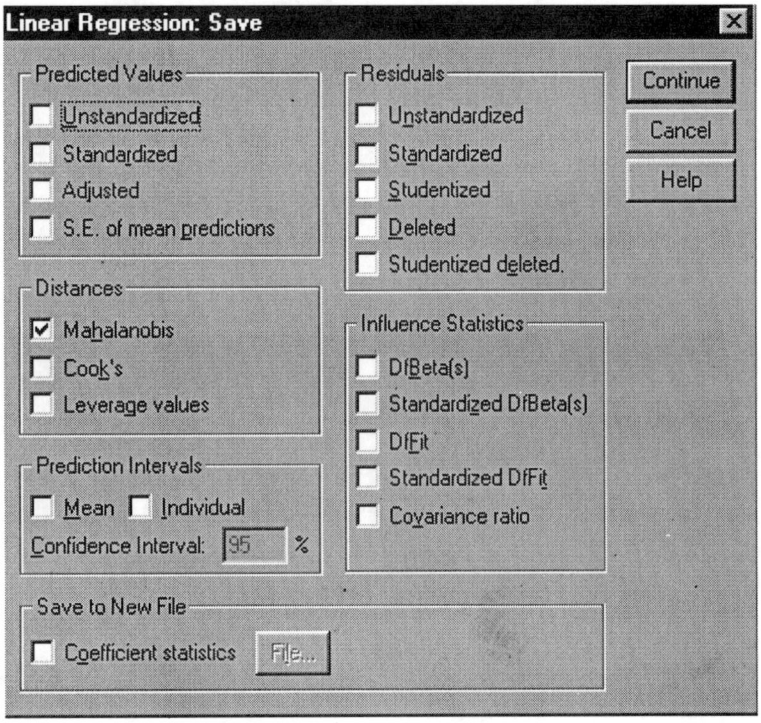

Linear Regression Save Dialogue Box (see Figure 3.26)

Once in this box, check **Mahalanobis** under Distances. Although this procedure does not produce output that is especially helpful in identifying multivariate outliers, a new variable (*mah_1*) is created for Mahalanobis distances, which is tested using chi-square (χ^2) criteria. Outliers are indicated by chi square values that are significant at *p*<.001 with the respective degrees of freedom. The number of variables being examined for outliers is used as the degrees of freedom. To determine the critical value for χ^2, one must utilize a table of critical values for chi square, available in most introductory statistics textbooks. For our example, the critical value of χ^2 at *p*<.001 and *df*=4 is 18.467. Consequently, cases with a Mahalanobis distance greater than 18.467 are considered multivariate outliers for the variables of *rincom_4*, *age*, *educ* and *hrs1*. Identification of the outlying cases can now be easily achieved using the **Explore** procedure for the variable *mah_1*. Within **Explore**, all that is necessary is checking **Outliers** within the **Statistics** Dialogue Box, as previously demonstrated. The Table of Extreme Values (see Figure 3.27) generated lists the five highest and lowest values for *mah_1*. Four cases (#222, #24, #616, #208) are identified as outliers as they exceed 18.467. With only four multivariate outliers in the entire data set, these cases are most appropriately deleted.

Figure 3.27 Extreme Values for Mahalanobis Distance.

Extreme Values

			Case Number	Value
Mahalanobis Distance	Highest	1	222	29.93848
		2	24	22.03483
		3	616	18.71248
		4	208	18.52947
		5	729	18.15252
	Lowest	1	292	.23228
		2	146	.24275
		3	550	.30986
		4	126	.32204
		5	443	.33112

Only cases (222, 24, 616, 208) with values that exceed the critical value of chi square are considered outliers.

Normality, Linearity, and Homoscedasticity

Since our example includes several quantitative variables, univariate normality should be examined for each individual variable; however, multivariate normality will need to be assessed as well. To assess univariate normality, the **Explore** procedure is conducted for each of these variables. Histograms, normal Q-Q plots, and descriptive statistics reveal the following: *age* has moderate, positive skewness; *hrs1* and *educ* are fairly normal but very peaked. *Age* will be transformed into *age_2* by taking the square root of *age*. *Hrs1* and *educ* will not be transformed at this point.

The next step is to analyze for multivariate normality and linearity. The most common method evaluating multivariate normality is creating scatterplots of all variables in relation to one another. If variable combinations are normal, scatterplots will display elliptical shapes. To create scatterplots of the four variables, open the following menus:

Graphs
      Scatter...

Scatterplot Dialogue Box (see Figure 3.28)

Since scatterplots will be created for several combinations of variables, click **Matrix**, then **Define**.

Figure 3.28 Scatterplot Dialogue Box.

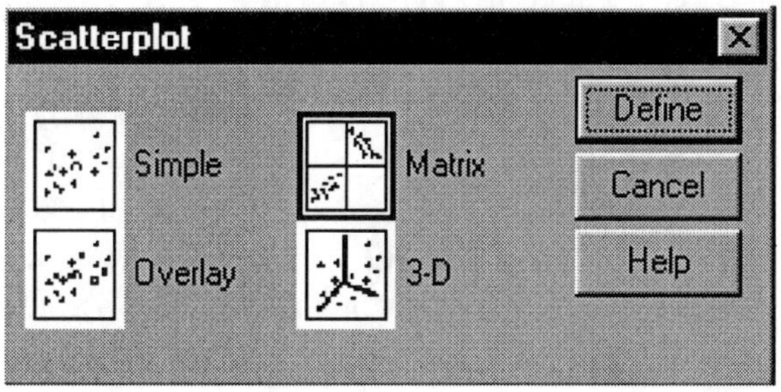

Scatterplot Matrix Dialogue Box (see Figure 3.29)

Identify the variables to be analyzed and move them to Matrix Variables.

Figure 3.29 Scatterplot Matrix Dialogue Box.

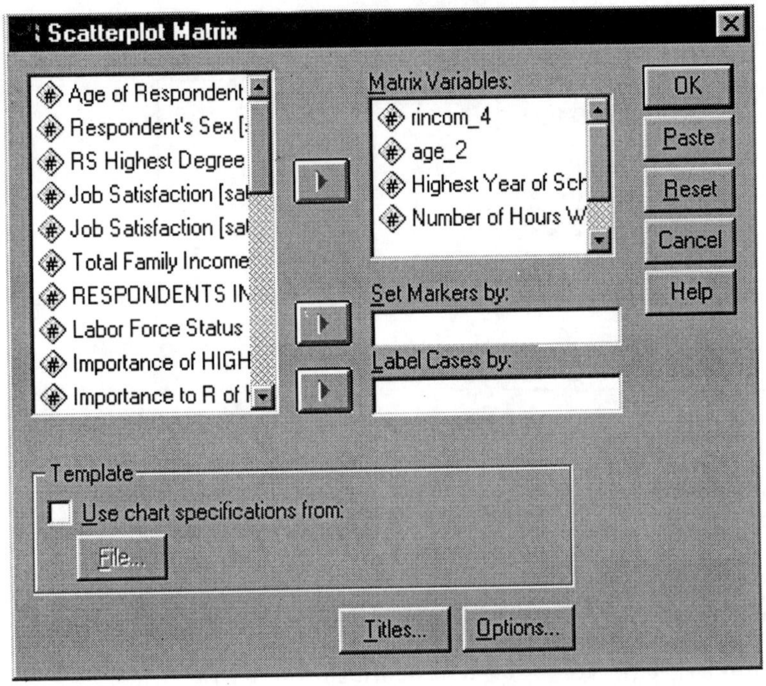

The output (see Figure 3.30) for our example displays nonelliptical shapes for all combinations, which implies failure of normality and linearity. For such a situation, two options are available: (1) re-check univariate normality for each variable; and (2) only utilize variables for univariate analyses.

Figure 3.30 Scatterplot matrix of *rincom_4, age_2, educ* and *hrs1*.

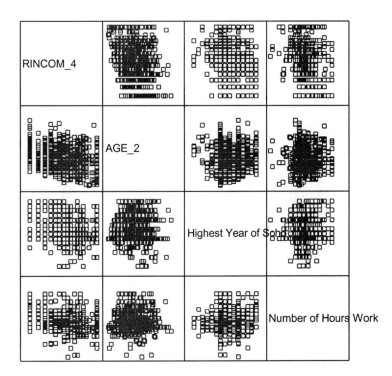

Since the use of bivariate scatterplots is fairly subjective in examining linearity, we recommend a more sophisticated method that compares standardized residuals to the predicted values of the DV. This method also provides some information regarding homoscedasticity. To create the residual plot for these variables, open the following menus:

Regression
        Linear

Linear Regression Dialogue Box (see Figure 3.31)

Move *rincom_4* to the Dependent Box. Select the IVs and move to Independent(s) Box. Then click **Plot**.

Linear Regression Plot Dialogue Box (see Figure 3.32)

Within this menu, select the standardized residuals (ZRESID) for the *Y*-axis. Select the standardized predicted values (ZPRED) for the *X*-axis. When the assumptions of linearity, normality, and homoscedasticity are met, residuals will create an approximate rectangular distribution with a concentration of scores along the center. Figure 3.33 displays fairly consistent scores throughout the plot with concentration in the center. When assumptions are not met, residuals may be clustered on the top or bottom of the plot (non-normality), may be curved (nonlinearity), or may be clustered on the right or left side (heteroscedasticity). Since such extreme clustering is not displayed, we will conclude that the assumptions of normality, linearity, and homoscedasticity are met for these variables.

Figure 3.31 Linear Regression Dialogue Box.

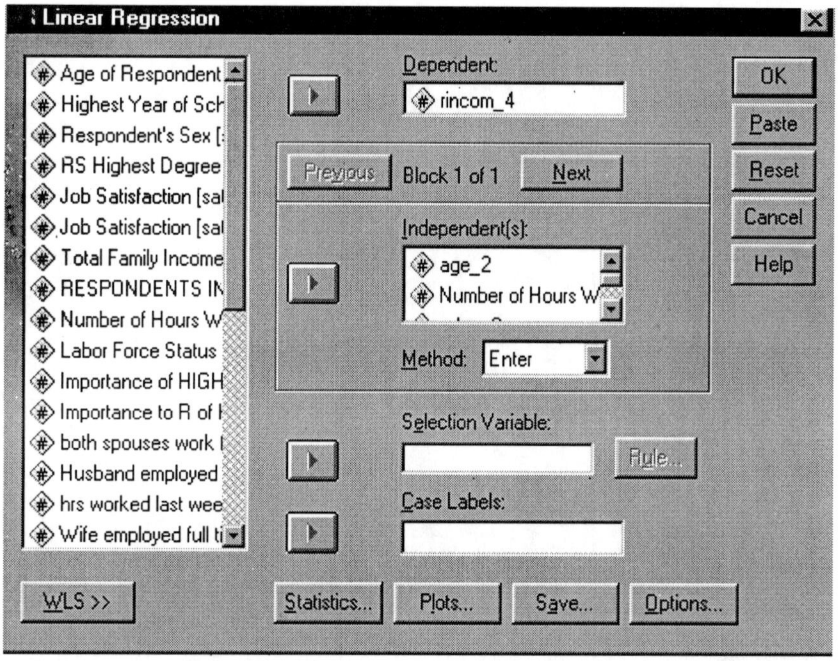

Figure 3.32 Linear Regression Plots Dialogue Box.

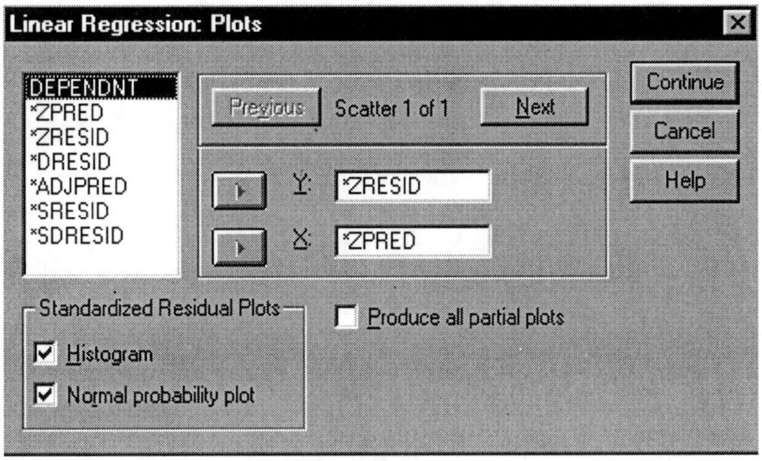

Figure 3.33 Scatterplot of Standardized Predicted Values by Standardized Residuals.

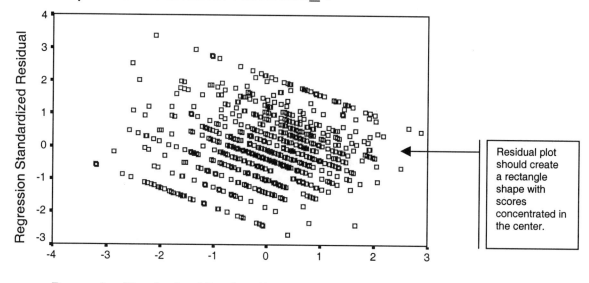

Scatterplot

Dependent Variable: RINCOM_4

Residual plot should create a rectangle shape with scores concentrated in the center.

SECTION 3.8   USING SPSS TO EXAMINE GROUPED DATA FOR MULTIVARIATE ANALYSIS

The following example describes the process of examining missing values, outliers, normality, linearity, and homoscedasticity for grouped multivariate data. A non-grouped multivariate scenario would typically follow the univariate non-grouped example previously presented. For our example, we will again utilize the data set *gssft.sav*, as we are interested in investigating group differences (*satjob2*) in *rincom_4, age_2,* and *educ*.

Missing Data and Outliers

Missing data would be assessed for each variable. Multivariate outliers would be examined using Mahalanobis distances within **Regression** for *each group*. Please refer to the previous example on methods for conducting this procedure. Tables of Extreme Values (see Figure 3.34) present chi square values for each possible outlier within each group. The critical value at *p*<.001 for chi square is again 18.467 with *df*=4. Thus, the satisfied group has two outliers (#222, #24), while the unsatisfied group has three outliers (#545, #551, #575). Notice the particular cases identified as outliers are slightly different from the previous example where data was ungrouped. Identified outliers will be deleted from further analysis.

Figure 3.34 Tables of Extreme Values.

Extreme Values for *Satjob2=1*

Extreme Values

			Case Number	Value	
Mahalanobis Distance	Highest	1	222	28.50217	Cases 222 & 24 are outliers since their values exceed chi square critical.
		2	24	19.54314	
		3	616	16.06943	
		4	208	15.98786	
		5	661	14.89095	
	Lowest	1	126	.21779	
		2	550	.22604	
		3	741	.25174	
		4	146	.25229	
		5	331	.27349	

Extreme Values for *Satjob2=2*

Extreme Values

			Case Number	Value	
Mahalanobis Distance	Highest	1	545	19.17850	Cases 545, 551 & 575 are outliers since their values exceed chi square critical.
		2	551	18.76909	
		3	575	18.49591	
		4	427	18.23602	
		5	729	15.20968	
	Lowest	1	292	.15434	
		2	619	.37817	
		3	637	.38071	
		4	677	.40415	
		5	527	.46097	

Normality, Linearity, and Homoscedasticity

Because groups are being compared, assumptions of normality, linearity and homoscedasticity must be examined for all the quantitative variables together by each group. Prior to multivariate examination, univariate examination should take place for each variable within each group. Although these variables have been assessed for assumptions in the previous example, examination was with ungrouped data. Consequently, assessment of normality and homoscedasticity will need to be conducted for each variable within each group. Using the **Explore** procedure provides the histograms, tests of normality, descriptive statistics, and normal Q-Q plots, all of which indicate that the four quantitative variables are fairly normal. Homoscedasticity (homogeneity of variance) will be assessed using the Levene's Test within the *t*-test of independent samples. These results indicate equality of variance for each variable between groups.

Figure 3.35 Scatterplot Matrices of *rincom_4, age_2, educ,* and *hrs1* by *satjob2.*

satjob2=1

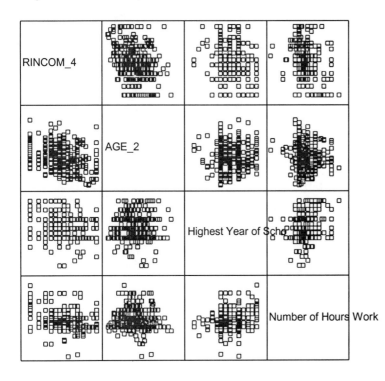

satjob2=2

Multivariate normality, linearity, and homoscedasticity can now be assessed. Multivariate normality and linearity are examined with a matrix of scatterplots for each group (see Figure 3.35). The results are quite similar to the previously produced scatterplot matrix of the same variables but with ungrouped data (see Figure 3.30). Although some plots display enlarged oval shapes, multivariate normality and linearity are questionable.

Homogeneity of variance-covariance matrices is evaluated within MANOVA by calculating Box's Test of Equality of Covariance. To do so, open the following menus:

Analyze
> **General Linear Model**
>> **Multivariate...**

Multivariate Dialogue Box (see Figure 3.36)

Move the DVs into the Dependent Variables box. Identify the IV and move to the Fixed Factor(s) box. Once variables have been identified, click **Options**.

Figure 3.36 Multivariate Dialogue Box.

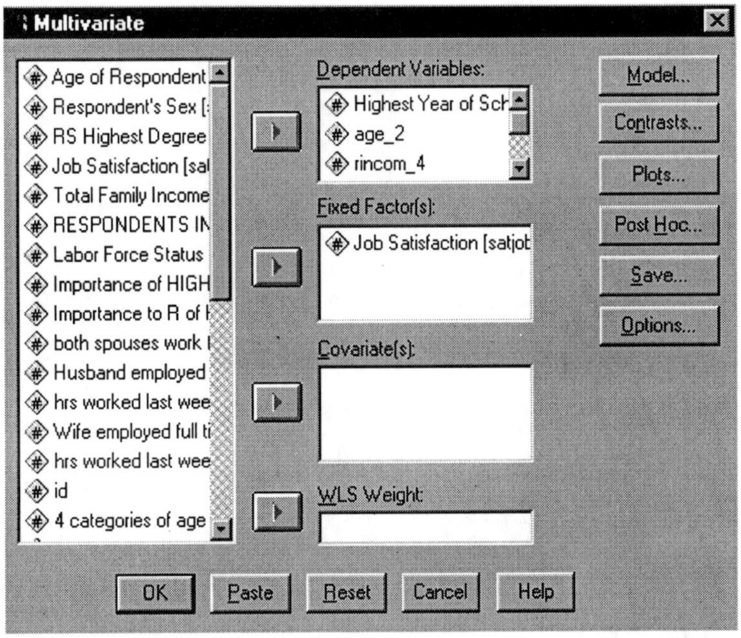

Multivariate Options Dialogue Box (see Figure 3.37)

Check Homogeneity Tests.

Since tests of homogeneity of variance-covariance matrices are quite strict, a more stringent critical value of .025 or .01 is often used rather than .05. Thus, when interpreting the results from the Box's Test (see Figure 3.38), the probability value was calculated at .044, which at the .025 level of significance would lead us to conclude that the covariance matrices for the dependent variable are fairly equivalent.

Figure 3.37 Multivariate Options Dialogue Box.

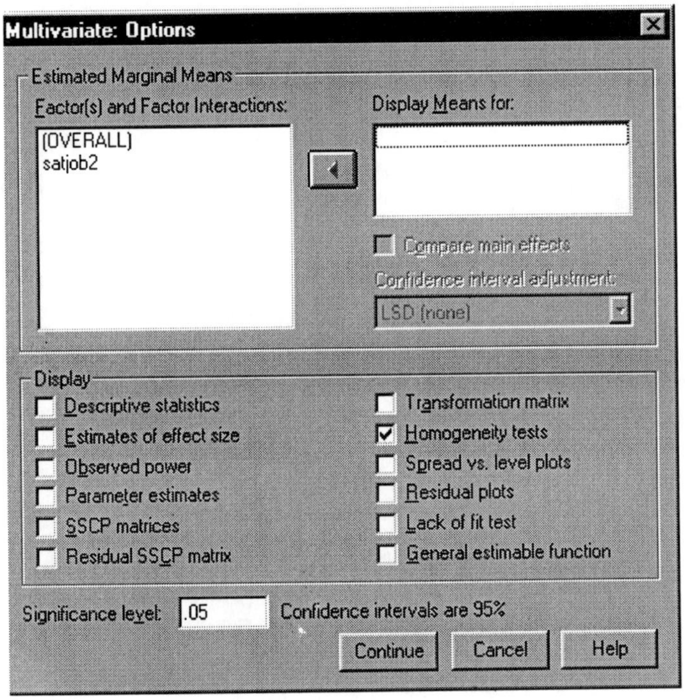

Figure 3.38 Box's Test of Equality of Covariance.

Box's Test of Equality of Covariance Matrices[a]

Box's M	18.852
F	1.873
df1	10
df2	2053031
Sig.	.044

Tests the null hypothesis that the observed covariance
matrices of the dependent variables are equal across groups.

a. Design: Intercept+SATJOB2

> Significance is NOT found at .025 or .01. Equality of covariance is concluded.

Summary

Screening data for missing data, outliers, and the assumptions of normality, linearity, and homoscedasticity is an important task prior to conducting statistical analyses. If data are not screened, conclusions drawn from statistical results may be erroneous. Figure 3.39 presents the steps for univariate and multivariate examination of grouped data, while Figure 3.40 presents the process of univariate and multivariate examination of ungrouped data.

Figure 3.39 Steps for Screening Grouped Data.

Examination & Process	SPSS Procedure	Technique for "Fixing"
Missing Data • Examine missing data for each variable.	• Run **Frequency** for categorical variables 1. ⌐ **Analyze...Descriptive Statistics...Frequencies** 2. Move IVs to Variables box. 3. ⌐ **OK.** • Run **Descriptive** for quantitative variables 1. ⌐ **Analyze...Descriptive Statistics...Frequencies** 2. Move quantitative variables to Variables box. 3. ⌐ **Options.** 4. Check **Mean, Standard Deviation, Kurtosis,** and **Skewness.** 5. ⌐ **Continue.** 6. ⌐ **OK.**	• Less than 5% missing cases→use Listwise default. • 5-15% missing cases→replace missing values with estimated value by conducting **Transform.** 1. ⌐ **Transform...Replace Missing Values.** 2. Identify variable to be transformed and move to New Variable box. 3. Identify new variable name (this occurs automatically). 4. Select method of replacement (e.g., mean, median). 5. ⌐ **OK.** • More than 15% missing cases→delete variable from analysis.
Univariate Outliers • Examine outliers for quantitative variable within each group.	• Run **Explore.** 1. ⌐ **Analyze...Descriptive Statistics...Explore** 2. Move DVs to Dependent Variable box. 3. Move IVs to Factor List box. 4. ⌐ **Statistics.** 5. Check **Descriptives** and **Outliers.** 6. ⌐ **Continue.** 7. ⌐ **Plots.** 8. Check **Boxplots,** and **Stem-and-leaf.** 9. ⌐ **Continue.** 10. ⌐ **OK.**	• More than 90-10 split between categories→ delete variable from analysis. • Small # of outliers→delete severe outliers. • Small to moderate # of outliers→replace with accepted minimum or maximum value by conducting **Recode.** 1. ⌐ **Transform...Recode...Into a different variable.** 2. Select variable to be transformed and move to Input→Output Variable box. 3. Type in new variable name under Output Variable Name box. 4. ⌐ **Change.** 5. ⌐ **Old and New Values.** 6. Identify value to be changed under Old Value. 7. Under New Value, identify appropriate new value. 8. ⌐ **Add.** 9. After all necessary values have been recoded, check **All Other Values** under Old Value. 10. Check **Copy Old Value(s)** under New Value. 11. ⌐ **Add.** 12. ⌐ **Continue.** 13. ⌐ **OK.**
Univariate Normality • Examine normality for quantitative variable within each group.	• Run **Explore.** 1. ⌐ **Analyze...Descriptive Statistics...Explore** 2. Move DVs to Dependent Variable box. 3. Move IVs to Factor List box. 4. ⌐ **Statistics.** 5. Check **Descriptives.** 6. ⌐ **Continue.** 7. ⌐ **Plots.** 8. Check **Histograms** and **Normality plots with tests.** 9. ⌐ **Continue.** 10. ⌐ **OK.**	• Transform variable accordingly (see Figure 3.3) using **Compute.** 1. ⌐ **Transform...Compute.** 2. Under Target, identify new variable name. 3. Identify appropriate function and move to Numeric Expression(s) box. 4. Identify variable to be transformed and move within the function equation (in place of ?). 5. ⌐ **OK.**

Figure 3.39 Steps for Screening Grouped Data. (*Continued*)

Univariate Homoscedasticity • Examine homogeneity of variances between/among groups	• Conduct t-test or ANOVA using **Compare Means** to run Levene's Test.	• *p* value is significant at .05 → reevaluate univariate normality and consider transformations.
Multivariate Outliers • Examine quantitative variables together by group for outliers.	• Conduct **Regression** to test Mahalanobis' Distance. 1. ⏏ **Analyze...Regression...Linear.** 2. Identify a variable that serves as a case number and move to Dependent Variable box. 3. Identify all appropriate quantitative variables and move to Independent(s) box. 4. ⏏ **Save.** 5. Check **Mahalanobis'.** 6. ⏏ **Continue.** 7. ⏏ **OK.** 8. Determine chi square χ^2 critical value at *p*<.001. • Conduct **Explore** to test outliers for Mahalanobis chi square χ^2. 1. ⏏ **Analyze...Descriptive Statistics...Explore** 2. Move *mah_1* to Dependent Variable box. 3. Leave Factor box empty. 4. ⏏ **Statistics.** 5. Check **Outliers.** 6. ⏏ **Continue.** 7. ⏏ **OK.**	• Delete outliers for subjects when χ^2 exceeds critical χ^2 at *p*<.001.
Multivariate Normality, Linearity • Examine normality and linearity of variable combinations by group.	• Create **Scatterplot Matrix** 1. ⏏ **Graphs...Scatter** 2. ⏏ **Matrix.** 3. ⏏ **Define.** 4. Identify appropriate quantitative variables and move to Matrix Variables. 5. ⏏ **OK.**	• Scatterplot shapes are not close to elliptical shapes → reevaluate univariate normality and consider transformations.
Multivariate Homogeneity of Variance-Covariance • Examine homogeneity of variance-covariance between/among groups.	• Conduct MANOVA using **Multivariate** to run homogeneity tests (see Chapter 6 for SPSS steps).	• *p* value is significant at .025 or .01 → reevaluate univariate normality and consider transformations.

See the next page for screening ungrouped data.

Figure 3.40 Steps for Screening Ungrouped Data.

Examination & Process	SPSS Procedure	Technique for "Fixing"
Missing Data • Examine missing data for each variable.	• Run **Descriptive** for quantitative variables. 1. 🖱 **Analyze...Descriptive Statistics...Frequencies** 2. Move quantitative variables to Variables box. 3. 🖱 **Options**. 4. Check **Mean, Standard Deviation, Kurtosis,** and **Skewness**. 5. 🖱 **Continue**. 6. 🖱 **OK**.	• Less than 5% missing cases→use Listwise default. • 5-15% massing cases→replace missing values with estimated value by conducting **Transform**. 1. 🖱 **Transform...Replace Missing Values**. 2. Identify variable to be transformed and move to New Variable box. 3. Identify new variable name (this occurs automatically). 4. Select method of replacement (e.g., mean, median). 5. 🖱 **OK**. • More than 15% missing cases→delete variable from analysis.
Univariate Outliers • Examine outliers for each quantitative variable.	• Run **Explore**. 1. 🖱 **Analyze...Descriptive Statistics...Explore** 2. Move DVs to Dependent Variable box. 3. Move IVs to Factor List box 4. 🖱 **Statistics**. 5. Check **Descriptives** and **Outliers**. 6. 🖱 **Continue**. 7. 🖱 **Plots**. 8. Check **Boxplots,** and **Stem-and-leaf**. 9. 🖱 **Continue**. 10. 🖱 **OK**.	• Small # of outliers→delete severe outliers. • Small to moderate # of outliers→replace with accepted minimum or maximum value by conducting **Recode**. 1. 🖱 **Transform...Recode...Into a different variable**. 2. Select variable to be transformed and move to Input→Output Variable box. 3. Type in new variable name under Output Variable Name box. 4. 🖱 **Change**. 5. 🖱 **Old and New Values**. 6. Identify value to be changed under Old Value. 7. Under New Value, identify appropriate new value. 8. 🖱 **Add**. 9. After all necessary values have been recoded, check **All Other Values** under Old Value. 10. Check **Copy Old Value(s)** under New Value. 11. 🖱 **Add**. 12. 🖱 **Continue**. 13. 🖱 **OK**.
Univariate Normality • Examine normality for each quantitative variable.	• Run **Explore**. 1. 🖱 **Analyze...Descriptive Statistics...Explore** 2. Move DVs to Dependent Variable box. 3. Move IVs to Factor List box. 4. 🖱 **Statistics**. 5. Check **Descriptives**. 6. 🖱 **Continue**. 7. 🖱 **Plots**. 8. Check **Histograms** and **Normality plots with tests**. 9. 🖱 **Continue**. 10. 🖱 **OK**.	• Transform variable accordingly (see Figure 3.3) using **Compute**. 1. 🖱 **Transform...Compute**. 2. Under Target, identify new variable name. 3. Identify appropriate function and move to Numeric Expression(s) box. 4. Identify variable to be transformed and move within the function equation (in place of ?). 5. 🖱 **OK**.

Figure 3.40 Steps for Screening Ungrouped Data. (*Continued*)

Multivariate Outliers • Examine quantitative variables together for outliers.	• Conduct **Regression** to test Mahalanobis' Distance. 1. ⁀ᵗ **Analyze...Regression...Linear.** 2. Identify a variable that serves as a case number and move to Dependent Variable box. 3. Identify all appropriate quantitative variables and move to Independent(s) box. 4. ⁀ᵗ **Save.** 5. Check **Mahalanobis'.** 6. ⁀ᵗ **Continue.** 7. ⁀ᵗ **OK.** 8. Determine chi square χ^2 critical value at $p<.001$. • Conduct **Explore** to test outliers for Mahalanobis chi square χ^2. 1. ⁀ᵗ **Analyze...Descriptive Statistics...Explore** 2. Move *mah_1* to Dependent Variable box. 3. Leave Factor box empty. 4. ⁀ᵗ **Statistics.** 5. Check **Outliers.** 6. ⁀ᵗ **Continue.** 7. ⁀ᵗ **OK.**	• Delete outliers for subjects when χ^2 exceeds critical χ^2 at $p<.001$.
Multivariate Normality, Linearity • Examine normality and linearity of variable combinations.	• Create **Scatterplot Matrix** 1. ⁀ᵗ **Graphs...Scatter** 2. ⁀ᵗ **Matrix.** 3. ⁀ᵗ **Define.** 4. Identify appropriate quantitative variables and move to Matrix Variables. 5. ⁀ᵗ **OK.**	• Scatterplot shapes are not close to elliptical shapes→reevaluate univariate normality and consider transformations.
Multivariate Homogeneity of Variance-Covariance • Examine standardized residuals to predicted values.	• Create residual plot using **Regression**. 1. ⁀ᵗ **Analyze...Regression...Linear** 2. Move DV to Dependent Variable box. 3. Move IVs to Independent(s) Variable box. 4. ⁀ᵗ **Plot.** 5. Select ZRESID for y-axis. 6. Select ZPRED for x-axis. 7. ⁀ᵗ **Continue.** 8. ⁀ᵗ **OK.**	• Residuals are clustered at the top, bottom, left, or right area in plot→ reevaluate univariate normality and consider transformations.

Exercises for Chapter 3

This exercise utilizes the data set *schools.sav*, which can be downloaded at the SPSS Web site. Open the URL: **www.spss.com/tech/DataSets.html** in your Web browser. Scroll down until you see "Data Used in SPSS Guide to Data Analysis—8.0 and 9.0" and click on the link "dataset.exe." When the "Save As" dialogue appears, select the appropriate folder and save the file. Preferably, this should be a folder created in the SPSS folder of your hard drive for this purpose. Once the file is saved, double-click the "dataset.exe" file to extract the data sets to the folder.[3]

---

[3]These directions have been tested on a number of computer platforms and have worked. However, it is possible that some platforms are configured in such a way that adjustments will need to be made to download the data.

1. You are interested in investigating if being above or below the median income (*medloinc*) impacts ACT means (*act94*) for schools. Complete the necessary steps to examine univariate grouped data in order to respond to the questions below. Although deletions and/or transformations may be implied from your examination, all steps will examine original variables.

    a. How many subjects have missing values for *medloinc* and *act94*?

    b. Is there a severe split in frequencies between groups?

    c. What are the cutoff values for outliers in each group?

    d. Which outlying cases should be deleted for each group?

    e. Analyzing histograms, normal Q-Q plots, and tests of normality, what is your conclusion regarding normality? If a transformation is necessary, which one would you use?

    f. Do the results from Levene's Test of Equal Variances indicate homogeneity of variance? Explain.

2. Examination of the variable of *scienc93* indicates a substantial to severe positively skewed distribution. Transform this variable using the two most appropriate methods. After examining the distributions for these transformed variables, which produced the best alteration?

3. You are interested in studying predictors (*math94me, loinc93,* and *read94me*) of the % graduating in 1994 (*grad94*).

    a. Examine univariate normality for each variable. What are your conclusions about the distributions? What transformations should be conducted?

    b. After making the necessary transformations, examine multivariate outliers using Mahalanobis' distance. What cases should be deleted?

    c. After deleting the multivariate outliers, examine multivariate normality and linearity by creating a Scatterplot Matrix.

    d. Examine the variables for homoscedasticity by creating a residuals plot (standardized vs. predicted values). What are your conclusions about homoscedasticity?

CHAPTER 4

FACTORIAL ANALYSIS OF VARIANCE

Many students enrolled in an introductory statistics course find one of the most challenging topics to be univariate analysis of variance. This is probably due to the "complex" nature of the formulae and the related hand calculations. After all, compared to a simple *t*-test, the nine separate equations and subsequent calculations that make up a univariate analysis of variance seem absolutely overwhelming! The advice we provide to our students is that they should not "get lost" in the equations and/or the numbers as they proceed through the calculations; rather, they should try to remain focused on the overall purpose of the test itself. (By the way, we also provide the same advice as we progress through our discussions of multivariate statistical techniques!) With that in mind, a brief review of univariate analysis of variance (ANOVA) is undoubtedly warranted.

I. UNIVARIATE ANALYSIS OF VARIANCE

The univariate case of ANOVA is a hypothesis testing procedure that simultaneously evaluates the significance of mean differences on a dependent variable (DV) between two or more treatment conditions or groups (Agresti & Finlay, 1997). The treatment conditions or groups are defined by the various levels of the independent variable (IV), or *factor* in ANOVA terminology. We will limit our discussion here to a *one-way ANOVA*, which studies the effect that one factor has on one dependent variable. For example, we might be interested in examining mean differences in achievement test scores (DV) for various school settings (IV), namely urban, suburban, and rural. Our research question might be stated, "Are there significant differences in achievement test scores for urban, suburban, or rural schools?" A second example could involve an investigation of mean differences in student performance in a college course based on four different styles of instruction. Here, the research question would be, "Do college students enrolled in the same college course perform differently based on the type of instruction they receive?"

The null hypothesis in a one-way ANOVA states that there is no difference among the treatment conditions or groups. Using the first example above, the null hypothesis would state that there is no difference in achievement test scores between urban, suburban, or rural schools. Restating this hypothesis using proper statistical notation would give us

$$H_0: \mu_{Urban} = \mu_{Suburban} = \mu_{Rural}$$

where μ represents the various group means.

Similarly, the alternative, or research, hypothesis for this scenario says that at least one of the group or treatment means is significantly different from the others. It is not necessary to state the alternative hypothesis using proper notation, since that would require accounting for every pairwise comparison (e.g., $\mu_{Urban} = \mu_{Suburban}$, $\mu_{Urban} = \mu_{Rural}$, $\mu_{Suburban} = \mu_{Rural}$), which can become extremely cumbersome when there are more than three levels of the IV. Additionally, in order to disprove the null hypothesis, we need only to find one pair that is significantly different and, at this point, we are not concerned with which pair that might be. Thus, restating the alternative hypothesis would give us

H_1:   At least one group mean is different from the others.

Therefore, this leaves the researcher with two possible interpretations of the results of a one-way ANOVA:

- There really are no differences between the treatment conditions or groups. Any observed differences are due only to chance, or sampling error. (*Fail to reject H_o*)
- Any observed differences between the conditions or groups represent real differences in the population. (*Reject H_o*)

In order to decide between these two possible outcomes, the researcher must rely on the specific and appropriate test statistic. The test statistic for ANOVA is the *F* ratio and has the following structure:

$$F = \frac{\text{variance between subjects}}{\text{variance expected due to chance (error)}} \qquad \text{(Equation 4.1)}$$

Recall that the *F* ratio is based on *variances* as opposed to mean *differences* (Gravetter & Wallnau, 1999) because we are now dealing with more than two groups. In a research situation involving two groups, for which a *t*-test would be appropriate, it is relatively simple to determine the size of the differences between two sample means; one simply subtracts the mean of the first group from the mean of the second group. However, if a third group is added, it becomes much more difficult to "describe" the difference among the sample means. The solution is to compute the variance for the set of sample means. If the three sample means cluster closely together (i.e., small differences), the variance will be small; if the means are spread out (i.e., large differences), the variances will be larger. Analysis of variance actually takes the total variability and analyzes, or *partitions*, it into two separate components. The variance calculated in the numerator in the equation for the *F* ratio, as indicated above, provides a singular value that describes the differences between the three sample means. Thus, the numerator is referred to as the **between-groups variability**. The variance in the denominator of the *F* ratio, often referred to as the **error variance** or **within-groups variability**, provides a measure of the variance that the researcher could expect due simply to chance factors, such as sampling error.

There are two possible causes or explanations for the differences that occur between groups or treatments (Gravetter & Wallnau, 1999). They are:

(1) The differences due to the *treatment effect*. In this case, the various treatments, or group characteristics, actually cause the differences. For example, students in the three different school settings (i.e., urban, suburban, and rural) performed differently on their achievement tests due to the setting in which they received instruction. Therefore, changing the school locations results in an increase or decrease in achievement test performance.

(2) The differences occur *simply due to chance*. Since all possible random samples from a population consist of different individuals with different scores, and even if there is no treatment effect at all, you would still expect to see some differences – most likely *random* differences – in the scores.

Therefore, when the between-groups variability is measured, we are actually measuring differences due to the effect of the treatment or to chance. In contrast, the within-groups variability is caused only by chance differences. Within each treatment or group, all subjects in that sample have been exposed to the same treatment or share the same characteristic. The researcher has not done anything that would result in different scores. Obviously, individuals within the same group will likely have different scores, but the reader should realize that these differences are due to random effects. The within-groups variability provides a measure of the differences that exist when there is no treatment effect that could have caused those differences (Gravetter & Wallnau, 1999).

If we substitute these explanations for variability into Equation 4.1, expressing each component in terms of its respective sources, we obtain the following equation:

$$F = \frac{\text{treatment effect} + \text{differences due to chance}}{\text{differences due to chance}} \qquad \text{(Equation 4.2)}$$

Thus, the F ratio becomes a comparison of the two partitioned components of the total variability. The value of the ratio will be used to determine whether differences are large enough to be attributed to a treatment effect, or if they are due simply to chance effects. When the treatment has no effect, any differences between treatment groups are due only to chance. In this situation, the numerator and denominator of the F ratio are measuring the same thing and should be roughly equivalent (Gravetter & Wallnau, 1999). That is,

$$F = \frac{0 + \text{differences due to chance}}{\text{differences due to chance}}$$

In this case, we would expect the value of the F ratio to be approximately equal to 1.00. A value near 1.00 would indicate that the differences between the groups are roughly the same as would be expected due to chance. There is no evidence of a treatment effect; therefore, our statistical decision would be to "fail to reject H_o" (i.e., the null hypothesis cannot be rejected; it remains a viable explanation for the differences). However, if there exists a treatment effect, the numerator of the F ratio would be larger than the denominator (due to the nonzero value of the "treatment effect"), which would result in an F-value greater than 1.00. If that value is substantially larger than 1.00 (i.e., larger than the critical value for F), a statistically significant treatment effect is indicated. In this situation, we would conclude that at least one of the sample means is significantly different from the others, permitting us to "reject H_o" (i.e., the null hypothesis is not true).

Earlier, in our discussions of the alternative hypothesis, we stated that we were concerned with which groups were different from which other groups. However, since we have now rejected the null hypothesis, it is important for us to be able to discuss specifically where the differences lie. Are students in urban schools different from those of rural schools? Or from students in suburban *and* rural schools? Or, do the differences occur between suburban and rural schools? In this situation, a *post hoc* comparison is the appropriate procedure to address these questions.

Post hoc tests, also known as *multiple comparisons*, enable the researcher to compare individual treatments two at a time, a process known as **pairwise comparisons**. If the goal of the analysis was to determine whether any groups differed from each other, one might ask why we did not just do a series of independent-samples *t*-tests. The reason is fairly simple: Conducting a series of hypothesis tests within a single research or analysis situation results in the accumulation of the risk of Type I error. For example, in our school setting analysis problem, there would be three pairwise comparisons — urban vs. suburban, urban vs. rural, and suburban vs. rural. If each test was conducted at an α level equal to .05, the overall risk of Type I error for this analysis would be equal to three times the pre-established α level. In other words, the **experimentwise alpha level** would be equal to .15. Now we have increased our probability of making an error in judgment from 5% of the time to making an error 15% of the time.

Most of us would not be willing to assume a risk of that magnitude. Several techniques, which basically involve the simultaneous testing of all pairwise comparisons, exist for addressing this problem (Agresti & Finlay, 1997). These include the Scheffé test, the Bonferroni test, and Tukey's Honest Significance Difference (HSD). An excellent discussion of the use, advantages, and disadvantages of each of these multiple comparison tests is provided by Harris (1998, pp. 371-398).

As with any inferential statistical test, assumptions must be met in order for proper use of the test and interpretation of the subsequent results. You will recall that the assumptions for a one-way analysis of variance are identical to those for an independent-samples t-test. Specifically, these assumptions are:

1. The observations within each sample must be independent of one another.
2. The populations from which the samples were selected must be normal.
3. The populations from which the samples were selected must have equal variances (also known as *homogeneity of variance*).

Generally speaking, the one-way analysis of variance is robust to violations of the normality and homogeneity of variances assumptions (Harris, 1998).

II. FACTORIAL ANALYSIS OF VARIANCE

The goal of most research studies is to determine the extent to which variables are related and, perhaps more specifically, to determine if the various levels of one variable differ with respect to values on a given dependent variable. We have just finished examining one of the simplest cases of the latter situation — a one-way analysis of variance. However, examining variables in relative isolation (i.e., two variables at a time) is seldom very informative. When conducting social science research, human beings are the subjects of the research study nearly all the time. When studying human behavior, it would be naïve to think that only one variable could influence another. Behaviors are usually influenced by "a variety of different variables acting and interacting simultaneously" (Gravetter & Wallnau, 1999). Research designs that include more than one factor are called *factorial designs*. In the remainder of this chapter, we will investigate the simplest of factorial designs — a *two-way analysis of variance*. A two-way analysis of variance, as the name suggests, consists of two IVs and one DV. We will limit our discussion here to designs where there is a separate sample for each treatment condition or characteristic; specifically, this is referred to as a *two-factor, independent-measures design*.

Section 4.1 Practical View

Purpose

The purpose of factorial analysis of variance is to test for mean differences with respect to some DV, similar to a simple one-way analysis of variance. However, in the case of a two-way analysis of variance design, there are now two IVs. Factorial analysis of variance allows the researcher not only to test the significance of group differences (based on levels of the two IVs), but also to test for any interaction effects between levels of IVs. The two-way ANOVA actually tests three separate hypotheses simultaneously in one analysis. Two hypotheses test the significance of the levels of the two IVs separately, and the third tests the significance of the interaction of the levels of the two IVs.

For example, assume we have a research design composed of two IVs (Factor A and Factor B, each with two levels) and one DV. Furthermore, we are interested in evaluating the mean differences on the DV produced by either Factor A or Factor B, or by Factor A and Factor B in combination. The structure of this two-factor design is shown in Figure 4.1.

Any differences produced by either Factor A or B, independent of the other, are called ***main effects***. In Figure 4.1, there are two main effects: (1) if there are differences between Level 1 and Level 2 of Factor A, there is a main effect due to Factor A; (2) if there are differences between Level 1 and Level 2 of Factor B, there is a main effect due to Factor B. Remember that we still do not know if these main effects are statistically significant—we must still conduct hypothesis tests in order to determine significance. These main effects provide us with two of the three hypotheses in a two-factor design. The main effect due to Factor A involves the comparison of its two levels. The null hypothesis for this main effect states that there is no difference in the scores due to the level of A. Symbolically, the null hypothesis would be:

$$H_o: \ \mu_{A_1} = \mu_{A_2}$$

The alternative hypothesis states that there is a difference in scores due to the level of A. The alternative hypothesis would appear as:

$$H_1: \ \mu_{A_1} \neq \mu_{A_2}$$

In similar fashion, the hypotheses corresponding to the main effects for Factor B would be represented by:

$$H_o: \ \mu_{B_1} = \mu_{B_2}$$
$$H_1: \ \mu_{B_1} \neq \mu_{B_2}$$

The test statistic for the hypothesis tests of the main effects is again the F ratio. The formula used for the calculation of the F ratio in this case is identical to Equation 4.1.

In addition to testing for the effects of each factor individually, a two-factor ANOVA allows the researcher to evaluate mean differences that are the result of unique combinations of levels of the two factors. This is accomplished by comparing all of the cell means. A ***cell*** is defined as each combination of a particular level of one factor and a particular level of the second factor. In the example shown in Figure 4.1, there are four cells. The ***cell means*** are simply the averages of all of the scores that fall into each cell. When mean differences exist in a given design that are not explained by the main effects, that design contains an interaction between factors. ***Interaction between factors*** occurs when the effect of one factor depends on different levels of the other factor (Gravetter & Wallnau, 1999).

In order to evaluate whether a significant interaction effect exists, the ANOVA procedure first computes any mean differences that cannot be explained by the main effects. After any additional mean differences are identified, they are also evaluated by using an F ratio. However, the nature of the F ratio changes somewhat:

$$F = \frac{\text{variance not explained by main effects}}{\text{variance expected due to chance (error)}} \qquad \text{(Equation 4.3)}$$

The null hypothesis for the test of an interaction effect states that there is no interaction between Factor A and Factor B. In other words, there are no differences between any of the cell means that cannot be explained by simply summing the individual effects of Factor A and Factor B (Harris, 1998). The alternative hypothesis states that there is a significant interaction effect; that is, there are mean differences that cannot be explained by Factor A or by Factor B alone.

Figure 4.1 Basic Structure for a Two-Factor Design, with Each IV Having Two Levels.

Factor *B*

	Level 1	Level 2
Level 1	Scores for *n* subjects in A_1B_1	Scores for *n* subjects in A_1B_2
Level 2	Scores for *n* subjects in A_2B_1	Scores for *n* subjects in A_2B_2

Factor *A* (left side label: Level 1 / Level 2)

Another way of identifying interaction effects between factors is to examine line plots of the cell means. Essentially, whenever there is an interaction, the lines on the graph will not be parallel (Aron & Aron, 1997). Line plots are a graphical method of depicting a pattern of differences among different levels of the factors. Realize that the lines do not have to cross in order to be significantly nonparallel. For example, Figure 4.2 shows two possible situations that could occur when examining interaction effects.

If there is an interaction between the levels of Factor *A* and Factor *B*, the distance between B_1 and B_2 on A_1 will be significantly different from the distance between B_1 and B_2 on A_2 (Newman & Newman, 1994). Clearly in Figure 4.2(a), these two distances are not significantly different; therefore, there is no interaction in this case. However, in Figure 4.2(b), a substantial discrepancy between the two distances exists and represents an interaction between Factors *A* and *B*.

There are two types of interaction that can be displayed graphically—ordinal and disordinal interactions. An interaction is said to be ***ordinal*** when the lines plotted on the graph do not cross within the values of the graph (Newman & Newman, 1994; Kennedy & Bush, 1985). In contrast, an interaction is ***disordinal*** when the lines plotted on the graph actually cross within the values of the graph. Figure 4.2(b) shows an interaction that is ordinal. An example of a disordinal interaction plot is presented in Figure 4.3.

When interpreting the results of a two-way analysis of variance, one should always examine the significance of the interaction first. If the interaction is significant, it does not make much sense to interpret any main effects. Knowing that two IVs in combination result in a significant effect on the DV is more informative than determining that one and/or the other IV have individual effects. If the interaction is significant, there may be situations when it is appropriate to evaluate the main effects; specifically, it may make logical sense to evaluate the main effects when the interaction is ordinal, simply because the interaction is not as pronounced as in a disordinal interaction (Newman & Newman, 1994). If the interaction is not significant, then the researcher should proceed to the evaluation of the main effects. This is done separately for each factor. Whenever significant group differences are identified, it is appropriate to conduct follow-up post hoc tests in order to determine where specific differences lie.

Another indicator of the strength of relation is effect size, represented by *eta* squared. Effect size is calculated for each factor and for the interaction, and indicates the amount of total variance that is explained by the IVs. An effect size of .50 or greater indicates significant importance.

Figure 4.2 Cell Mean Plots Show (a) No Interaction and (b) Interaction Between Factors.

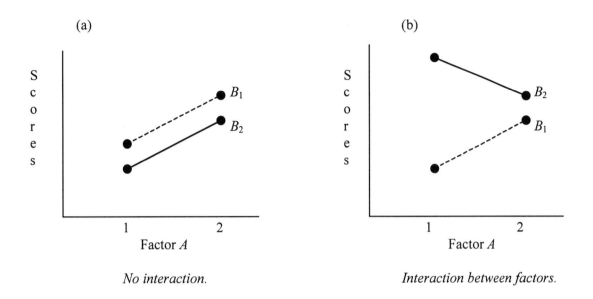

No interaction.   Interaction between factors.

Figure 4.3 Cell Mean Plot Showing a Disordinal Interaction Between Factors.

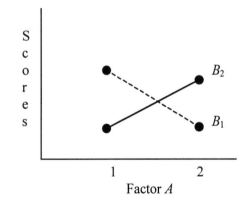

Sample Research Questions

Suppose we were interested in analyzing the effect of gender and age category on income. Eight different groups of employees are created; one for each combination of gender and age category (see Figure 4.4). The independent variables for this study would consist of gender and age category, and the dependent variable would be income.

Research questions should parallel the hypotheses that would be tested by this two-way ANOVA. Thus, this study would address the following research questions:

1. Are there significant mean differences for income between male and female employees?
2. Are there significant mean differences for income by age category among employees?
3. Is there a significant interaction on income between gender and age category?

Figure 4.4 Structure of a 4 × 2, Two Factor Design.

Gender

	Female	*Male*
18-29	Income for *n* subjects *female—(18-29)*	Income for *n* subjects *male—(18-29)*
30-39	Income for *n* subjects *female—(30-39)*	Income for *n* subjects *male—(30-39)*
40-49	Income for *n* subjects *female—(40-49)*	Income for *n* subjects *male—(40-49)*
50+	Income for *n* subjects *female—(50+)*	Income for *n* subjects *male—(50+)*

Age Category

SECTION 4.2 ASSUMPTIONS AND LIMITATIONS

The validity of the results of a factorial ANOVA are dependent upon three assumptions with which the reader should be familiar after having studied previous independent-measures designs (e.g., independent-samples *t*-test, one-way ANOVA, etc.). These assumptions are:
(1)   The observations within each sample must be randomly sampled and must be independent of one another.
(2)   The distributions of scores on the dependent variable must be normal in the populations from which the data were sampled.
(3)   The distributions of scores on the dependent variable must have equal variances.
The assumption of independence is primarily a design issue, as opposed to a statistical one. Provided the researcher has randomly sampled and, more important, randomly assigned subjects to treatments, it is usually safe to believe that this particular assumption has not been violated.

Methods of Testing Assumptions

Generally speaking, analysis of variance is robust to violations of the normality assumption, especially when the sample size is relatively large, and should not be a cause for substantial concern (Gravetter & Wallnau, 1999). Slight departures from normality are to be expected, but even larger deviations will seldom have much effect on the interpretation of results (Kennedy & Bush, 1985). However, if the researcher wishes to have a greater degree of confidence concerning this assumption, there are methods of testing for violations of it. Initially, one can simply "eye ball" the distributions of the data in each cell by obtaining histograms and boxplots of the dependent variable. If the distribution appears to be relatively normal (i.e., there are no marked departures from normality in the form of extreme values), it is safe to assume that the assumption of normality has not been violated. If, however, the shape of the distribution is clearly nonnormal or there appear to be cases with extreme values, the researcher will probably want to submit the data to a more stringent test. For example, the researcher may want to obtain numerical values for skewness and kurtosis and test their significance (as discussed in Chapter 3). Whether or not a distribution is normally distributed can also be tested by means of a chi-square goodness-of-fit test (Kennedy & Bush, 1985). Recall that a goodness-of-fit test purports to test the null hypothesis that a population has a specific shape or proportions. If one specifies in the null that the distribution is normal, then the results of the chi-square test would indicate statistically whether or not the distribution was normal. A decision to "reject the null hypothesis" would lead the researcher to the conclusion that the assumption of normality had been violated and, thus, could possibly call into question the results of the analysis of variance. On the other hand, a decision of "fail to reject the null hypothesis" would permit the researcher to conclude that there was not enough evidence to support the claim that the distribution is not normal.

Violation of the assumption of homogeneity of variance is more crucial than a violation of the other assumptions. Initially, one should examine the standard deviations in the boxplots for each cell. If one suspects that the data do not meet the homogeneity requirement, a specific test should be conducted prior to beginning the analysis of variance. Hartley's F-max test allows the researcher to use sample variances to determine if there are any differences among population variances. Hartley's test is simply a ratio of the largest group variance to the smallest group variance. The $F(\text{max})$ statistic is then compared to a critical value obtained from a table of the distribution of $F(\text{max})$ statistics. Another possible statistical test of homogeneity is Cochran's test, which is also a test of the ratio between variances. A third option is Levene's test, which essentially consists of performing a one-way analysis of variance on data that has been transformed (Kennedy & Bush, 1985). All three of these methods are statistical tests of inference and involve the evaluation of a significance level (if calculated by means of a computer program) or of a test statistic by comparing the obtained value to a critical value (if calculated by hand). One disadvantage of these three tests is that they are only appropriate for situations that involve groups with equal sample sizes. Bartlett's test is an alternative appropriate for unequal n's, but can provide misleading results since it is also extremely sensitive to nonnormality (Kennedy & Bush, 1985).

SECTION 4.3 PROCESS AND LOGIC

The Logic Behind Factorial ANOVA

The logic of a two-way analysis of variance is quite similar to the logic behind its one-way counterpart, with the addition of a few more computational complexities (Harris, 1998). Recall that in a one-way ANOVA, two estimates of variance are computed. One of these consists of only random error (i.e., the variance within groups or MS_W); the other consists of random error as well as group differences (i.e., the variance between groups or MS_B). If there are no differences between the population means of the various groups in the design, the estimates of the two variances will be approximately equal, resulting in the ratio of one (MS_B) to the other (MS_W) approximately equal to 1.00. This ratio, of course, is the F ratio.

In a two-way ANOVA, the total variance is again partitioned into separate components. The within-groups variability remains the same (SS_W, and subsequently, MS_W); however, the between-groups variability is further partitioned into three sub-components: variability due to Factor A (SS_A), variability due to Factor B (SS_B), and variability due to the interaction of Factor A and Factor B (SS_{AxB}). This subsequent partitioning of sums of squares is shown in Figure 4.5.

Figure 4.5 Partitioning of Sums of Squares Variability in a Two-Way ANOVA.

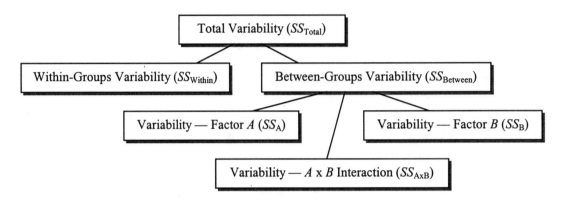

In a two-way ANOVA, all between-groups variability components (i.e., MS_A, MS_B, and MS_{AxB}) are compared to the within-groups variability (i.e., MS_W) individually. If the group means in the population are different for the various levels of Factor A (after removing or controlling for the effects of Factor B), then MS_A will be greater than MS_W; that is, F_A will be significantly greater than 1.00. Similarly, if the group means in the population are different for the various levels of Factor B (after removing the effects of Factor A), then MS_B will be greater than MS_W; that is, F_B will be significantly > 1.00. Finally, if the group means in the population are different for the various combinations of levels for Factor A and Factor B (after removing the individual effects of Factor A and Factor B), then MS_{AxB} will be greater than MS_W; that is, F_{AxB} will be significantly greater than 1.00.

The F ratio is clearly a measure of the statistical significance of the differences between group means or differences between combinations of levels of the IVs. There is, however, another measure that one can obtain that more directly examines the magnitude of the relationship between the independent and dependent variables. This measure is called ***eta squared*** (η^2) and is commonly viewed as the proportion of variance in the dependent variable explained by the independent variable(s) in the sample. Eta squared is viewed as a descriptive statistic and is interpreted as a measure of effect size (Harris, 1998).

Interpretation of Results

Since groups in factorial ANOVA are created by two or more factors or independent variables, it is important to determine if factors are interacting (working together) to effect the dependent variable. Typically a line plot is created to graphically display any factor interaction. If lines overlap and crisscross, factor interaction is present. Although a line plot may reveal some factor interaction, the ANOVA results may show that the interaction is not statistically significant. Consequently, it is important to determine interaction significance using the F ratio and p level for interaction generated from the factorial ANOVA test. If factors significantly interact such that they are working together to effect the dependent variable, we cannot determine the effect that each separate factor has on the dependent variable by looking at the main effects for each factor. Although factor main effects may be significant even while factor interaction is significant, caution should be used when drawing inferences about factor main effects. Effect size should also be analyzed to determine the strength of such effects.

In summary, the first step in interpreting the factorial ANOVA results is to determine if an interaction is present among factors by looking at the F ratio and its level of significance for the interaction. If no interaction is present, then look at the F ratio and its level of significance for each factor's main effect. This will indicate the degree that each factor effects the dependent variable. If there is an interaction among factors, then you cannot discuss how each factor effects the dependent variable; instead, you can only draw conclusions about how the factors work together to effect the dependent variable.

Figure 4.6 Visual Representation of Interaction Between Factors.

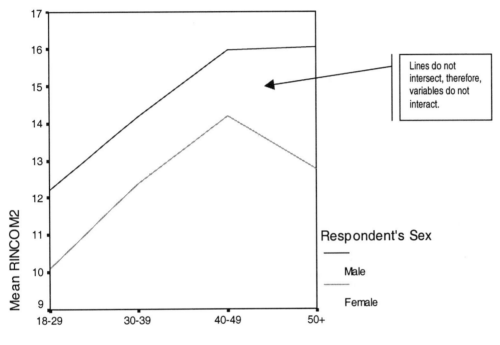

Continuing with the example of gender (*sex*) and age category (*agecat4*) differences in income (*rincom93*), data was first screened for missing data and outliers and then examined for test assumptions. Data screening led to the transformation of *rincom93* to *rincom2*, which eliminated all cases with income greater than or equal to 22. Although group distributions indicate moderate negative skewness, no further transformations were conducted since ANOVA is not highly sensitive to nonnormality if group sample sizes are fairly large. Homogeneity of variance was tested within the ANOVA. A line plot was then created of income means for each group. The nonoverlapping lines in Figure 4.6 demonstrate the lack of factor interaction between gender and age category. Figures 4.7–4.9 present the output from conducting the **Univariate ANOVA**. Levene's test for equal variances is presented in Figure 4.7 and indicates homogeneity of variance among groups, $F(7,701)=1.06$, $p=.387$. The ANOVA summary table (see Figure 4.8) indicates no significant factor interaction, $F(3,701)=.97$, $p=.408$, partial $\eta^2=.004$. Significant group differences were found in gender ($F(7,701)=40.48$, $p<.001$, partial $\eta^2=.055$) and age category ($F(3,701)=21.64$, $p<.001$, partial $\eta^2=.085$). Although significant group differences were found, one should note the small effect size for each factor. Effect size indicates that a very small proportion of variance in income is accounted for by the IVs.

Since the ANOVA results can indicate only group differences and not identify which groups are different, the Scheffé post hoc test was conducted to compare all group combinations and identify any significantly different pairs. Figure 4.8 presents the output from the Scheffé post hoc test for age category. A post hoc test was not conducted for gender since it has only two categories. Results indicate that the age category of 18–29 significantly differs in income from all other age categories. In addition, those 30–39 are significantly different in income from those 40–49 years of age.

Figure 4.7 Levene's Test for Equality of Error Variances.

Levene's Test of Equality of Error Variances[a]

Dependent Variable: RINCOM2

F	df1	df2	Sig.
1.061	7	701	.387

Tests the null hypothesis that the error variance of the dependent variable is equal across groups.

a. Design: Intercept+AGECAT4+SEX+AGECAT4 * SEX

Nonsignficance indicates homogeneity of variance.

Figure 4.8 ANOVA Summary Table for Interaction and Main Effects.

Tests of Between-Subjects Effects

Dependent Variable: RINCOM2

Source	Type III Sum of Squares	df	Mean Square	F	Sig.	Eta Squared
Corrected Model	2236.798[a]	7	319.543	15.360	.000	.133
Intercept	121426.954	1	121426.954	5836.953	.000	.893
AGECAT4	1350.516	3	450.172	21.640	.000	.085
SEX	842.073	1	842.073	40.478	.000	.055
AGECAT4 * SEX	60.316	3	20.105	.966	.408	.004
Error	14583.002	701	20.803			
Total	149966.000	709				
Corrected Total	16819.800	708				

a. R Squared = .133 (Adjusted R Squared = .124)

Main effects for each factor.

Interaction between factors

F-ratios and p-values show no significant interaction between factors. Age category and gender are significant.

Effect sizes are very small.

Figure 4.9 Scheffé Multiple Comparisons for Student Level.

Multiple Comparisons

Dependent Variable: RINCOM2

Scheffe

(I) 4 categories of age	(J) 4 categories of age	Mean Difference (I-J)	Std. Error	Sig.	95% Confidence Interval	
					Lower Bound	Upper Bound
18-29	30-39	-2.1172*	.4933	.000	-3.4997	-.7348
	40-49	-3.9165*	.5101	.000	-5.3461	-2.4870
	50+	-3.2930*	.5455	.000	-4.8216	-1.7643
30-39	18-29	2.1172*	.4933	.000	.7348	3.4997
	40-49	-1.7993*	.4421	.001	-3.0381	-.5605
	50+	-1.1757	.4824	.116	-2.5277	.1762
40-49	18-29	3.9165*	.5101	.000	2.4870	5.3461
	30-39	1.7993*	.4421	.001	.5605	3.0381
	50+	.6235	.4996	.669	-.7765	2.0236
50+	18-29	3.2930*	.5455	.000	1.7643	4.8216
	30-39	1.1757	.4824	.116	-.1762	2.5277
	40-49	-.6235	.4996	.669	-2.0236	.7765

Based on observed means.

*. The mean difference is significant at the .05 level.

Asterisk indicates which groups are significantly different.

Writing Up Results

Since several assumptions must be fulfilled when conducting a factorial ANOVA, the summary of results should first include the steps taken to screen the data. Group means and standard deviations of the dependent variable should first be reported in a table. A line plot of group means, as seen in Figure 4.6, should be created to provide a visual representation of any interaction among factors. ANOVA results (F ratios, degrees of freedom for the particular factor and error, levels of significance, and effect sizes) are then presented in a narrative format in the following order: main effects of each factor, interaction of factors, effect size of factors and interaction, and post hoc results. The following results statement applies the results from Figures 4.6 – 4.9.

The two-way analysis of variance was conducted to investigate income differences in gender and age category among employees. ANOVA results, presented in Table 1, showed a significant main effect for gender, ($F(1,701)=40.48$, $p<.001$, partial $\eta^2=.055$) and age category ($F(3,701)=21.64$, $p<.001$, partial $\eta^2=.085$). Interaction between factors was not significant, $F(3,701)=.97$, $p=.408$, partial $\eta^2=.004$. However, calculated effect size for each factor indicates a small proportion of income variance is accounted for by each factor. The Scheffé post hoc test was conducted to determine which age categories were significantly different. Results revealed that the age category of 18-29 significantly differed in income from all other age categories. In addition, those 30-39 were significantly different in income from those 40-49 years of age.

Table 1   Two-way ANOVA Summary Table.

Source	SS	df	MS	F	p	ES
Between treatments	2236.80	7	399.54			
Age Category	1350.52	3	450.17	21.64	<.001	.085
Gender	842.07	1	842.07	40.48	<.001	.055
Age Category x Gender	60.32	3	20.10	.966	.408	.004
Within treatments	14583.00	701	5.12			
Total	149966.00	709				

Recall that each of the obtained F ratios is reported with its degrees of freedom and the degrees of freedom for error in parentheses. If space allows, an ANOVA summary table as shown in Table 1 should be created. In addition, a narrative results statement should be presented.

SECTION 4.4 SAMPLE STUDY AND ANALYSIS

This section provides a complete example that applies the entire process of conducting a factorial ANOVA: development of research questions and hypotheses, data screening methods, test methods, interpretation of output, and a presentation of results. This example utilizes the data set *gssft.sav* from the SPSS Web site.

Problem

Employers and employees may be interested in determining if income is different for those with differing degrees and whether the differences are the same for people who are very satisfied or not very satisfied with their jobs. The following research questions and respective null hypotheses address the main effects for each factor and the possible interaction between factors.

Research Questions

Null Hypotheses

RQ1: Does income differ among employees with different degrees?

$H_0 1$: Income will not differ among employees with different degrees.

RQ2: Does income differ for satisfied and unsatisfied employees?

$H_0 2$: Income will not differ between satisfied and unsatisfied employees.

RQ3: Does the relationship between income and degree differ for satisfied and unsatisfied employees?

$H_0 3$: The relationship between income and degree will not differ for satisfied and unsatisfied employees.

The independent variables are nominal and include highest degree attained (*degree*) and job satisfaction (*satjob2*). As seen in Figure 4.10, these two factors create ten different groups for comparison on the dependent variable of respondent's income in 1991 (*rincom91*), which is an interval/ratio variable. The reader should note that the variable of *rincom91* was previously transformed to *rincom2* to eliminate subjects who reported income greater than or equal to a value of 22. This transformed variable will be utilized in the following example.

Method

The first step in conducting a factorial ANOVA is to examine the data for missing subjects and outliers and to ensure that test assumptions are fulfilled. The **Explore** procedure was conducted to evaluate outliers and normality. Stem-and-leaf plots reveal that several groups within *degree* and *sat-job2* have outlying values of 3 or less. Therefore, *rincom2* was transformed again to *rincom3* in order to eliminate subjects who reported income of 3 or less. Figure 4.11 displays the stem-and-leaf plots for *satjob2*. **Explore** was conducted again with *rincom3*. Tests of Normality indicate nonnormal distributions for the majority groups. Group histograms reveal slight to substantial negative skewness. However, *rincom3* will not be transformed again since ANOVA is not heavily dependent upon fulfilling the normality assumption as long as group sample sizes are adequate. Homogeneity of variance will be examined within the ANOVA.

The next step is to create a line plot that displays any interaction between the factors (see Figure 4.12). Then **Univariate ANOVA** is conducted along with Bonferroni's post hoc test for highest degree attained. See Section 4.5 in this chapter on SPSS "How To" for a more detailed explanation of steps taken to generate the following output.

Figure 4.10 2 x 5 Factor Design of Respondents Income by Degree and Job Satisfaction.

Highest Degree Attained

	Less than High School	High School	Junior College	Bachelor	Graduate
Very Satisfied with Job	Income for *n* employees who are very satisfied with jobs and have less than a high school diploma	Income for *n* employees who are very satisfied with jobs and have a high school diploma	Income for *n* employees who are very satisfied with jobs and have junior college degree	Income for *n* employees who are very satisfied with jobs and have a bachelors degree	Income for *n* employees who are very satisfied with jobs but have a graduate degree
Not Very Satisfied with Job	Income for *n* employees who are not very satisfied with jobs and have less than a high school diploma	Income for *n* employees who are not very satisfied with jobs and have a high school diploma	Income for *n* employees who are not very satisfied with jobs and have junior college degree	Income for *n* employees who are not very satisfied with jobs and have a bachelors degree	Income for *n* employees who are not very satisfied with jobs and have a graduate degree

Figure 4.11 Stem-and-Leaf Plots for Job Satisfaction.

```
RINCOM2 Stem-and-Leaf Plot for SATJOB2= Very satisfied

 Frequency     Stem &  Leaf

    16.00 Extremes      (=<3.0)
     1.00         4 .  0
     2.00         5 .  00
     2.00         6 .  00
     3.00         7 .  000
     7.00         8 .  0000000
     9.00         9 .  000000000
    11.00        10 .  00000000000
    22.00        11 .  0000000000000000000000
    13.00        12 .  0000000000000
    17.00        13 .  00000000000000000
    21.00        14 .  000000000000000000000
    36.00        15 .  000000000000000000000000000000000000
    41.00        16 .  00000000000000000000000000000000000000000
    25.00        17 .  0000000000000000000000000
    30.00        18 .  000000000000000000000000000000
    23.00        19 .  00000000000000000000000
     8.00        20 .  00000000
    26.00        21 .  00000000000000000000000000

 Stem width:      1.00; Each leaf:        1 case(s)



RINCOM2 Stem-and-Leaf Plot for SATJOB2= Not very satisfied

 Frequency     Stem &  Leaf

    22.00 Extremes      (=<3.0)
     8.00         4 .  00000000
     4.00         5 .  0000
     4.00         6 .  0000
     4.00         7 .  0000
     9.00         8 .  000000000
    24.00         9 .  000000000000000000000000
    24.00        10 .  000000000000000000000000
    33.00        11 .  000000000000000000000000000000000
    29.00        12 .  00000000000000000000000000000
    27.00        13 .  000000000000000000000000000
    35.00        14 .  00000000000000000000000000000000000
    41.00        15 .  00000000000000000000000000000000000000000
    36.00        16 .  000000000000000000000000000000000000
    24.00        17 .  000000000000000000000000
    27.00        18 .  000000000000000000000000000
    15.00        19 .  000000000000000
    14.00        20 .  00000000000000
    17.00        21 .  00000000000000000

 Stem width:      1.00; Each leaf:        1 case(s)
```

Figure 4.12 Line Plot of Interaction Between Degree and Job Satisfaction.

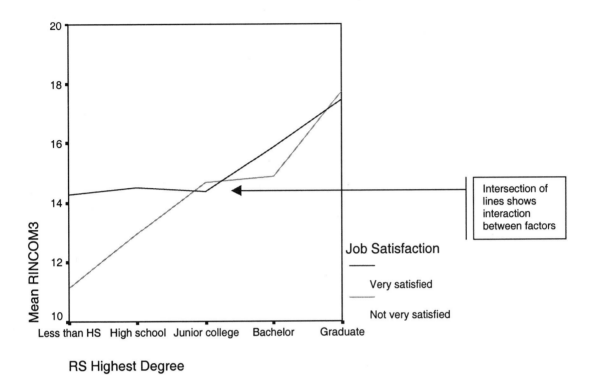

Figure 4.13 Levene's Test for Equality of Variances.

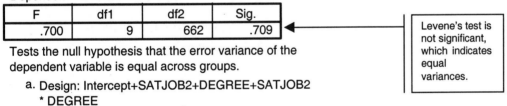

Levene's Test of Equality of Error Variances[a]

Dependent Variable: RINCOM3

F	df1	df2	Sig.
.700	9	662	.709

Tests the null hypothesis that the error variance of the dependent variable is equal across groups.

a. Design: Intercept+SATJOB2+DEGREE+SATJOB2 * DEGREE

Levene's test is not significant, which indicates equal variances.

Output and Interpretation of Results

Levene's test of equality of variances was conducted within ANOVA and indicates homogeneity of variance within groups (see Figure 4.13). Although the line plot of degree and job satisfaction shows interaction between factors, Figure 4.14 reveals that the factor interaction is not statistically significant. Consequently, we can then determine if the main effects of each factor are significant. *F* ratios and levels of significance reveal that income is significantly different with respect to the levels of job satisfaction and degree. However, effect size (eta squared) is quite small for both factors. Figure 4.15 displays the results of the Bonferroni's post hoc test for highest degree attained. This table presents mean differences for every possible paired combination of highest degree attained. Significant mean differences at the .05 level are indicated by an asterisk. Results reveal that income of individuals with a graduate degree is significantly different from all other degree groups. Those with a high school diploma are not significantly different from those with no high school diploma or those with a junior college degree. In addition, income is not significantly different between junior college and bachelor degree holders. A post hoc test on job satisfaction is not necessary, since this variable has only two categories.

Figure 4.14 Univariate ANOVA Table.

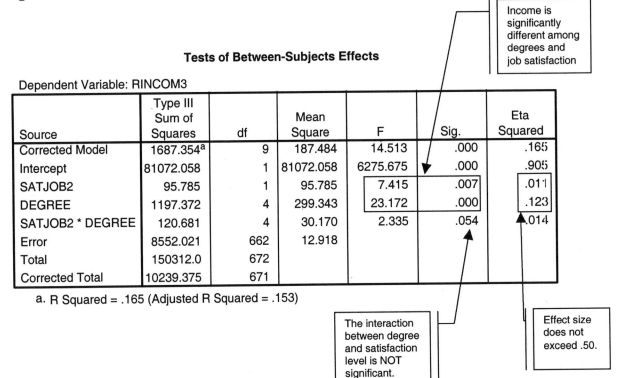

Presentation of Results

The following report briefly summarizes the two-way ANOVA results from the previous example. Notice that Table 1 was created using data from Figure 4.14 and that Figure 4.12 is used to display factor interaction.

Data were screened to ensure that the assumptions of factorial ANOVA were fulfilled. To eliminate outliers, subjects with income values of less than or equal to 3 or greater than or equal to 22 were removed. A Univariate ANOVA was conducted; a summary of results are presented in Table 1. Main effect results revealed that income was significantly different among employees with differing degrees, $F(4, 662) = 23.17$, $p < .001$, partial $\eta^2 = .123$. Income was also significantly different for employees who were very satisfied or not very satisfied with their jobs, $F(1, 662) = 7.42$, $p = .007$, partial $\eta^2 = .011$. Although the interaction between degree and job satisfaction was not statistically significant, $F(4, 662) = 2.34$, $p = .054$, Figure 4.12 reveals some interaction. Estimates of effect size revealed low strength in associations. Bonferroni's post hoc test was conducted to determine which degree groups were significantly different in income. Results reveal that income of individuals with a graduate degree is significantly different from all other degree groups. Those with a high school diploma are not significantly different from those with no high school diploma or those with a junior college degree. In addition, income is not significantly different between junior college and bachelor degree holders.

Table 1 Two-way ANOVA Summary Table.

Source	SS	df	MS	F	p	ES
Between treatments	1687.35	9	187.48			
Job Satisfaction	95.79	1	95.79	7.42	.007	.011
Degree	1197.37	4	1197.37	23.17	<.001	.123
Job Satisfaction x Degree	120.68	4	30.17	2.34	.054	.014
Within treatments	8552.02	662	12.92			
Total	150312.0	672				

Figure 4.15 Bonferroni Multiple Comparisons for Degree.

Multiple Comparisons

Dependent Variable: RINCOM3

Bonferroni

(I) RS Highest Degree	(J) RS Highest Degree	Mean Difference (I-J)	Std. Error	Sig.	95% Confidence Interval Lower Bound	Upper Bound
0 Less than HS	1 High school	-1.4122	.5752	.143	-3.0321	.2078
	2 Junior college	-2.3249*	.7395	.017	-4.4077	-.2420
	3 Bachelor	-3.0935*	.6158	.000	-4.8277	-1.3592
	4 Graduate	-5.3777*	.6761	.000	-7.2820	-3.4735
1 High school	0 Less than HS	1.4122	.5752	.143	-.2078	3.0321
	2 Junior college	-.9127	.5390	.909	-2.4308	.6054
	3 Bachelor	-1.6813*	.3504	.000	-2.6682	-.6944
	4 Graduate	-3.9656*	.4481	.000	-5.2275	-2.7036
2 Junior college	0 Less than HS	2.3249*	.7395	.017	.2420	4.4077
	1 High school	.9127	.5390	.909	-.6054	2.4308
	3 Bachelor	-.7686	.5821	1.000	-2.4081	.8709
	4 Graduate	-3.0529*	.6456	.000	-4.8712	-1.2345
3 Bachelor	0 Less than HS	3.0935*	.6158	.000	1.3592	4.8277
	1 High school	1.6813*	.3504	.000	.6944	2.6682
	2 Junior college	.7686	.5821	1.000	-.8709	2.4081
	4 Graduate	-2.2843*	.4991	.000	-3.6899	-.8786
4 Graduate	0 Less than HS	5.3777*	.6761	.000	3.4735	7.2820
	1 High school	3.9656*	.4481	.000	2.7036	5.2275
	2 Junior college	3.0529*	.6456	.000	1.2345	4.8712
	3 Bachelor	2.2843*	.4991	.000	.8786	3.6899

Based on observed means.

*. The mean difference is significant at the .05 level.

Asterisk indicates significant group differences.

SECTION 4.5 SPSS "HOW TO"

This section demonstrates the steps for conducting a factorial ANOVA using the **Univariate ANOVA** procedure with the preceding *gssft.sav* example. This procedure analyzes the independent effects of the factors and covariates as well as the interaction among factors. In addition, different models may be used to estimate the sum of squares. To open the **Univariate ANOVA** dialogue box as shown in Figure 4.16, select the following menus:

> **Analyze**
> > **General Linear Model**
> > > **Univariate**

Univariate Dialogue Box (see Figure 4.16)

Once in this dialogue box, identify one DV (*rincom3*) and move it into the Dependent Variable box. Then select the independent variables (*satjob2* and *degree*) that create the groups and move them into the Fixed Factor(s) box. After you have defined the variables, click the **Model** button.

Figure 4.16 Univariate ANOVA Dialogue Box.

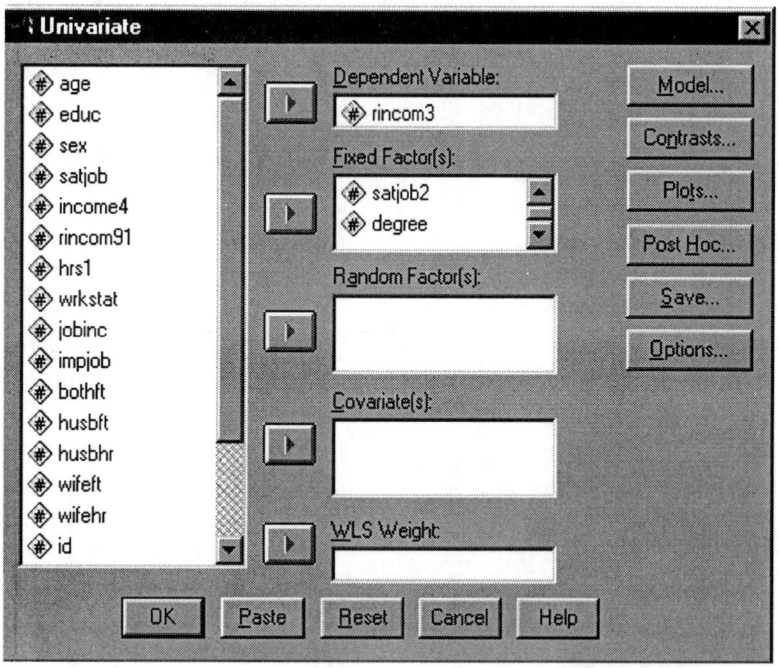

This box allows you to choose between a full factorial model or a custom model. The full factorial model is the default and usually the most appropriate since it will test all main effects and interactions. The custom model will test only the main effects and interactions you select. If building a custom model, be sure to select interaction variables together and click the **Build Term(s)** button. The **Model** dialogue box also allows you to identify the way in which Sum of Squares will be calculated. Type III is the default and the most commonly used since it is useful when subgroups vary in sample size but have no missing cells/subgroups. Type IV is used for models with empty cells/groups.

Figure 4.17 Univariate ANOVA Model Dialogue Box.

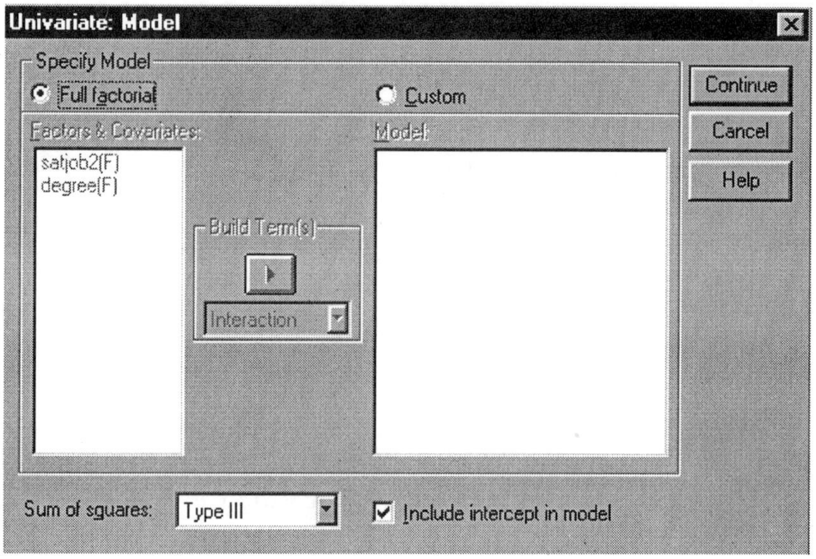

Figure 4.18 Univariate ANOVA Options Dialogue Box.

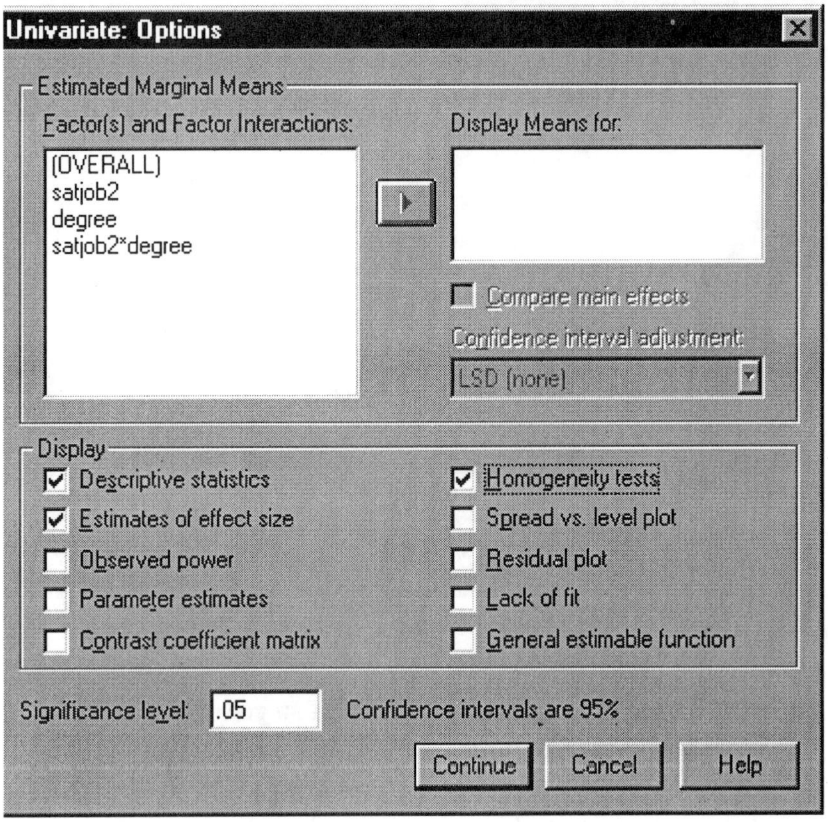

`Univariate Options` Dialogue Box (see Figure 4.18)

This dialogue box provides several options for display that can help you in examining your data. Three commonly used options are:

`Descriptive Statistics`—produces means, standard deviations, and counts for each subgroup.

`Estimates of Effect Size`—calculates eta squared; represents the amount of total variance that is explained by the independent variables.

`Homogeneity Tests`—calculates the Levene statistic for testing the equality of variance for all subgroups.

You can also change the level of significance within this box; the default is $\alpha = .05$.

`Univariate Post Hoc` Dialogue Box (see Figure 4.19)

Post hoc tests allow you to identify which groups are different from one another within a single factor. However, you can only test factors that have three or more groups. A variety of post hoc tests are provided when equal variances are assumed and not assumed. Scheffé and Bonferroni tests are the most conservative. For our example, we utilized Bonferroni.

Figure 4.19 Univariate ANOVA Post Hoc Dialogue Box.

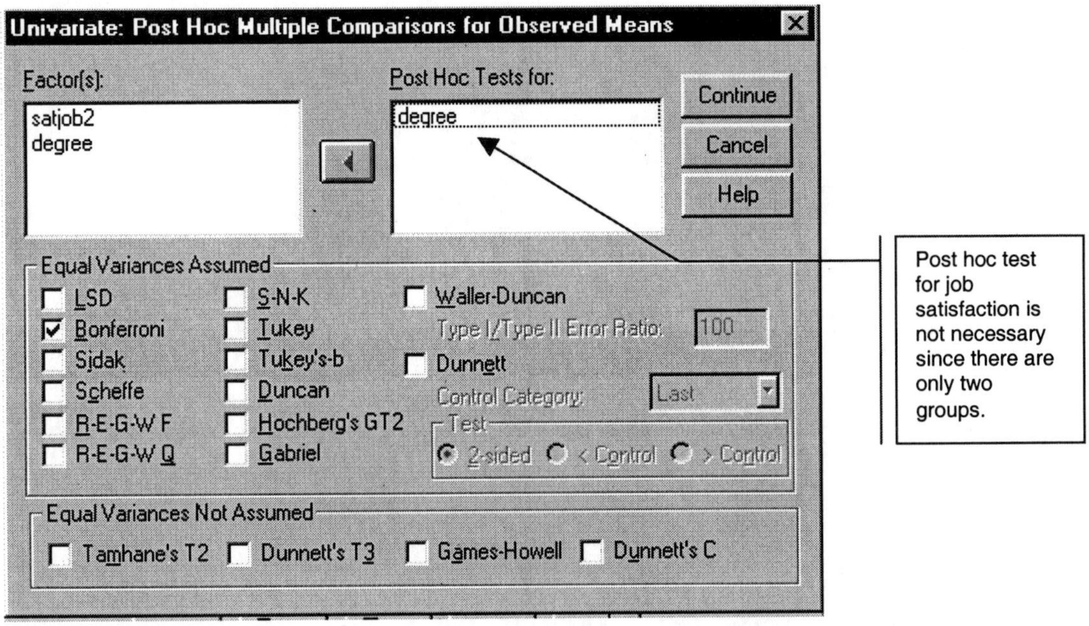

SECTION 4.6 VARIATIONS OF THE TWO-FACTOR DESIGN

The reader should be aware that there are many variations of this design that come under the heading of factorial designs. For example, measures taken on subjects may be *dependent* rather than independent. In other words, subjects may each have a score at each level of the factor(s) as in a design that involves a pretest, posttest, and an extended follow-up test. For this reason, these designs are often referred to as *within-groups*, *repeated-measures*, or *randomized-block* ANOVAs.

Another variation of the factorial ANOVA design involves fixed and random effects. A *fixed effect* is a factor in which all levels of interest of the IV have been included. For example, if our study of interest involved an examination of the effect of gender, we would most likely include both females and males. In contrast, a *random effect* is a factor for which the levels represent only a sample of all possible levels to which we hope to be able to generalize the results. For example, suppose we are interested in teachers' effectiveness using a new approach to classroom instruction. We might include five secondary science teachers as the five levels of the variable called *teacher* in our study. Since we are interested not only in these five teachers (obviously, we hope to be able to generalize our results to the entire population of secondary science teachers), the variable *teacher* and its subsequent levels would be considered a random effect.

Another important distinction in factorial designs is that of crossed and nested designs. A *crossed* design is one in which all combinations of the levels of the IVs are included, or represented, by at least one observation. For example, if we are concerned with how employees' scores on some motivation inventory are affected by gender (i.e., female or male) and education level (i.e., high school degree, college degree, or graduate degree), we would have at least one person in each of the six possible combined level categories. In contrast, a *nested* design is one in which some possible combinations of levels are missing. For example, we might study the effectiveness of three teachers using curriculum A and three different teachers using curriculum B. Notice that each teacher is paired with only one of the two curricula.

Finally, there are designs in which the *cells*, or combination of levels of IVs, do not contain equal numbers of subjects. This is often the case in nonexperimental research designs. Unequal n's are usually a concern only when some of the assumptions of factorial analysis of variance have been violated. In this case, the nature of the statistical analysis becomes much more complex (Harris, 1998).

If the reader is interested in a more detailed account of any of the previously mentioned variations on factorial designs, there are many excellent resources including, but not limited to, Agresti & Finlay (1997), Aron & Aron (1997), Gravetter & Wallnau (1999), Harris (1998), and Kennedy & Bush (1985).

Summary

The purpose of a factorial ANOVA is to determine group differences when two or more factors create these groups. Factorial ANOVA will test the main effect of each factor on the dependent variable as well as the interaction among factors. Usually, post hoc tests are conducted in conjunction with the ANOVA to determine which groups are significantly different. Prior to conducting the factorial ANOVA, data should be screened to ensure fulfillment of test assumptions—independence of observations, normal distributions of subgroups, and equal variances among subgroups. The SPSS *Univariate ANOVA* table provides F ratios and p values that indicate the significance of factor main effects and interaction. If factor interaction is significant, then conclusions about the main effects of each factor are limited since factors are working together to effect the dependent variable. Post hoc results will identify which group combinations are significantly different. Figure 4.20 provides a checklist for conducting a factorial ANOVA.

Figure 4.20 Checklist for Conducting a Factorial ANOVA.

I.   Screen Data
   a.   Missing Data?
   b.   Outliers?
      ❑   Run Outliers and review stem-and-leaf plots and boxplots within **Explore.**
      ❑   Eliminate or transform outliers if necessary.
   c.   Normality?
      ❑   Run Normality Plots with Tests within **Explore.**
      ❑   Review boxplots and histograms.
      ❑   Transform data if necessary.
   d.   Homogeneity of Variance?
      ❑   Run Levene's Test of Equality of Variances within **Univariate.**
      ❑   Transform data if necessary.
   e.   Factor Interaction?
      ❑   Create lineplot of DV by IVs.

II.   Conduct ANOVA
   a.   Run Factorial ANOVA with post hoc test
      1. ⟁ **Analyze...**⟁ **General Linear Model...**⟁ **Univariate**
      2. Move DV to Dependent Variable box.
      3. Move IVs to Fixed Factor box.
      4. ⟁ **Model.**
      5. ⟁ **Full Factorial.**
      6. ⟁ **Continue.**
      7. ⟁ **Options**
      8. Check **Descriptive statistics, Estimates of effect size,** and **Homogeneity Tests.**
      9. ⟁ **Continue**
      10. ⟁ **Post hoc**
      11. Select post hoc method.
      12. ⟁ **Continue.**
      13. ⟁ **OK.**
   b.   Interpret factor interaction.
   c.   If no factor interaction, interpret main effects for each factor.

III. Summarize Results
   a.   Describe any data elimination or transformation.
   b.   Present line plot of factor interaction.
   c.   Narrate main effects for each factor and interaction (F-ratio, p value, and effect size).
   e.   Draw conclusions.

Exercises for Chapter 4

1. The table below presents means for the number of hours worked last week (*hrs1*) for individuals by general happiness and job satisfaction (*satjob2*). Using the data below, draw a line plot. Use the line plot to complete the following steps to estimate the factor main effects on the dependent variable and the interaction between factors.

	Very Satisfied with Job	Not Very Satisfied with Job
Very Happy	43	40
Pretty Happy	42	42
Not Too Happy	38	40

a. Develop the appropriate hypotheses for main effects and interaction.

b. Do the factors interact? If so, do you think the interaction will be statistically significant? Explain.

c. Do you think that there will be a significant main effect for the factor of general happiness? If so, which groups do you think will be significantly different?

d. Do you think that there will be a significant main effect for the factor of job satisfaction?

2. The data in question 1 came from *gss.sav*, which can be downloaded at the SPSS Web site. Open the URL: www.spss.com/tech/DataSets.html in your Web browser. Scroll down until you see "Data Used in SPSS Guide to Data Analysis—8.0 and 9.0" and click on the link "dataset.exe." When the "Save As" dialogue appears, select the appropriate folder and save the file. Preferably, this should be a folder created in the SPSS folder of your hard drive for this purpose. Once the file is saved, double-click the "dataset.exe" file to extract the data sets to the folder. Use this data set and your hypotheses from question 1 to complete a factorial ANOVA analysis and Bonferroni's post hoc test. The variables used were: hours worked last week (*hrs1*), general happiness (*happy*), and job satisfaction (*satjob2*). Use the questions below to guide your interpretation of the output.

a. Is factor interaction significant? Explain.

b. Are main effects significant? Explain.

c. How do these results compare with your estimation in problem #1?

3. Use the *salary.sav* data file to determine if current salaries (*salnow*) are related to gender (*sex*) and minority status (*minority*).

      a.   Develop the appropriate hypotheses for main effects and interaction.

      b.   Evaluate your data to ensure that it meets the necessary assumptions.

      c.   Run the appropriate analyses and interpret your results.

      d.   Write a results statement.

CHAPTER 5

ANALYSIS OF COVARIANCE

As mentioned in the previous chapter, there are numerous variations, extensions, and elaborations of the one-way analysis of variance technique. In this chapter, we discuss one such technique, analysis of covariance, which can be used to improve research design efficiency by adjusting the effect of variables that are related to the dependent variable. We begin by examining a basic comparison between analysis of variance and analysis of covariance, followed by the specific details of analysis of covariance.

I. ANALYSIS OF VARIANCE VERSUS ANALYSIS OF COVARIANCE

One-way analysis of variance compares the means on some dependent variable for several groups, as defined by the various levels of the independent variable. In many situations, especially in the social sciences, it is difficult to imagine that differences on a dependent variable could be attributed only to the effect of *one* independent variable. It would likely take very little persuasion to convince the reader that *many* variables may in fact affect that particular dependent variable. Oftentimes, a researcher may be able to identify one or more variables that also demonstrate an effect on the DV. If one is still interested only in the effect of the original IV on the DV, the effects of these "accompanying" variables, also known as ***concomitant*** variables, can be controlled for, or ***partialed out***, of the results. In analysis of variance, the effects of any concomitant variables are simply ignored. The covariance analysis itself mirrors the ordinary analysis of variance, but only after the effect of the unwanted variable has been partialed out. The variable whose effects have been partialed out of the results is called the ***covariate***. The results of the analysis are then interpreted just like any other analysis of variance.

In the previous chapter, we presented an example that investigated income level differences. Here we will apply an analysis of covariance design to these same variables. Let us assume, for example, that we wanted to determine whether or not job satisfaction, gender, or the interaction between them has an effect on income level. However, our data set also includes a measure for level of education, and we would probably be justified in assuming that level of education has an effect on income level. In other words, income level differences may exist due to differences in the level of education of individuals in the sample; in general, people who have higher levels of education tend to have higher levels of income. If this is the case, we will never be able to accurately determine income differences due to job satisfaction or gender. Furthermore, the problem is that we are not concerned with the education level variable—it may not have been central to our research interests and/or research questions. Therefore, we would like to control for the effect of level of education on income level. Stated another way, level of education has been identified as the covariate for our analysis.

II. ANALYSIS OF COVARIANCE

While analysis of variance is similar to factorial analysis of variance, the use of analysis of covariance provides researchers with a technique that allows us to more appropriately analyze data collected in social science settings. The examination of the relationships among variables considered in relative isolation can be somewhat troubling, especially when dealing with human beings as the subjects in a research study. Oftentimes, there are extraneous variables present that may influence the dependent measures. Analysis of covariance is an extension of analysis of variance where the main effects and interactions are assessed *after* the effects of some other concomitant variable has been removed. The effects of the covariate are removed by adjusting the scores on the DV in order to reflect initial differences on the covariate. In this chapter, we will discuss the purposes, proper applications, and interpretations of analysis of covariance.

SECTION 5.1 PRACTICAL VIEW

Purpose

The use of analysis of covariance essentially has three major purposes (Tabachnick & Fidell, 1996). The first of these purposes is to increase the sensitivity of the *F*-tests of main effects and interactions by reducing the error variance, primarily in experimental studies. This is accomplished by removing from the error term (i.e., the within-groups variability) unwanted, *predictable* variance associated with the covariate(s). Recall, for a moment, the generic equations for an *F*-test as discussed in Chapter 4. The error term corresponds to the denominator of the formula for the calculation of an *F*-statistic:

$$F = \frac{\text{variance between subjects}}{\text{variance due to chance (error)}} = \frac{\text{variance between subjects}}{\text{variance within subjects}}$$

This predictable variance (or **systematic bias**) results from the use of intact groups that differ systematically on several variables and is best addressed through means of random assignment of subjects to groups (Stevens, 1992). When random assignment is not possible, due to constraints within the research setting, the inclusion of a covariate in the analysis can be helpful in reducing the error variance. The covariate is used to assess any undesirable variance in the DV. This variance is actually estimated by scores on the covariate. If the covariate has a substantial effect on the DV, a portion of the within-variability will be statistically removed, resulting in a smaller error term (i.e., a smaller denominator). This ultimately produces a larger value for the *F*-statistic, therefore producing a more sensitive test (Stevens, 1996). In other words, we are now more likely to reject the null hypotheses concerning the main and interaction effects. An example of a source of this undesirable variance would be individual differences possessed by the subjects prior to entry into the research study.

The second purpose of analysis of covariance involves a statistical adjustment procedure (Tabachnick & Fidell, 1996). This is most appropriately used in nonexperimental situations where subjects cannot be randomly assigned to treatments. In this situation, analysis of covariance is used as a statistical matching procedure where the means of the DV for each group are adjusted to what they would be if all groups had scored equally on the covariate. In this manner, instead of the ideal situation consisting of subjects randomly assigned to groups/treatments, the researcher now has two (or more) groups containing subjects that have been matched based on the covariate scores. Although this condition is certainly not as advantageous as having random assignment, it does improve the research design when random assignment is not feasible. Analysis of covariance is also used primarily for descriptive purposes in nonexperimental studies: The covariate improves the predictability of the DV, but there is no implication of causality. Tabachnick & Fidell (1996) warn that "…if the research question to be answered involves causality, [the use of] ANCOVA is no substitute for running an experiment" (p. 322).

The third purpose of analysis of covariance is to interpret differences in levels of the IV when several DVs are included in the analysis. This procedure is known as *multivariate analysis of covariance* (MANCOVA) and will be discussed in greater detail in the next chapter. Briefly, it is often the goal of a research study to assess the contribution of each DV to the significant differences in the IVs. One method of accomplishing this is to remove the effects of all other DVs by treating them as covariates in the analysis.

The most common use of analysis of covariance is its use for the first purpose described above (i.e., increase the sensitivity of the *F*-tests of main effects and interactions by reducing the error variance, primarily in experimental studies). A generic example of the classical application of ANCOVA would involve the random assignment of subjects to various levels of one or more IVs. The subjects are then measured on one or more covariates. A commonly used covariate consists of scores on some pretest, measured in identical fashion as the DV (i.e., a posttest) but prior to the manipulation of the levels of the IV(s). This pretest measure is followed by manipulation of the IV (i.e., implementation of the treatment), which is then followed by the measurement of the DV. It is important to keep in mind that the covariate need not consist of a pretest; demographic characteristics (e.g., level of education, socio-economic level, gender, IQ, etc.) that differ from, but are related to, the DV can serve as covariates (Tabachnick & Fidell, 1996). Covariates can be incorporated in all variations of the basic ANOVA design, including factorial between-subjects, within-subjects, crossed, and nested designs, although analyses of these designs are not readily available in most computer programs (Tabachnick & Fidell, 1996).

Let us now apply this generic example to our current, concrete example. The example that we began discussing at the beginning of this chapter was actually a factorial analysis of covariance design. Recall that in factorial analysis of *variance*, the researcher not only tests the significance of group differences (based on levels of the two IVs), but also tests for any interaction effects between levels of IVs. The additional component of the design of a factorial analysis of *covariance* is that the significance of group differences as well as the interaction effects are tested only *after* the effects of some covariate have been removed. Similar to the two-way ANOVA, the two-way ANCOVA actually tests three separate hypotheses simultaneously in one analysis. Two hypotheses test the significance of the levels of the two IVs separately, after removing the effects of the covariate, and the third tests the significance of the interaction of the levels of the two IVs, also after removing the effects of the covariate. Specifically, these three null hypotheses would be stated as the following:

H_o: $\mu_{A_1} = \mu_{A_2}$ (main effect of variable A)

H_o: $\mu_{B_1} = \mu_{B_2}$ (main effect of variable B)

H_o: $\mu_{A_1 B_1} = \mu_{A_1 B_2} = \mu_{A_2 B_1} = \mu_{A_2 B_2}$ (interaction effect of variables A and B)

where μ refers to the income level group means, and the subscripts "1" and "2" indicate the two levels of the independent variables, or factors, A and B.

In our example, recall that we wanted to determine whether or not job satisfaction, gender, or the interaction between them has an effect on income level, after controlling for education level. The DV is the actual income as reported by each subject in our data set. Our two IVs—gender and job satisfaction—each has two levels. Therefore, we have a four-cell design as depicted in Figure 5.1. Since we have identified education level as a covariate in our analysis, the variability in income levels which is attributable to level of education has been partialed out; in other words, the effect of education level on income has been controlled. The reader should be reminded that controlling for the effects of a covariate is a statistical procedure accomplished only by the computer during the analysis procedure.

A critical issue that needs to be addressed by the researcher in any analysis of covariance design is the choice of covariate(s). Generally speaking, any variables that should theoretically correlate with the DV or that have been shown to correlate with the DV on similar types of subjects should be considered as possible covariates (Stevens, 1992). Ideally, if quantitative variables are being considered, one should choose as covariates those variables that are significantly correlated with the DV and that have low correlations among themselves (in cases where more than one covariate is being used). If there exists a weak correlation between two covariates, they will each be removing from the DV relatively unique portions of the error variance, which is advantageous since we want to obtain the greatest amount of total error reduction through the inclusion of covariates. On the other hand, if there exists a strong correlation between two covariates (e.g., $r > .80$), then those two covariates are removing essentially the same error variance from the DV; in other words, the inclusion of the second covariate has contributed very little to improving the design and resultant analysis (Stevens, 1992). If categorical variables are being considered as covariates, the degree of relationship with the DV or with other co-variates is not an issue.

Figure 5.1   Structure for a Two-Factor Analysis of Covariance Design.

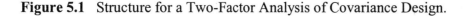

Gender

		Female	*Male*
Job Satisfaction	*Very Satisfied with Job*	Income* for *n*=140 subjects *female–very satisfied*	Income* for *n*=173 subjects *male–very satisfied*
	Not Very Satisfied with Job	Income* for *n*=180 subjects *female–not very satisfied*	Income* for *n*=217 subjects *male–not very satisfied*

* indicates that degree (education level) has been
partialed out of income level means.

The *number* of covariates to be included in an analysis is also a decision that should not be taken lightly by the researcher. Huitema (1980, p.161) has provided a formula that can be used to provide guidance in determining the number of covariates to be included in a study, based on the number of groups and subjects. The formula recommends limiting the number of covariates such that

$$\frac{C + (J - 1)}{N} < .10$$

where C is the number of covariates, J is the number of groups, and N is the total number of subjects in the study. For example, if we wanted to conduct a research study consisting of three groups and a total of 45 subjects, then $(C + 2)/45 < .10$, or $C < 2.5$. Thus, we should probably include fewer than 2.5 (i.e., 1 or 2) covariates in this study. If the ratio is greater than .10, the adjusted means resulting from the analysis of covariance are likely to be unstable (i.e., the adjusted means for our sample would be substantially different from other samples drawn from the same population) (Stevens, 1992).

Sample Research Questions

In our study, as it is represented in Figure 5.1, we have two IVs with two levels each, creating four groups. After controlling for the effect of education level, we are concerned with investigating three components of the design:

- ◆ the main effect of gender on income level;
- ◆ the main effect of job satisfaction on income level; and
- ◆ the interaction effect between gender and job satisfaction on income level.

Remember that our research questions should parallel our null hypotheses as we have previously stated them. Therefore, this study would address the following research questions:

(1)   Are there significant mean differences for income level between males and females, after controlling for education level?

(2)   Are there significant mean differences for income level between individuals who are very satisfied with their jobs and those who are not very satisfied with their jobs, after controlling for education level?

(3)   Is there a significant interaction on income level between gender and job satisfaction, after controlling for education level?

SECTION 5.2   ASSUMPTIONS AND LIMITATIONS

The results of an analysis of covariance are valid only to the extent to which the assumptions are not violated. Analysis of covariance is subject to the same assumptions as analysis of variance. As a reminder to the reader, these assumptions are:

(1)   The observations within each sample must be randomly sampled and must be independent of one another.

(2)   The distributions of scores on the dependent variable must be normal in the population from which the data were sampled.

(3)   The distributions of scores on the dependent variable must have equal variances.

In addition, analysis of covariance also rests on three additional assumptions. These assumptions are:

(4)   A linear relationship exists between the dependent variable and the covariate(s).

(5)   The regression slopes for a covariate are homogeneous (i.e., the slope for the regression line is the same for each group).

(6)   The covariate is reliable and is measured without error.

Methods of Testing Assumptions

Potential violations of the first three ANCOVA assumptions are examined in the same manner as they are in an analysis of variance design. Recall that the assumption of random sampling and independent observations really constitutes issues of design and should be addressed prior to any collection of data. The assumption of normally distributed scores on the dependent variable can be assessed initially by inspection of histograms, boxplots, and normal Q-Q plots, but is probably best tested statistically by examining the values (and the associated significance tests) for skewness and kurtosis and through the use of the Kolmogorov-Smirnov test. Finally, the assumption of homogeneity of variances is best tested using Box's Test or one of three different statistical tests, namely Hartley's F-max test, Cochran's test, or Levene's test.

The three additional assumptions, unique to analysis of covariance, address two issues with which the reader is familiar and one issue new to discussions of screening data. The first of these additional assumptions, that of a linear relationship between the covariate(s) and the dependent variable, is analogous to the general multivariate assumption of linearity as discussed in Chapter 3. The researcher need only be concerned with this assumption when the covariate is quantitative. If a nonlinear relationship exists between the quantitative covariate and DV, the error terms are not reduced as fully as they could be and, therefore, the group means are incompletely adjusted (Tabachnick & Fidell, 1996). A violation of this assumption could ultimately result in errors in statistical decision making. This assumption of linearity is roughly assessed by inspecting the bivariate scatterplots between the covariate and the DV. More precisely, however, one should test this assumption by obtaining and examining residuals plots comparing the standardized residuals to the predicted values for the dependent variable. If it is apparent that serious curvilinearity is present, one should then examine the within-cells scatterplots of the DV with each covariate (Tabachnick & Fidell, 1996). If curvilinear relationships are indicated, they may be corrected by transforming some or all of the variables. If transforming the variables would create difficulty in interpretations, one might consider eliminating the covariate that appears to produce nonlinearity and replacing it with another appropriate covariate (Tabachnick & Fidell, 1996).

Similar to the ANOVA/ANCOVA assumption of random selection and independent observations, a second ANCOVA assumption, that of the inclusion of a reliable covariate, is an issue of research design and is most appropriately addressed prior to data collection. Steps should be taken at the outset of a research study to ensure that covariates—as should all variables—be measured as error-free as possible. Failure of covariates to be measured reliably can again result in a less sensitive F-test, which could cause errors in statistical decision-making.

The final ANCOVA assumption presents a new issue to the reader. This assumption states that the regression slopes for a covariate are homogeneous (i.e., that the slope for the regression line is the same for each group). This assumption is also referred to as *homogeneity of regression*. Regression, which is discussed in detail in Chapter 7, is a statistical technique in which the relationship between two variables can be used so that the DV can be predicted from the IV. Briefly, the method used in the prediction process is often referred to as the *regression line*, or the best-fitting line through a series of points, as defined by a mathematical equation. If the values for n subjects on two variables are plotted as scatterplots, those series of points would form a unique shape. There will be an equally unique line that will mathematically satisfy the requirements to be the best-fitting line through those points. Since the selection of a covariate in ANCOVA is based upon the extent to which there exists a predictable relationship (i.e., a strong relationship) between the covariate and the DV, regression is incorporated into ANCOVA in order to predict scores on the DV based on knowledge of scores on the covariate (Kennedy & Bush, 1985). The homogeneity of regression slopes assumption states that the regression slopes (i.e., those best-fitting lines between the covariate and the DV) are equal for each group in the analysis. In a design consisting of three groups, each containing 15 subjects, the assumption of homogeneous regression slopes is met if the slopes appear like those pictured in Figure 5.2(a). Figure 5.2(b) depicts a violation of the assumption of homogeneous regression slopes.

A violation of the assumption of homogeneous regression slopes is a crucial one with respect to the validity of the results of an analysis of covariance. If the slopes are unequal, it is implied that there is a different DV-covariate slope in some cells of the design or that there is an interaction between the IV and the covariate (Tabachnick & Fidell, 1996). If an IV-covariate interaction exists, the relationship between the covariate and the DV is different at different levels of the IV(s). If one again examines Figure 5.2(b), it should be clear that, at low scores on the covariate, Group 2 outscores Group 1 on the DV; however, at higher covariate scores, Group 1 outscores Group 2 (and does so at an increasing rate). Therefore, the covariate adjustment needed for the various cells is different.

Various computer programs will provide a statistical test (specifically, an F-test) of the assumption of homogeneity of regression slopes. The null hypothesis being tested in this case is that all regression slopes are equal. If the researcher is to continue in the use of analysis of covariance, he/she would hope to fail to reject that particular null hypothesis, thus indicating that the assumption is tenable and that analysis of covariance is an appropriate technique to apply. In SPSS, this is determined by examining the results of the F-test for the interaction of the IV(s) by the covariate(s). If the F-test is significant, then ANCOVA should *not* be conducted.

SECTION 5.3   PROCESS AND LOGIC

The Logic Behind ANCOVA

The logic of analysis of covariance is nearly identical to the logic behind analysis of variance. Recall that in an ANOVA, two estimates of variance are computed. One of these consists of only random error (i.e., the variance within groups or MS_W); the other consists of random error as well as group

Figure 5.2   Regression Lines Between a Dependent Variable and Covariate for Three Groups Depicting (a) Homogeneous Regression Slopes and (b) Heterogeneous Regression Slopes.

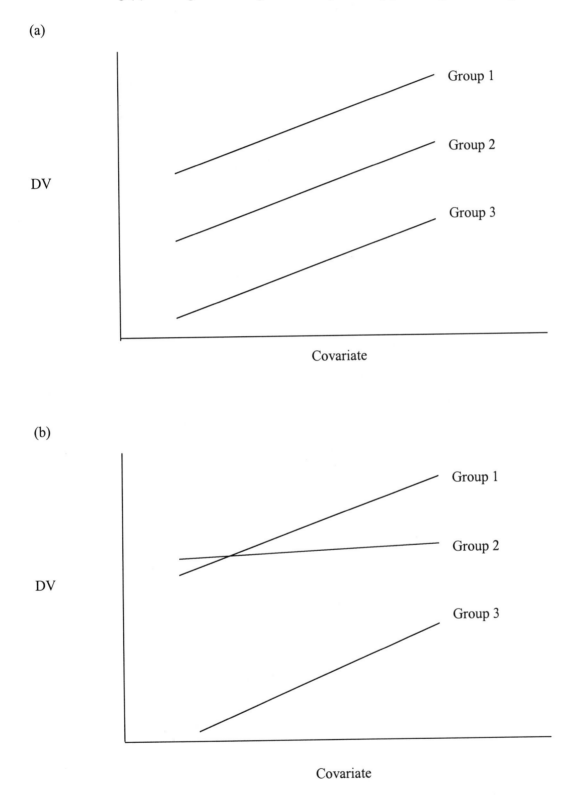

differences (i.e., the variance between groups or MS_B). If there are no differences between the population means of the various groups in the design, the estimates of the two variances will be approximately equal, resulting in the ratio of one (MS_B) to the other (MS_W) approximately equal to 1.00. This ratio, of course, is the F ratio.

Analysis of covariance parallels the above procedure, with one additional component: the adjustment of DV scores. Initially, the covariate is measured prior to the manipulation of the independent variable (i.e., implementation of the treatments). Following the implementation of the treatments, the dependent measures are collected. The initial phase of the analysis involves the statistical adjustment of the DV group means in order to control for the effects of the covariate on the DV. From this point on, the logic behind analysis of covariance is the same as that behind analysis of variance. A ratio of random errors plus treatment effect (MS_B) to random error (MS_W—reduced due to the adjustment of DV group means) is obtained. The magnitude of the resultant F ratio is then evaluated in order to determine the significance of the effect of group differences, after controlling for the covariate, on the DV.

As an extension of this one-way ANCOVA design, a factorial ANCOVA again mirrors the factorial ANOVA design, following the adjustment of DV means. The total variance is partitioned into separate components. The within-groups variability remains the same (SS_W, and subsequently, MS_W) but has been reduced due to the inclusion of the covariate; the between-groups variability is partitioned into the variability due to Factor A (SS_A), variability due to Factor B (SS_B), and variability due to the interaction of Factor A and Factor B (SS_{AxB}). The reader may want to refer to Figure 4.5 in Chapter 4.

Then, similar to the two-way ANOVA, all between-groups variability components (i.e., MS_A, MS_B, and MS_{AxB}) are compared to the reduced within-groups variability (i.e., MS_W) individually. If the group means in the population are different for the various levels of Factor A (after removing or controlling for the effects of Factor B), then MS_A will be greater than MS_W; that is, F_A will be significantly greater than 1.00. The same logic applies to testing the effects of Factor B (after removing the effects of Factor A) and for testing the various combinations of levels for Factor A and Factor B (after removing the individual effects of Factor A and Factor B).

In addition to the F ratio, we can again obtain a measure of effect size. Recall that eta squared (η^2) is a measure of the magnitude of the relationship between the independent and dependent variables and is interpreted as the proportion of variance in the dependent variable explained by the independent variable(s) in the sample, after partialing out the effects of the covariate(s).

Interpretation of Results

Interpretation of ANCOVA results is similar to that of ANOVA; however, with the inclusion of covariates, interpretation of a preliminary or custom ANCOVA is necessary in order to test the assumption of homogeneity of regression slopes. Basically, this preliminary analysis tests for the interaction between the factors (IVs) and covariates. In addition, homogeneity of variance will also be tested in this preliminary analysis. If the F-test of factor-covariate interaction is significant, then the full ANCOVA should not be conducted. If factor-covariate interaction is not significant, then one can proceed with interpreting the Levene's Test as well as proceeding with the full ANCOVA. If the Levene's test is not significant, then homogeneity of variance is assumed. Once homogeneity of regression slopes and homogeneity of variance have been established, then the full factorial can be conducted.

Interpretation of the full ANCOVA results must also take into account the interaction among the factors (IVs). Consequently, line graphs are typically created to graphically represent any interaction between/among factors based upon the DV. Overlapping lines indicate factor interaction, which may limit the inferences drawn from the analysis. Fortunately, the ANCOVA procedure also calculates the significance of such interaction. If interaction is statistically significant, the main effect for each factor on the DV is not a valid indicator of effect since factors are working together to affect the DV. Effect size (η^2) for each factor and interaction is also a good indicator of the strength of these effects.

Since ANCOVA is adjusting group means as if subjects scored equally on the covariate(s), the utility of each covariate is also analyzed within ANCOVA. *F* ratios and *p* values for each covariate indicate the degree to which the covariate significantly influences the DV. If the covariate substantially influences the DV, the subsequent group adjustments will typically lead to lower *F*-values and higher *p* values for main effects and interactions. Although a significant covariate decreases the likelihood of significant factor main effects, the results should present a more accurate picture of group differences when adjusted for some covariate(s). Effect size for the covariate(s) should also be examined to determine the amount of variance accounted for by each covariate.

In summary, the first step in interpreting the results of ANCOVA is to determine if factor interaction is present by examining the *F* ratio and *p* value for the interaction. If no interaction is present, then each factor's main effect can be reliably interpreted. *F* ratios, *p* values, and effect sizes are examined for each factor's main effect and indicate the degree to which each adjusted factor affects the DV. A comparison of adjusted group means can also indicate which groups differ from one another. This is often helpful since *post hoc* analyses are not available in *SPSS ANCOVA*. The influence of covariate(s) should also be assessed by examining *F* ratios, *p* values, and η^2.

Continuing with our example that seeks to determine the effect of gender (*sex*) and job satisfaction (*satjob2*) on respondent's income (*rincom91*) while controlling for highest degree attained (*degree*), data was first screened for missing data and outliers and then examined for testing assumptions. Data screening led to the transformation of *rincom91* to *rincom2* in order to eliminate all cases with income equal to or exceeding 22. Although group distributions of *rincom2* are slightly negatively skewed, no further transformation was done. Linearity of the covariate and DV was not assessed since *degree* is categorical. Factor interaction was then investigated by creating a line plot of income for gender and job satisfaction (see Figure 5.3). The line plot reveals no factor interaction. Next, homogeneity of regression slopes and homogeneity of variance were evaluated by conducting a preliminary **Univariate** ANOVA. The ANOVA summary table indicates no significant interaction between the factors and covariate, $F(3, 703)=2.18$, $p=.089$ (see Figure 5.4). In addition, the Levene's test reveals equal variances among groups, $F(3, 706)=.602$, $p=.614$ (see Figure 5.5). With test assumptions fulfilled, **Univariate** ANOVA was conducted and produced the output in Figures 5.6 – 5.8. Figures 5.6 and 5.7 present the unadjusted and adjusted means for each group. The ANCOVA summary table is presented in Figure 5.8 and indicates no factor interaction. Main effects for gender ($F(1, 705)=37.68$, $p<.001$, partial $\eta^2=.051$) and job satisfaction ($F(1, 705)=10.87$, $p=.001$, partial $\eta^2=.015$) are significant. However, one should be cautious in drawing inferences regarding the effect of each IV since effect sizes are very small.

Writing Up Results

ANCOVA requires the fulfillment of several assumptions. If data screening leads to any subject elimination and/or variable transformation, these should be reported in the summary of results. The narrative of results should also include ANCOVA results (*F* ratios, degrees of freedom, *p* values, and effect sizes) for the main effect of each factor and covariate as well as the interaction of factors. Figures and tables may also be generated to support results (e.g., line graph of factor interactions, ANCOVA summary table, and a table comparing unadjusted and adjusted group means). The following results statement applies the results from Figures 5.3 – 5.8.

A 2 x 2 analysis of covariance was conducted to determine the effect of gender and job satisfaction on income when controlling for highest degree attained. Initial data screening led to the transformation of income by eliminating all values greater than or equal to 22. Interaction of factors was first analyzed by creating a line plot, which demonstrates no interaction (see Figure 5.3). ANCOVA results (see Table 1) indicate a significant main effect for gender, $F(1, 705)=37.68$, $p<.001$, partial $\eta^2=.051$, and a significant main effect for job satisfaction, $F(1, 705)=10.87$, $p=.001$, partial $\eta^2=.015$. Interaction between gender and job satisfaction was not significant, $F(1, 705)=0.00$, $p=.958$. The covariate of degree significantly influenced the dependent variable of income, $F(1, 705)=74.39$, $p<.001$, partial $\eta^2=.095$. Table 2 presents the adjusted means for gender and job satisfaction, which indicate that males ($M=14.73$) have significantly higher income than females ($M=12.63$) and that satisfied ($M=14.24$) employees earn more than unsatisfied ($M=13.11$).

Table 1 ANCOVA Summary Table

Source	SS	df	MS	F	p	η^2
Between treatments	2602.19	4	650.55	32.14	<.001	.154
Degree	1505.81	1	1505.81	74.39	<.001	.095
Gender	762.68	1	762.68	37.68	<.001	.051
Job Satisfaction	220.07	1	220.07	10.87	.001	.015
Gender x Satisfaction	0.00	1	0.00	0.00	.958	.000
Error	424.77	705	0.60			
Total	5883.00	710				

Table 2 Adjusted and Unadjusted Group Means for Income

	Adjusted M	Unadjusted M
Males	14.73	14.67
Females	12.63	12.55
Very Satisfied	14.24	14.46
Not Very Satisfied	13.11	13.13

Figure 5.3 Line Plot of Gender and Job Satisfaction for Income.

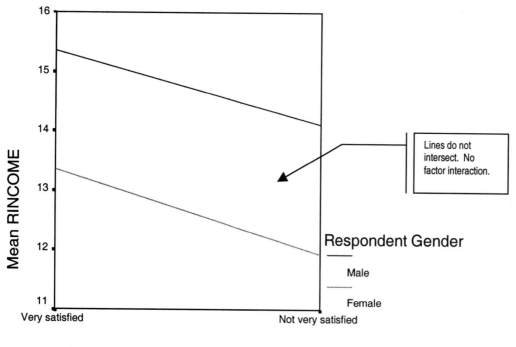

Figure 5.4 ANCOVA Summary Table for Homogeneity of Regression Slopes.

Tests of Between-Subjects Effects

Dependent Variable: RINCOM2

Source	Type III Sum of Squares	df	Mean Square	F	Sig.	Eta Squared
Corrected Model	2733.846[a]	6	455.641	22.655	.000	.162
Intercept	28091.978	1	28091.978	1396.740	.000	.665
SEX	108.004	1	108.004	5.370	.021	.008
SATJOB2	261.418	1	261.418	12.998	.000	.018
DEGREE	1295.667	1	1295.667	64.421	.000	.084
SEX * SATJOB2 * DEGREE	131.716	3	43.905	2.183	.089	.009
Error	14139.113	703	20.113			
Total	150407.000	710				
Corrected Total	16872.959	709				

a. R Squared = .162 (Adjusted R Squared = .155)

No significant interaction. Assumption of homogeneity of regression slopes is fulfilled.

Figure 5.5 Levene's Test of Homogeneity of Variance.

Levene's Test of Equality of Error Variances[a]

Dependent Variable: RINCOM2

F	df1	df2	Sig.
.602	3	706	.614

Tests the null hypothesis that the error variance of the dependent variable is equal across groups.

Not significant. Assumption of equal variances is fulfilled.

a. Design: Intercept+SEX+SATJOB2+DEGREE+SEX * SATJOB2 * DEGREE

Figure 5.6 Unadjusted Descriptive Statistics for Income by Gender and Job Satisfaction.

Descriptive Statistics

Dependent Variable: RINCOM2

Respondent's Sex	Job Satisfaction	Mean	Std. Deviation	N
Male	Very satisfied	15.3584	4.5250	173
	Not very satisfied	14.1198	4.9819	217
	Total	14.6692	4.8181	390
Female	Very satisfied	13.3500	4.9165	140
	Not very satisfied	11.9278	4.4433	180
	Total	12.5500	4.7022	320
Total	Very satisfied	14.4601	4.8018	313
	Not very satisfied	13.1259	4.8637	397
	Total	13.7141	4.8783	710

Figure 5.7 Adjusted Descriptive Statistics for Income by Job Satisfaction and Gender.

1. Respondent's Sex

Dependent Variable: RINCOM2

Respondent's Sex	Mean	Std. Error	95% Confidence Interval Lower Bound	Upper Bound
Male	14.726[a]	.229	14.276	15.177
Female	12.628[a]	.254	12.131	13.126

[a] Evaluated at covariates appeared in the model: RS Highest Degree = 1.79.

2. Job Satisfaction

Dependent Variable: RINCOM2

Job Satisfaction	Mean	Std. Error	95% Confidence Interval Lower Bound	Upper Bound
Very satisfied	14.242[a]	.256	13.740	14.745
Not very satisfied	13.113[a]	.227	12.667	13.558

[a] Evaluated at covariates appeared in the model: RS Highest Degree = 1.79.

Figure 5.8 Univariate ANCOVA Summary Table.

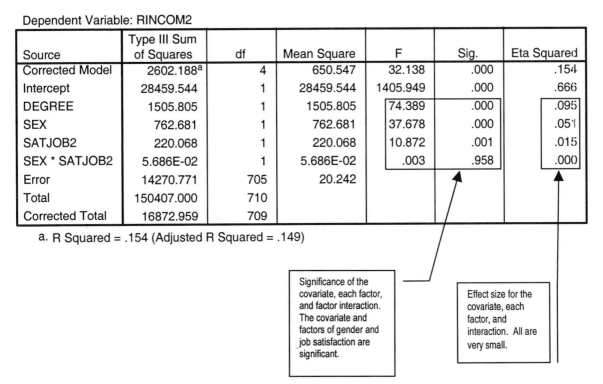

Tests of Between-Subjects Effects

Dependent Variable: RINCOM2

Source	Type III Sum of Squares	df	Mean Square	F	Sig.	Eta Squared
Corrected Model	2602.188ᵃ	4	650.547	32.138	.000	.154
Intercept	28459.544	1	28459.544	1405.949	.000	.666
DEGREE	1505.805	1	1505.805	74.389	.000	.095
SEX	762.681	1	762.681	37.678	.000	.051
SATJOB2	220.068	1	220.068	10.872	.001	.015
SEX * SATJOB2	5.686E-02	1	5.686E-02	.003	.958	.000
Error	14270.771	705	20.242			
Total	150407.000	710				
Corrected Total	16872.959	709				

a. R Squared = .154 (Adjusted R Squared = .149)

Significance of the covariate, each factor, and factor interaction. The covariate and factors of gender and job satisfaction are significant.

Effect size for the covariate, each factor, and interaction. All are very small.

SECTION 5.4 SAMPLE STUDY AND ANALYSIS

This section provides a complete example that applies the entire process of conducting ANCOVA: development of research questions and hypotheses, data screening methods, test methods, interpretation of output, and presentation of results. The SPSS data set of *gssft.sav* is utilized.

Problem

Suppose you are interested in determining if hours worked per week is different by gender and for those who are satisfied or not satisfied with their jobs, while equalizing these groups on income. The following research questions and respective null hypotheses are generated for the main effects of each factor and the possible interaction between factors.

Research Questions

RQ1: Does hours worked per week differ by gender among employees when controlling for income differences?

RQ2: Does hours worked per week differ for satisfied and unsatisfied employees when controlling for income differences?

Null Hypotheses

H_01: Hours worked per week will not differ by gender among employees when controlling for income differences.

H_02: Hours worked per week will not differ for satisfied and unsatisfied employees when controlling for income differences.

RQ3: Does the relationship between hours worked per week and gender differ for satisfied and unsatisfied employees when controlling for income differences? ⟶ H₀3: The relationship between hours worked per week and gender will not differ for satisfied and unsatisfied employees when controlling for income differences.

The IVs are categorical and include gender (*sex*) and job satisfaction *(satjob2)*. The DV is hours worked per week (*hrs1*); the covariate is *rincom2*. Both the DV and covariate are quantitative variables. The IVs create a 2 x 2 factor design for hours worked per week as seen in Figure 5.9.

Figure 5.9 Structure of 2 x 2, Two-Factor Design for Hours Worked per Week.

Gender

	Female	*Male*
Very Satisfied with Job	Hours worked per week* for *n*=140 *females—very satisfied*	Hours worked per week* for *n*=173 *males—very satisfied*
Not Very Satisfied with Job	Hours worked per week* for *n*=180 *females—not very satisfied*	Hours worked per week* for *n*=217 *males—not very satisfied*

Job Satisfaction (row label, left of table)

*Indicates that income has been partialed out of group means for hours worked per week.

Method

Before ANCOVA can be conducted, examination of data for missing cases, outliers, and fulfillment of test assumptions must occur. The **Explore** procedure was conducted to identify missing values, outliers, and evaluate normality. Stem-and-leaf plots indicate extreme values in *hrs1* for each group in job satisfaction (see Figure 5.10) and gender (see Figure 5.11). Since the cutoff for extreme values differs for each group, the most conservative criteria (<=16 and >= 80) will be used in eliminating outliers. Thus, *hrs1* will be transformed into *hrs2* so that cases less than or equal to 16 will be transformed to 17 and cases greater than or equal to 80 will be transformed to 79. This procedure will reduce only the number of outliers, not eliminate all outliers.

Because the DV has been transformed, the **Explore** procedure will be conducted again to examine normality of *hrs2*. Although the tests of normality indicate nonnormal distributions for all groups (see Figure 5.12), ANCOVA is not highly sensitive to nonnormality as long as group sizes are large and fairly equivalent. Another assumption to test is the linear relationship between the covariate and the DV. This is appropriate only if the covariate is quantitative. Since income is quantitative, a scatterplot will be created to determine the linearity of the relationship between *hrs2* and *rincom2*. The scatterplot (see Figure 5.13) indicates a linear trend.

Figure 5.10 Stem-and-Leaf Plots for Hours Worked by Job Satisfaction.

```
Number of Hours Worked Last Week Stem-and-Leaf Plot for
SATJOB2= Very satisfied

 Frequency     Stem &  Leaf

     7.00 Extremes      (=<25)
     3.00         3 .  0
     2.00         3 .  2
    10.00         3 .  555
     6.00         3 .  67
     2.00         3 .  8
   114.00         4 .  0000000000000000000000000000000000000&
    10.00         4 .  2223
    32.00         4 .  44555555555
     6.00         4 .  66&
    13.00         4 .  8888&
    41.00         5 .  00000000000001
     3.00         5 .  2
    13.00         5 .  5555
     4.00         5 .  6
      .00         5 .
    25.00         6 .  00000000
    33.00 Extremes      (>=65)

Stem width:    10
Each leaf:     3 case(s) & denotes fractional leaves.
```

Extreme values.

```
Number of Hours Worked Last Week Stem-and-Leaf Plot for
SATJOB2= Not very satisfied

 Frequency     Stem &  Leaf

    10.00 Extremes      (=<25)
     2.00         2 .  &
    15.00         3 .  002&
    33.00         3 .  55567888&
   177.00         4 .  00000000000000000000000000000000000000001223444
    34.00         4 .  55555688&
    55.00         5 .  0000000000024&
    36.00         5 .  55555568&
    29.00         6 .  0000000&
    26.00 Extremes      (>=65)

Stem width:    10
Each leaf:     4 case(s) & denotes fractional leaves.
```

Extreme values.

Figure 5.11 Stem-and-Leaf Plots for Hours Worked by Gender.

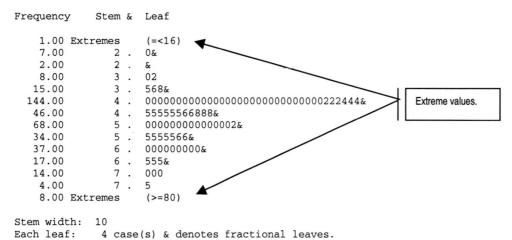

```
Number of Hours Worked Last Week Stem-and-Leaf Plot for
SEX= Male

 Frequency     Stem &  Leaf

     1.00 Extremes     (=<16)
     7.00        2 .  0&
     2.00        2 .  &
     8.00        3 .  02
    15.00        3 .  568&
   144.00        4 .  000000000000000000000000000222444&
    46.00        4 .  55555566888&
    68.00        5 .  000000000000002&
    34.00        5 .  5555566&
    37.00        6 .  000000000&
    17.00        6 .  555&
    14.00        7 .  000
     4.00        7 .  5
     8.00 Extremes     (>=80)

Stem width:  10
Each leaf:   4 case(s) & denotes fractional leaves.
```

```
Number of Hours Worked Last Week Stem-and-Leaf Plot for
SEX= Female

 Frequency     Stem &  Leaf

     9.00 Extremes     (=<28)
     7.00        3 .  00
     4.00        3 .  2
    18.00        3 .  5555&
     7.00        3 .  67
    12.00        3 .  88&
   150.00        4 .  000000000000000000000000000000000000001
     6.00        4 .  23
    28.00        4 .  455555
     4.00        4 .  6&
     8.00        4 .  88
    26.00        5 .  000000
     3.00        5 .  2
    17.00        5 .  5555&
     1.00        5 .  &
     3.00        5 .  &
    33.00 Extremes     (>=60)

Stem width:  10
Each leaf:   4 case(s) & denotes fractional leaves.
```

Figure 5.12 Tests of Normality for Job Satisfaction and Gender.

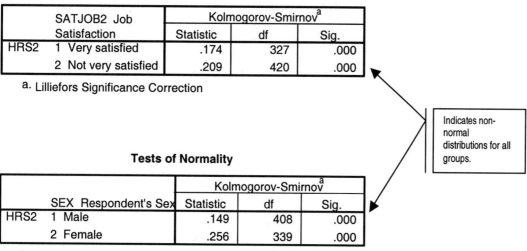

Tests of Normality

	SATJOB2 Job Satisfaction	Kolmogorov-Smirnov[a]		
		Statistic	df	Sig.
HRS2	1 Very satisfied	.174	327	.000
	2 Not very satisfied	.209	420	.000

a. Lilliefors Significance Correction

Tests of Normality

	SEX Respondent's Sex	Kolmogorov-Smirnov[a]		
		Statistic	df	Sig.
HRS2	1 Male	.149	408	.000
	2 Female	.256	339	.000

a. Lilliefors Significance Correction

> Indicates non-normal distributions for all groups.

Figure 5.13 Scatterplot of Income by Hours Worked per Week.

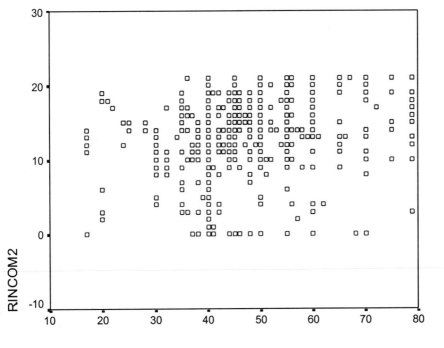

The next assumption to be tested is homogeneity of regression slopes to determine if significant interaction between the covariate and the factors is present. If significant interaction is found, ANCOVA results are not meaningful since the interaction implies that differences on the dependent variable among groups vary as a function of the covariate. In such a situation, ANCOVA should not be conducted. To evaluate homogeneity of regression slopes, a preliminary ANCOVA is conducted using **Univariate** analysis. This preliminary ANCOVA can also be used to evaluate homogeneity of variance.   For our example, results (see Figure 5.14) indicate that the interaction of gender, job satisfaction, and income is not significant, $F(3,703)=1.72$, $p=.161$, partial $\eta^2=.007$. In addition, the Levene's test for equal variances (see Figure 5.15) indicates that variances between groups are fairly equivalent, $F(3,706)=2.27$, $p=.079$). Since interaction between the factors and covariate was not found, factor interaction can now be analyzed by creating a line plot (see Figure 5.16). Next, **Univariate** ANOVA was conducted with *rincom2* as a covariate. See Section 5.5 on SPSS "How To" for an explanation on the steps necessary to generate the following output.

Figure 5.14 ANCOVA Summary Table for Homogeneity of Regression Slopes.

Tests of Between-Subjects Effects

Dependent Variable: HRS2

Source	Type III Sum of Squares	df	Mean Square	F	Sig.	Eta Squared
Corrected Model	7025.503[a]	6	1170.917	9.731	.000	.077
Intercept	121798.226	1	121798.226	1012.219	.000	.590
SEX	65.529	1	65.529	.545	.461	.001
SATJOB2	12.804	1	12.804	.106	.744	.000
RINCOM2	2419.342	1	2419.342	20.106	.000	.028
SEX * SATJOB2 * RINCOM2	621.925	3	207.308	1.723	.161	.007
Error	84590.527	703	120.328			
Total	1634449.000	710				
Corrected Total	91616.030	709				

a. R Squared = .077 (Adjusted R Squared = .069)

Indicates interaction between factors and covariate is NOT significant.

Figure 5.15 Levene's Test of Homogeneity of Variance.

Levene's Test of Equality of Error Variances[a]

Dependent Variable: HRS2

F	df1	df2	Sig.
2.268	3	706	.079

Indicates equal variances among groups.

Tests the null hypothesis that the error variance of the dependent variable is equal across groups.

a. Design: Intercept+SEX+SATJOB2+RINCOM2+SEX * SATJOB2 * RINCOM2

Figure 5.16 Line Plot of Hours Worked by Gender and Job Satisfaction.

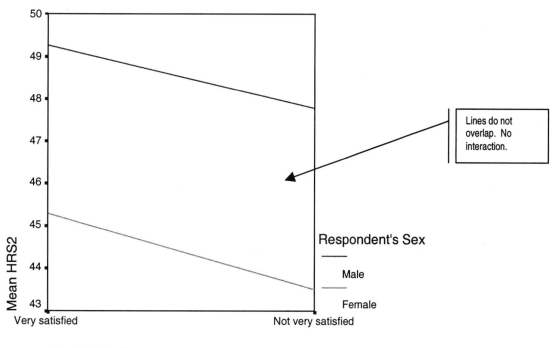

Output and Interpretation of Results

Figure 5.17 presents the means and standard deviations for each group prior to adjustment. Figure 5.18 displays the adjusted means. The summary table of ANCOVA results (see Figure 5.19) indicate that factor interaction is not significant, therefore the main effects of each factor and covariate can be more accurately interpreted. *F* ratios and *p* values indicate that after adjustment for income, hours worked per week is significantly different for males and females. However, the effect size ($\eta^2=.021$) for gender is quite small in that gender accounts for only 2.1% of the variance in hours worked per week. No significant differences in hours worked per week were found in job satisfaction after adjustment for income. Results also indicate that the covariate (*rincom2*) significantly adjusted the DV.

Figure 5.17 Unadjusted Descriptive Statistics for Hours Worked by Gender and Job Satisfaction.

Descriptive Statistics

Dependent Variable: HRS2

Respondent's Sex	Job Satisfaction	Mean	Std. Deviation	N
Male	Very satisfied	49.4624	11.7020	173
	Not very satisfied	47.7465	11.3747	217
	Total	48.5077	11.5377	390
Female	Very satisfied	45.2071	11.5357	140
	Not very satisfied	43.6111	10.0381	180
	Total	44.3094	10.7310	320
Total	Very satisfied	47.5591	11.8011	313
	Not very satisfied	45.8715	10.9713	397
	Total	46.6155	11.3674	710

Figure 5.18 Adjusted Descriptive Statistics for Hours Worked by Gender and Job Satisfaction.

1. Respondent's Sex

Dependent Variable: HRS2

Respondent's Sex	Mean	Std. Error	95% Confidence Interval	
			Lower Bound	Upper Bound
Male	48.171[a]	.567	47.057	49.285
Female	44.864[a]	.627	43.634	46.094

a. Evaluated at covariates appeared in the model: RINCOM2 = 13.7141.

2. Job Satisfaction

Dependent Variable: HRS2

Job Satisfaction	Mean	Std. Error	95% Confidence Interval	
			Lower Bound	Upper Bound
Very satisfied	47.064[a]	.627	45.832	48.296
Not very satisfied	45.971[a]	.557	44.876	47.065

a. Evaluated at covariates appeared in the model: RINCOM2 = 13.7141.

Figure 5.19 ANCOVA Summary Table.

The factor of gender is significant, but effect size is quite small.

Tests of Between-Subjects Effects

Dependent Variable: HRS2

Source	Type III Sum of Squares	df	Mean Square	F	Sig.	Eta Squared
Corrected Model	6405.268[a]	4	1601.317	13.249	.000	.070
Intercept	124454.483	1	124454.483	1029.687	.000	.594
RINCOM2	2823.075	1	2823.075	23.357	.000	.032
SEX	1807.177	1	1807.177	14.952	.000	.021
SATJOB2	203.103	1	203.103	1.680	.195	.002
SEX * SATJOB2	1.690	1	1.690	.014	.906	.000
Error	85210.761	705	120.866			
Total	1634449.000	710				
Corrected Total	91616.030	709				

a. R Squared = .070 (Adjusted R Squared = .065)

Interaction is NOT significant.

Presentation of Results

The following report summarizes the two-way ANCOVA results from the previous example.

A 2 x 2 analysis of covariance was conducted on hours worked per week. Independent variables consisted of gender and job satisfaction (very satisfied versus not very satisfied). The covariate was income. Data screening for outliers led to the transformation of hours worked per week. Subjects with hours less than or equal to 16 were recoded 17, while subjects with hours greater than or equal to 80 were recoded 79. After significant adjustment by the covariate of income, hours worked per week varied significantly with gender, $F(1, 705)=14.95$, $p<.001$, partial $\eta^2=.021$. Table 1 presents a summary of the ANCOVA results. Comparison of adjusted group means, as displayed in Table 2, reveals that males work significantly more hours per week than females. No statistically significant difference was found for job satisfaction.

Table 1 ANCOVA Summary Table

Source	SS	df	MS	F	p	η^2
Between Treatments	6405.27	4	1601.32	13.25	<.001	.049
Income	2823.08	1	2823.08	23.36	<.001	.032
Gender	1807.18	1	1807.18	14.95	<.001	.021
Job Satisfaction	203.10	1	203.10	1.68	.195	.002
Gender x Satisfaction	1.69	1	1.69	0.01	.906	.000
Error	85210.76	705	120.87			
Total	1642559.00	709				

Table 2 Adjusted and Unadjusted Group Means for Hours Worked per Week

	Adjusted M	Unadjusted M
Males	48.18	48.51
Females	44.86	44.31
Very Satisfied	47.06	47.56
Not Very Satisfied	45.97	45.87

SECTION 5.5 SPSS "HOW TO"

This section presents the steps for examining homogeneity of regression slopes and conducting analysis of covariance using **Univariate** procedure with the preceding *gssft.sav* example. To open the **Univariate** dialogue box as shown in Figure 5.20, select the following:

> **Analyze**
> > **General Linear Model**
> > > **Univariate**

Univariate Dialogue Box (see Figure 5.20)

Once in this box, click the DV (*hrs2*) and move it to the Dependent Variable box. Click the IVs (*sex* and *satjob2*) and move each to the Fixed Factor(s) box. Click the covariate (*rincom2*) and move it to the Covariate(s) box. Then click **Model.**

Univariate Model Dialogue Box (see Figure 5.21)

Under specify model, click **Custom**. Move each IV and covariate(s) to the Model box. Then hold down the Ctrl Key and highlight all IVs and covariate(s). Once highlighted, continue to hold down the Ctrl Key and move this selection to the Model box. This should create the interaction between all IVs and covariate(s) (e.g., *rincom2*satjob2*sex*). Also check to make sure that Interaction is specified in the Build Terms box. Click **Continue** and then **Options**.

Figure 5.20 Univariate ANOVA Dialogue Box.

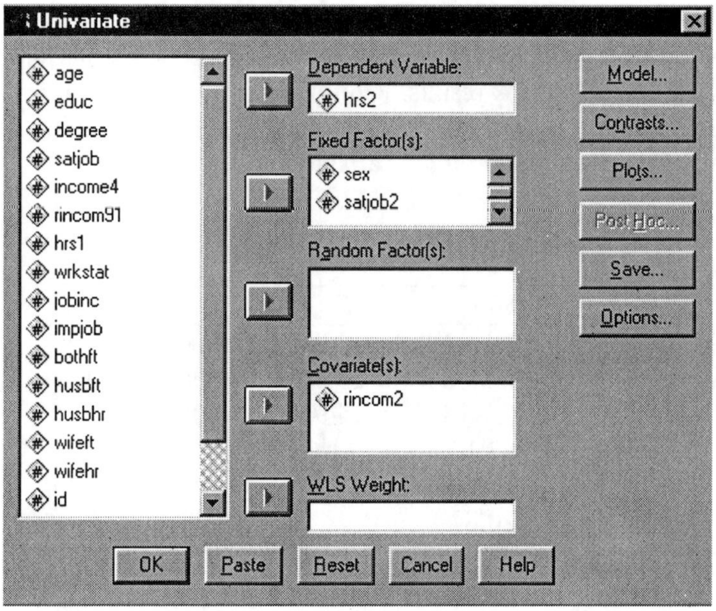

Figure 5.21 Univariate ANOVA Model Dialogue Box for Homogeneity-of-Slopes-Test.

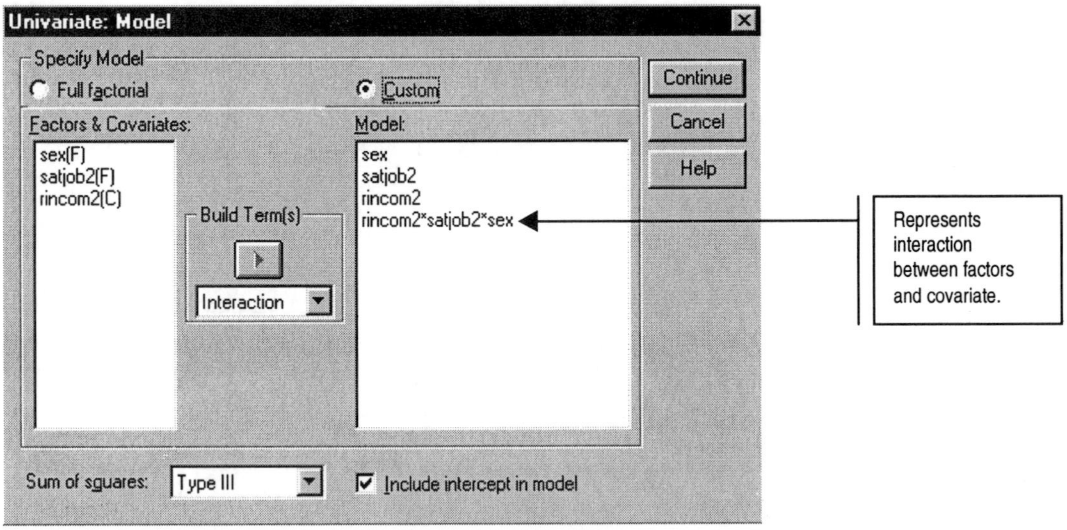

Univariate Options Dialogue Box (see Figure 5.22)

Click each IV (*sex* and *satjob2*) in the Factor(s) and Factor Interaction box and move each to the Display Means box. Select **Descriptive Statistics**, **Estimates of Effect Size**, and **Homogeneity Tests** in the Display box. The reader should note that these options are described in the previous chapter on ANOVA. Click **Continue**. Back in the Univariate Dialogue box, click **OK**.

Figure 5.22 Univariate ANOVA Options Dialogue Box.

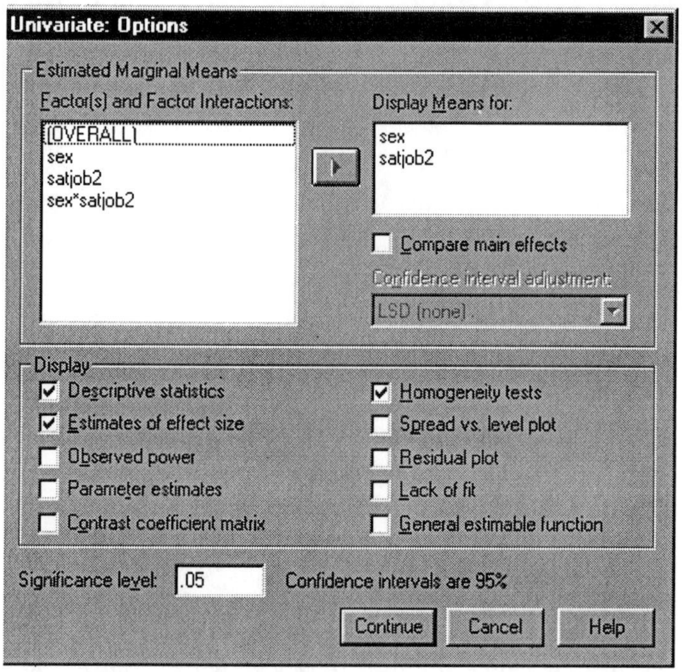

Figure 5.23 Univariate ANOVA Model Dialogue Box for Homogeneity-of-Slopes Test.

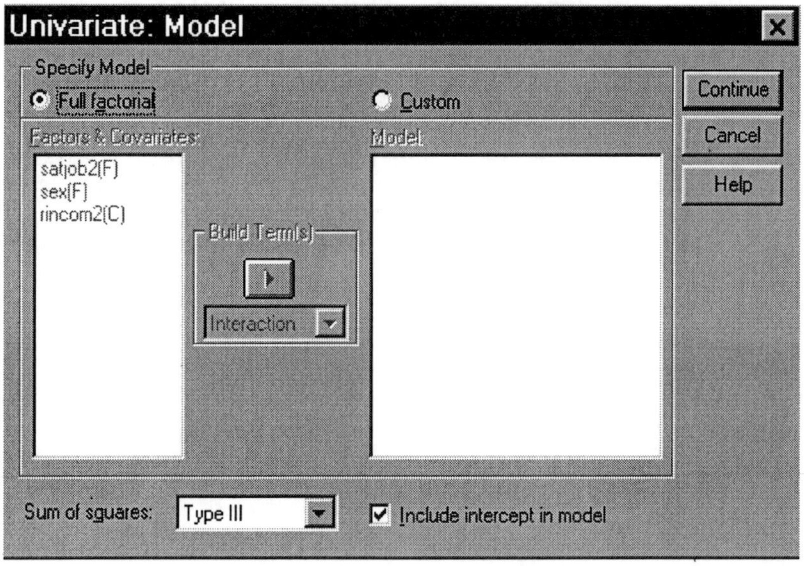

This process will create the output to evaluate homogeneity of regression slopes. If no interaction between IVs and covariate(s) has been found, then the following steps for ANCOVA can be conducted. Note that the steps within the Univariate Dialogue Box and Options Dialogue Box are the same as when conducting the homogeneity of regression slopes test.

```
Analyze
      General Linear Model
            Univariate
```

Univariate Dialogue Box (see Figure 5.20)

Once in this box, click the DV (*hrs2*) and move to the Dependent Variable box. Click the IVs (*sex* and *satjob2*) and move to the Fixed Factor(s) box. Click the covariate (*rincom2*) and move to the Covariate(s) box. Then click **Model.**

Univariate Model Dialogue Box (see Figure 5.23)

Under specify model, click **Full Factorial**. Within this box, you can also identify the method of calculating the Sum of Squares. Type III is the default and the most commonly used since it is best used when subgroups vary in sample size but have no missing cells. Back in the Univariate Dialogue Box, click **Options**.

Univariate Options Dialogue Box (see Figure 5.21)

Click each IV (*sex* and *satjob2*) in the Factor(s) and Factor Interaction box and move to the Display Means box. Select **Descriptive Statistics, Estimates of Effect Size,** and **Homogeneity Tests** in the Display box. Click **Continue**. Back in the Univariate Dialogue box, click **Model**.

Summary

Analysis of covariance allows the researcher to determine group differences while controlling for the effect of one or more variables. Essentially, the influence that the covariate(s) has on the DV is partitioned out before groups are compared; thus, ANCOVA compares group means that have been adjusted to control for any influence the covariate(s) may have on the DV. Prior to conducting ANCOVA, data should be screened for missing data, outliers, normality of subgroups, homogeneity of variance, and homogeneity of regression slopes. The SPSS Univariate procedure generates an ANCOVA summary table that provides F ratios, p values, and effect sizes, which indicate the significance of factor and covariate main effects as well as factor interaction. If factor interaction is significant, then conclusions are limited about the main effects for each factor. Figure 5.24 provides a checklist for conducting analysis of covariance. SPSS steps for screening missing data, outliers, normality, and homogeneity of variance are presented in Chapter 3. Steps for creating a line plot are presented in Chapter 4.

Figure 5.24 Checklist for Conducting ANCOVA.

I.   **Screen Data**
   a.   Missing Data?
   b.   Outliers?
      ❏   Run Outliers and review stem-and-leaf plots and boxplots within **Explore**.
      ❏   Eliminate or transform outliers if necessary.
   c.   Normality?
      ❏   Run Normality Plots with Tests within **Explore**.
      ❏   Review boxplots and histograms.
      ❏   Transform data if necessary.
   d.   Homogeneity of Variance?
      ❏   Run Levene's Test of Equality of Variances within **Univariate**.
      ❏   Transform data if necessary.
   e.   Homogeneity of regression slopes?
      ❏   Run Univariate ANOVA for Homogeneity of regression slopes.
         1. ⌐ᵗ **Analyze...**⌐ᵗ **General Linear Model...**⌐ᵗ **Univariate**.
         2. Move DV to Dependent Variable box.
         3. Move IVs to Fixed Factor box.
         4. Move covariate(s) to Covariate box.
         5. ⌐ᵗ **Model.**
         6. ⌐ᵗ **Custom.**
         7. Move each IV and covariate to the Model box.
         8. Hold down Ctrl key and highlight all IVs and covariate(s), ⌐ᵗ ▶ while still holding down the Ctrl key in order to move interaction to Model box.
         9. ⌐ᵗ **Continue.**
         10. ⌐ᵗ **Options.**
         11. Check **Descriptive statistics, Estimates of effect size,** and **Homogeneity Tests.**
         12. ⌐ᵗ **Continue.**
         13. ⌐ᵗ **OK.**
      ❏   If factors and covariates interact, do not conduct ANCOVA.
   f.   Factor Interaction
      ❏   Create lineplot of DV by IVs.

II.   **Conduct ANCOVA**
   a.   Run ANCOVA using Univariate ANOVA
         1. ⌐ᵗ **Analyze...**⌐ᵗ **General Linear Model...**⌐ᵗ **Univariate**
         2. Move DV to Dependent Variable box.
         3. Move IVs to Fixed Factor box.
         4. Move covariate(s) to Covariate box.
         5. ⌐ᵗ **Model.**
         6. ⌐ᵗ **Full Factorial.**
         7. ⌐ᵗ **Continue.**
         8. ⌐ᵗ **Options**
         9. Check **Descriptive statistics, Estimates of effect size,** and **Homogeneity Tests.**
         10. ⌐ᵗ **Continue**
         11. ⌐ᵗ **OK.**
   b.   Interpret factor interaction.
   c.   If no factor interaction, interpret main effects for each factor.

III.   **Summarize Results**
   a.   Describe any data elimination or transformation.
   b.   Present line plot of factor interaction.
   c.   Present table of adjusted and unadjusted group means.
   d.   Narrate main effects for each factor and interaction (F-ratio, p value, and effect size).
   e.   Draw conclusions.

Exercises for Chapter 5

The exercises below utilize the data set *gssft.sav,* which can be downloaded at the SPSS Web site. Open the URL: **www.spss.com/tech/DataSets.html** in your Web browser. Scroll down until you see "Data Used in SPSS Guide to Data Analysis—8.0 and 9.0" and click on the link "dataset.exe." When the "Save As" dialogue appears, select the appropriate folder and save the file. Preferably, this should be a folder created in the SPSS folder of your hard drive for this purpose. Once the file is saved, double-click the "dataset.exe" file to extract the data sets to the folder.

Address the following research problem: You are interested in evaluating the effect of gender (*sex*) and highest degree attained (*degree*) on respondent's income (*rincom1*) while controlling for hours worked per week (*hrs1*).

1. Develop the appropriate research questions and/or hypotheses for main effects and interaction.

2. Screen data for missing data and outliers. What steps, if any, are necessary for reducing missing data and outliers?

3. Test the assumptions of normality, homogeneity of regression slopes, and homogeneity of variance.

   a.   What steps, if any, are necessary for increasing normality?

   b.   Do the covariate and factors interact? Can you conclude homogeneity of regression slopes?

   c.   Can you conclude homogeneity of variance?

4. Create a line plot of the factors. Do factors interact?

5. Conduct ANCOVA.

   a.   Is factor interaction significant? Explain.

   b.   Are main effects significant? Explain.

   c.   Does the covariate significantly influence the DV? Explain.

   d.   What can you conclude from the effect size for each main effect?

6. Write a results statement.

CHAPTER 6

MULTIVARIATE ANALYSIS OF VARIANCE AND COVARIANCE

All of the statistical analysis techniques discussed to this point have involved only one dependent variable. In this chapter, for the first time, we consider *multivariate statistics*—statistical procedures that involve more than one dependent variable. The focus of this chapter is on two of the most widely used multivariate procedures: the multivariate variations of analysis of variance and analysis of covariance. These versions of analysis of variance and covariance are designed to handle two or more dependent variables within the standard ANOVA/ANCOVA designs. We begin by discussing multivariate analysis of variance in detail, followed by a discussion of the application of covariance analysis in the multivariate setting.

I. MANOVA

Like ANOVA, multivariate analysis of variance (MANOVA) is designed to test the significance of group differences. The only substantial difference between the two procedures is that MANOVA can include several dependent variables, whereas ANOVA can handle only one DV. Oftentimes, these multiple dependent variables consist of different measures of essentially the same thing (Aron & Aron, 1999), but this need not always be the case. At a minimum, the DVs should have some degree of linearity and share a common conceptual meaning (Stevens, 1992). They should "make sense" as a group of variables. As you will soon see, the basic logic behind a MANOVA is essentially the same as in a univariate analysis of variance.

SECTION 6.1 PRACTICAL VIEW

Purpose

The clear advantage of a multivariate analysis of variance over a univariate analysis of variance is the inclusion of multiple dependent variables. Stevens (1992) provides two reasons why a researcher should be interested in using more than one DV when comparing treatments or groups based on differing characteristics:

(1) Any worthwhile treatment or substantial characteristic will likely affect subjects in more than one way; hence, the need for additional criterion (dependent) measures.

(2) The use of several criterion measures permits the researcher to obtain a more "holistic" picture, and therefore a more detailed description, of the phenomenon under investigation (pp. 151-152). This stems from the idea that it is extremely difficult to obtain a "good" measure of a trait (e.g.,

math achievement, self-esteem, etc.) from one variable; multiple measures on variables representing a common characteristic are bound to be more representative of that characteristic.

ANOVA tests whether mean differences among k groups on a single DV are significant, or likely to have occurred by chance. However, when we move to the multivariate situation, the multiple DVs are treated in combination. In other words, MANOVA tests whether mean differences among k groups on a *combination of DVs* are likely to have occurred by chance. As part of the actual analysis, a "new" DV is created. This new DV is, in fact, a linear combination of the original measured DVs, combined in such a way as to maximize the group differences (i.e., separate the k groups as much as possible). The new DV is created by developing a linear equation where each measured DV has an associated weight and, when combined and summed, creates maximum separation of group means with respect to the new DV:

$$Y_{\text{new}} \; = \; a_1 Y_1 \; + \; a_2 Y_2 \; + \; a_3 Y_3 \; + ... + \; a_n Y_n, \qquad \text{(Equation 6.1)}$$

where Y_n is an original DV, a_n is its associated weight, and n is the total number of original measured DVs. An ANOVA is then conducted on this newly created variable.

Let us consider the following example: Assume we wanted to investigate the differences in worker productivity, as measured by income level (DV_1) and hours worked (DV_2), for individuals of different age categories (IV). Our analysis would involve the creation of a new DV, which would be a linear combination (DV_{new}) of our subjects' income levels and numbers of hours worked that maximizes the separation of our age category groups. Our new DV would then be subjected to a univariate ANOVA by comparing variances on DV_{new} for the various groups as defined by age category.

One could also have a *factorial MANOVA*—a design that would involve multiple IVs as well as multiple DVs. In this situation, a different linear combination of DVs is formed for each main effect and each interaction (Tabachnick & Fidell, 1996). For example, we might consider investigating the effects of gender (IV_1) and job satisfaction (IV_2) on employee income (DV_1) and years of education (DV_2). Our analysis would actually provide three new DVs—the first linear combination would maximize the separation between males and females (IV_1), the second linear combination would maximize the separation among job satisfaction categories (IV_2), and the third would maximize the separation among the various cells of the interaction between gender and job satisfaction.

At this point, one might be inclined to question why a researcher would want to engage in a multivariate analysis of variance, as opposed to simply doing a couple of comparatively simple analyses of variance. MANOVA has several advantages over its simpler univariate counterpart (Tabachnick & Fidell, 1996). First, as previously mentioned, by measuring several DVs instead of only one, the chances of discovering what actually changes as a result of the differing treatments or characteristics (and any interactions) improves immensely. If we wanted to know what measures of work productivity are affected by gender and age, we improve our chances of uncovering these effects by including hours worked as well as income level.

There are also several statistical reasons for preferring a multivariate analysis over a univariate one (Stevens, 1992). A second advantage is that, under certain conditions, MANOVA may reveal differences not shown in separate ANOVAs (Tabachnick & Fidell, 1996; Stevens, 1992). Assume we have a one-way design, with two levels on the IV and two DVs. If separate ANOVAs are conducted on two DVs, the distributions for each of the two groups (and for each DV) might overlap sufficiently, such that a mean difference probably would not be found. However, when the two DVs are considered in combination with each other, the two groups may differ substantially and could result in a statistically significant difference between groups. Therefore, a MANOVA may sometimes be more powerful than separate ANOVAs.

Third, the use of several univariate analyses leads to a greatly inflated overall Type I error rate. Consider a simple design with one IV (with two levels) and five DVs. If we assume that we wanted to test for group differences on each of the DVs (at $\alpha = .05$ level of significance), we would have to conduct five univariate tests. Recall that at an α-level of .05, we are assuming a 95% chance of no Type I errors. Because of the assumption of independence, we can multiply the probabilities. The effect of these error rates is compounded over all of the tests such that the overall probability of *not* making a Type I error becomes:

$$(.95)(.95)(.95)(.95)(.95) = .77$$

In other words, the probability of at least one false rejection (i.e., Type I error) becomes

$$1 - .77 = .23$$

which, as we all know, is an unacceptably high rate of possible statistical decision error (Stevens, 1992). Therefore, using this approach of fragmented univariate tests results in an overall error rate which is entirely too risky. The use of MANOVA includes a condition that maintains the overall error rate at the .05 level, or whatever α-level is pre-selected (Harris, 1998).

Finally, the use of several univariate tests ignores some very important information. Recall that if several DVs are included in an analysis, they should be correlated to some degree. A multivariate analysis incorporates the intercorrelations among DVs into the analysis (this is essentially the basis for the linear combination of DVs).

The reader should keep in mind, however, that there are disadvantages in the use of MANOVA. The main disadvantage is the fact that MANOVA is substantially more complicated than ANOVA (Tabachnick & Fidell, 1996). In the use of MANOVA, there are several important assumptions that need to be met. Furthermore, the results are sometimes ambiguous with respect to the effects of IVs on individual DVs. Finally, situations in which MANOVA is more powerful than ANOVA, as discussed a few paragraphs ago, are quite limited; often the multivariate procedure is much *less* powerful than ANOVA (Tabachnick & Fidell, 1996). It has been recommended that one carefully consider the need for additional DVs in an analysis in light of the added complexity (Tabachnick & Fidell, 1996).

In the univariate case, the null hypothesis stated that the population means are equal:

$$H_0: \ \mu_1 = \mu_2 = \mu_3 = \ldots = \mu_k$$

The calculations for MANOVA, however, are based on matrix algebra (as opposed to scalar algebra). The null hypothesis in MANOVA states that the population *mean vectors* are equal:

$$H_0: \ \mu_1 = \mu_2 = \mu_3 = \ldots = \mu_k$$

For the univariate analysis of variance, recall that the F-statistic is used to test the tenability of the null hypothesis. This test statistic is calculated by dividing the variance between the groups by the variance within the groups. There are several available test statistics for multivariate analysis of variance, but the most commonly used criterion is Wilks' Lambda (Λ). (Other test statistics for MANOVA include Pillai's Trace, Hotelling's Trace, and Roy's Largest Root.) Without going into great detail, Wilks' Lambda is obtained by calculating $|\mathbf{W}|$ (a measure of the within-groups sum-of-squares and cross-products matrix—a multivariate generalization of the univariate sum-of-squares within [SS_W]) and dividing it by $|\mathbf{T}|$ (a measure of the total sum-of-squares and cross-products matrix—also a multivariate generalization, this time of the total sum-of-squares [SS_T]). The obtained value of Wilks' Λ ranges from zero to one. It is important to note that Wilks' Λ is an *inverse criterion*; i.e., the smaller the value of Λ, the more evidence for treatment effects or group differences (Stevens, 1992). The reader should realize that this is the opposite relationship that F has to the amount of treatment effect.

In conducting a MANOVA, one first tests the overall multivariate hypothesis (i.e., that all groups are equal on the combination of DVs). This is accomplished by evaluating the significance of the test associated with Λ. If the null hypothesis is retained, it is common practice to stop the interpretation of the analysis at this point and conclude that the treatments or conditions have no effect on the DVs. However, if the overall multivariate test is significant, the researcher then would likely wish to discover which of the DVs is being affected by the IV(s). To accomplish this, one conducts a series of univariate analyses of variance on the individual DVs. This will undoubtedly result in multiple tests of significance, which will result in an inflated Type I error rate.

To counteract the potential of an inflated error rate due to multiple ANOVAs, an adjustment must be made to the alpha level used for the tests. This *Bonferroni-type* adjustment involves setting a more stringent alpha level for the test of each DV so that the alpha for the *set* of DVs does not exceed some critical value (Tabachnick & Fidell, 1996). That critical value for testing each DV is usually the overall α-level for the analysis (e.g., $\alpha = .05$) divided by the number of DVs. For example, if one had three DVs and wanted an overall α equal to .05, each univariate test could be conducted at $\alpha = .016$, since $.05/3 = .0167$. One should note that rounding down is necessary to create an overall alpha less than .05. The following equation may be used to check adjustment decisions:

$$\alpha = 1 - [(1 - \alpha_1)(1 - \alpha_2)...(1 - \alpha_p)]$$

where the overall error rate (α) is based on the error rate for testing the first DV (α_1), the second DV (α_2), and all others to the pth DV (α_p). All alphas can be set at the same level, or more important DVs can be given more liberal alphas (Tabachnick & Fidell, 1996).

Finally, for any univariate test of a DV that results in significance, one then conducts univariate post hoc tests (as discussed in Chapter 4) in order to identify where specific differences lie (i.e., which levels of the IV are different from which other levels). To summarize the analysis procedure for MANOVA, a researcher would follow these steps:

(1)   Examine the overall multivariate test of significance—if the results are significant, proceed to the next step; if not, stop.

(2)   Examine the univariate tests of individual DVs—if any are significant, proceed to the next step; if not, stop.

(3)   Examine the post hoc tests for individual DVs.

Sample Research Questions

In our first sample study in this chapter, we are concerned with investigating differences in worker productivity, as measured by income level (DV_1) and hours worked (DV_2), for individuals of different age categories (IV)—a one-way MANOVA design. Therefore, this study would address the following research questions:

(1)   Are there significant mean differences in worker productivity (as measured by the combination of income and hours worked) for individuals of different ages?

(2)   Are there significant mean differences in income levels for individuals of different ages?

(2a) If so, which age categories differ?

(3)   Are there significant mean differences in hours worked for individuals of different ages?

(3a) If so, which age categories differ?

Our second sample study will demonstrate a two-way MANOVA where we investigate the gender (IV_1) and job satisfaction (IV_2) differences in income level (DV_1) and years of education (DV_2). One should note the following questions address the MANOVA analysis, univariate ANOVA analyses, and post hoc analyses:

(1)   a.   Are there significant mean differences in the combined DV of income and years of education for males and females?

b.   Are there significant mean differences in the combined DV of income and years of education for different levels of job satisfaction? If so, which job satisfaction categories differ?

c.   Is there a significant interaction between gender and job satisfaction on the combined DV of income and years of education?

(2)   a.   Are there significant mean differences on income between males and females?

b.   Are there significant mean differences on income between different levels of job satisfaction? If so, which job satisfaction categories differ?

c.   Is there a significant interaction between gender and job satisfaction on income?

(3)   a.   Are there significant mean differences in years of education between males and females?

b.   Are there significant mean differences in years of education among different levels of job satisfaction? If so, which job satisfaction categories differ?

c.   Is there a significant interaction between gender and job satisfaction on years of education?

SECTION 6.2   ASSUMPTIONS AND LIMITATIONS

Since we are introducing our first truly multivariate technique in this chapter, we have a "new" set of statistical assumptions to discuss. They are new in that they apply to the multivariate situation; however, they are quite analogous to the assumptions for univariate analysis of variance, which we have already examined (see Chapter 4). For multivariate analysis of variance, these assumptions are:

(1)   The observations within each sample must be randomly sampled and must be independent of each other.

(2)   The observations on all dependent variables must follow a multivariate normal distribution in each group.

(3)   The population covariance matrices for the dependent variables in each group must be equal (this assumption is often referred to as the *homogeneity of covariance matrices* assumption or the assumption of homoscedasticity).

(4)   The relationships among all pairs of DVs for each cell in the data matrix must be linear.

As a reminder to the reader, the assumption of independence is primarily a design issue, not a statistical one. Provided the researcher has randomly sampled and assigned subjects to treatments, it is usually safe to believe that this assumption has not been violated. We will focus our attention on the assumptions of multivariate normality, homogeneity of covariance matrices, and linearity.

Methods of Testing Assumptions

As discussed in Chapter 3, multivariate normality implies that the sampling distribution of the means of each DV in each cell and all linear combinations of DVs are normally distributed (Tabachnick & Fidell, 1996). Multivariate normality is a difficult entity to describe and even more difficult to assess. Initial screening for multivariate normality consists of assessments for univariate normality (see Chapter 3) for all variables, as well as examinations of all bivariate scatterplots (see Chapter 3) to check that they are approximately elliptical (Stevens, 1996). Specific graphical tests for multivariate normality do exist, but are not available in standard statistical software packages (Stevens, 1996) and will not be discussed here.

It is probably most important to remember that both ANOVA and MANOVA are robust to moderate violations of normality, provided the violation is created by skewness and not by outliers (Tabachnick & Fidell, 1996). With equal or unequal sample sizes and only a few DVs, a sample size of about 20 in the smallest cell should be sufficient to ensure robustness to violations of univariate and multivariate normality. If it is determined that the data have substantially deviated from normal, transformations of the original data should be considered.

Recall that the assumption of equal covariance matrices (i.e., homoscedasticity) is a necessary condition for multivariate normality (Tabachnick & Fidell, 1996). The failure of the relationship between two variables to be homoscedastic is caused either by the nonnormality of one of the variables or by the fact that one of the variables may have some sort of relationship to the transformation of the other variable. Therefore, checking for univariate and multivariate normality is a good starting point for assessing possible violations of homoscedasticity. Specifically, possible violations of this assumption may be assessed by interpreting the results of Box's Test. The reader should note that a violation of the assumption of homoscedasticity, similar to a violation of homogeneity, will not prove fatal to an analysis (Tabachnick & Fidell, 1996; Kennedy & Bush, 1985). However, the results will be greatly improved if the heteroscedasticity is identified and corrected (Tabachnick & Fidell, 1996) by means of data transformations. On the other hand, if homogeneity of variance-covariance is violated, a more robust multivariate test statistic, Pillai's Trace, can be selected when interpreting the multivariate results.

Linearity is best assessed through inspection of bivariate scatterplots. If both variables in the pair are normally distributed and linearly related, the shape of the scatterplot should be elliptical. If one of the variables is not normally distributed, the relationship will not be linear and the scatterplot between the two variables will not appear oval shaped. As mentioned in Chapter 2, assessing linearity by means of bivariate scatterplots is an extremely subjective procedure. In situations where nonlinearity between variables is apparent, the data can once again be transformed in order to enhance the linear relationship.

SECTION 6.3 PROCESS AND LOGIC

The Logic Behind MANOVA

As previously mentioned, the calculations for MANOVA somewhat parallel those for a univariate ANOVA, although they exist in multivariate form (i.e., they rely on matrix algebra). Since several variables are involved in this analysis, calculations are based on a *matrix* of values, as opposed to the mathematical manipulations of a single value. Specifically, the matrix used in the calculations is the sum-of-squares and cross-products (SSCP) matrix, which you will recall is the precursor to the variance-covariance matrix (see Chapter 1).

In univariate ANOVA, recollect that the calculations are based on a partitioning of the total sum-of-squares into the sum-of-squares between the groups and the sum-of-squares within the groups:

$$SS_{\text{Total}} = SS_{\text{Between}} + SS_{\text{Within}}$$

In MANOVA, the calculations are based on the corresponding matrix analogue (Stevens, 1992), in which the total sum-of-squares and cross-products matrix (**T**) is partitioned into a between sum-of-squares and cross-products matrix (**B**) and a within sum-of-squares and cross-products matrix (**W**):

$$SSCP_{\text{Total}} = SSCP_{\text{Between}} + SSCP_{\text{Within}}$$

or

$$\mathbf{T} = \mathbf{B} + \mathbf{W} \qquad \text{(Equation 6.2)}$$

Wilks' Lambda (Λ) is then calculated by using the *determinants*—a sort of *generalized* variance for an entire set of variables—of the SSCP matrices (Stevens, 1992). The resulting formula for Λ becomes:

$$\Lambda = \frac{|\mathbf{W}|}{|\mathbf{T}|} = \frac{|\mathbf{W}|}{|\mathbf{B} + \mathbf{W}|} \qquad \text{(Equation 6.3)}$$

If there is no treatment effect or group differences, then $\mathbf{B} = 0$ and $\Lambda = 1$ indicating no differences between groups on the linear combination of DVs; whereas, if **B** were very large (i.e., substantially greater than 0), then Λ would approach 0, indicating significant group differences on the combination of DVs.

As in all of our previously discussed ANOVA designs, we can again obtain a measure of strength of association, or effect size. Recall that eta squared (η^2) is a measure of the magnitude of the relationship between the independent and dependent variables and is interpreted as the proportion of variance in the dependent variable explained by the independent variable(s) in the sample. For MANOVA, eta squared is obtained in the following manner:

$$\eta^2 = 1 - \Lambda$$

In the multivariate situation, η^2 is interpreted as the variance accounted for in the best linear combination of DVs by the IV(s) and/or interactions of Ivs.

Interpretation of Results

The MANOVA procedure generates several test statistics to evaluate group differences on the combined DV: Pillai's Trace, Wilks' Lambda, Hotelling's Trace, and Roy's Largest Root. When the IV has only two categories, the F test for Pillai's Trace, Wilks' Lambda, and Hotelling's Trace will be identical. When the IV has three or more categories, the F test for these three statistics will differ slightly but will maintain consistent significance or non-significance. Although, these test statistics may vary only slightly, Wilks' Lambda is the most commonly reported MANOVA statistic. Pillai's Trace is used when homogeneity of variance-covariance is in question. If two or more IVs are included in the analysis, factor interaction must be evaluated before main effects.

In addition to the multivariate tests, the output for MANOVA typically includes the test for homogeneity of variance-covariance (Box's Test), univariate ANOVAs, and univariate post hoc tests. Since homogeneity of variance-covariance is a test assumption for MANOVA and has implications in how to interpret the multivariate tests, the results of Box's Test should be evaluated first. Highly sensitive to the violation of normality, Box's Test should be interpreted with caution. Typically, if Box's Test is significant at $p<.001$ and group sample sizes are extremely unequal, then robustness cannot be assumed due to unequal variances among groups (Tabachnick & Fidell, 1996).   In such a situation, a more robust MANOVA test statistic, Pillai's Trace, is utilized when interpreting the MANOVA results. If equal variances are assumed, Wilks' Lambda is commonly used as the MANOVA test statistic.   Once the test statistic has been determined, factor interaction (F ratio and p value) should be assessed if two or more IVs are included in the analysis.   Like two-way ANOVA, if interaction is significant, then inferences drawn from the main effects are limited.   If factor interaction is *not* significant, then one should proceed to examine the F ratios and p values for each main effect.   When multivariate significance is found, the univariate ANOVA results can indicate the degree to which groups differ for each DV.   A more conservative alpha level should be applied using the Bonferroni adjustment.   Post hoc results can then indicate which groups are significantly different for the DV if univariate significance is found for that particular DV.

In summary, the first step in interpreting the MANOVA results is to evaluate the Box's Test.   If homogeneity of variance-covariance is assumed, utilize the Wilks' Lambda statistic when interpreting the multivariate tests.   If the assumption of equal variances is violated, use Pillai's Trace.   Once the multivariate test statistic has been identified, examine the significance (F ratios and p values) of factor interaction.   This is necessary only if two or more IVs are included.   Next evaluate the F ratios and p values for each factor's main effect.   If multivariate significance is found, interpret the univariate ANOVA results to determine significant group differences for each DV.   If univariate significance is revealed, examine the post hoc results to identify which groups are significantly different for each DV.

For our example that investigates age category (*agecat4*) differences in respondent's income (*rincom91*) and hours worked per week (*hrs1*), data were screened for missing data and outliers and then examined for fulfillment of test assumptions.   Data screening led to the transformation of *rincom91* to *rincom2* in order to eliminate all cases with income equal to zero and cases equal to or exceeding 22. *Hrs1* was also transformed to *hrs2* as a means of reducing the number of outliers; those less than or equal to 16 were recoded 17, and those greater than or equal to 80 were recoded 79.   Although normality of these transformed variables is still questionable, group sample sizes are quite large and fairly equivalent.   Therefore, normality will be assumed.   Linearity of the two DVs was then tested by creating a scatterplot and calculating the Pearson correlation coefficient.   Results indicate a linear relationship. Although the correlation coefficient is statistically significant, it is still quite low ($r=.253$, $p<.001$).   The last assumption, homogeneity of variance-covariance, will be tested within MANOVA.   Thus, MANOVA was conducted utilizing the **Multivariate** procedure.   The Box's Test (see Figure 6.1) reveals that equal variances can be assumed, $F(9, 2886561)=.766$, $p=.648$; therefore, Wilks' Lambda will be used as the test statistic.   Figure 6.2 presents the MANOVA results.   The Wilks' Lambda criteria indicates significant group differences in age category with respect to income and hours worked per week, Wilks' $\Lambda=.909$, $F(6,1360)=11.04$, $p<.001$, multivariate $\eta^2=.046$.   Univariate ANOVA results (see Figure 6.3) were interpreted using a more conservative alpha level ($\alpha=.025$).   Results reveal that age category significantly differs for only income ($F(3, 681)=21.00$, $p<.001$, partial $\eta^2=.085$) and not hours worked per week ($F(3, 681)=.167$, $p=.919$, partial $\eta^2=.001$).   Examination of post hoc results reveal that income of those 18-29 years of age significantly differs from all other age categories (see Figure 6.4). In addition, income for individuals 30-39 years differ from those 40-49 years.

Figure 6.1 Box's Test for Homogeneity of Variance-Covariance.

Box's Test of Equality of Covariance Matrices[a]

Box's M	6.936
F	.766
df1	9
df2	2886561
Sig.	.648

Tests the null hypothesis that the observed covariance
matrices of the dependent variables are equal across groups.

a. Design: Intercept+AGECAT4

Box's test is not
significant. Use
Wilks' Lambda
criteria.

Figure 6.2 Multivariate Tests for Income and Hours Worked by Age Category.

Multivariate Tests[c]

Effect		Value	F	Hypothesis df	Error df	Sig.	Eta Squared
Intercept	Pillai's Trace	.957	7507.272[a]	2.000	680.000	.000	.957
	Wilks' Lambda	.043	7507.272[a]	2.000	680.000	.000	.957
	Hotelling's Trace	22.080	7507.272[a]	2.000	680.000	.000	.957
	Roy's Largest Root	22.080	7507.272[a]	2.000	680.000	.000	.957
AGECAT4	Pillai's Trace	.091	10.791	6.000	1362.000	.000	.045
	Wilks' Lambda	.909	11.035[a]	6.000	1360.000	.000	.046
	Hotelling's Trace	.100	11.279	6.000	1358.000	.000	.047
	Roy's Largest Root	.099	22.457[b]	3.000	681.000	.000	.090

a. Exact statistic

b. The statistic is an upper bound on F that yields a lower bound on the significance level.

c. Design: Intercept+AGECAT4

Indicates that age
category
significantly differs
for the combined
DV.

Figure 6.3 Univariate ANOVA Summary Table.

Tests of Between-Subjects Effects

Source	Dependent Variable	Type III Sum of Squares	df	Mean Square	F	Sig.	Eta Squared
Corrected Model	RINCOM2	1029.016[a]	3	343.005	20.995	.000	.085
	HRS2	64.281[b]	3	21.427	.167	.919	.001
Intercept	RINCOM2	128493.5	1	128493.5	7864.965	.000	.920
	HRS2	1410954	1	1410954	10972.708	.000	.942
AGECAT4	RINCOM2	1029.016	3	343.005	20.995	.000	.085
	HRS2	64.281	3	21.427	.167	.919	.001
Error	RINCOM2	11125.807	681	16.337			
	HRS2	87568.119	681	128.588			
Total	RINCOM2	149966.0	685				
	HRS2	1575151	685				
Corrected Total	RINCOM2	12154.823	684				
	HRS2	87632.400	684				

a. R Squared = .085 (Adjusted R Squared = .081)

b. R Squared = .001 (Adjusted R Squared = -.004)

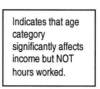

Indicates that age category significantly affects income but NOT hours worked.

Writing Up Results

Once again, any data transformations utilized to increase the likelihood of fulfilling test assumptions should be reported in the summary of results. The summary should then report the results from the multivariate tests by first indicating the test statistic utilized and its respective value and then reporting the F ratio, degrees of freedom, p value, and effect size for each IV main effect. If follow-up analysis was conducted using Univariate ANOVA, these results should be summarized next. Report the F ratio, degrees of freedom, p value, and effect size for the main effect on each DV. Utilize the post hoc results to indicate which groups were significantly different within each DV. Finally, you may want to create a table of means and standard deviations for each DV by the IV categories. In summary, the MANOVA results narrative should address the following:

(1) Subject elimination and/or variable transformation;
(2) MANOVA results (test statistic, F-ratio, degrees of freedom, p-value, and effect size);
    (a) Main effects for each IV on the combined DV;
    (b) Main effect for the interaction between IVs;
(3) Univariate ANOVA results (F-ratio, degrees of freedom, p-value, and effect size);
    (a) Main effect for each IV and DV;
    (b) Comparison of means to indicate which groups differ on each DV;
(4) Post hoc results (mean differences and levels of significance).

Utilizing our previous example, the following statement applies the results from Figures 6.1–6.4. A one-way multivariate analysis of variance (MANOVA) was conducted to determine age category differences in income and hours worked per week. Prior to the test, variables were transformed to eliminate outliers. Cases with income equal to zero or equal to or exceeding 22 were eliminated. Hours worked per week was also transformed; those less than or equal to 16 were recoded 17 and those greater than or equal to 80 were recoded 79. MANOVA results revealed significant differences among the age categories on the dependent variables, Wilks' $\Lambda=.909$, $F(6,1360)=11.04$, $p<.001$, multivariate $\eta^2=.046$. Analysis of variance (ANOVA) was conducted on each dependent variable as a follow-up test to MANOVA. Age category differences were significant for income, $F(3, 681)=21.00$, $p<.001$, partial $\eta^2=.085$. Differences in hours worked per week were not significant, $F(3, 681)=.167$, $p=.919$, partial $\eta^2=.001$. The Bonferroni post hoc analysis revealed that income of those 18-29 years significantly differs from all other age categories. In addition, income for individuals 30-39 years differs from those 40-49. Table 1 presents means and standard deviations for income and hours worked per week by age category.

Table 1 Means and Standard Deviations for Income and Hours Worked per Week by Age Category

Age	Income		Hours Worked per Week	
	M	**SD**	**M**	**SD**
18-29 years	11.87	4.14	46.32	10.32
30-39 years	14.03	3.88	47.03	11.42
40-49 years	15.32	3.87	46.49	11.75
50+ years	14.96	4.42	46.33	11.51

Figure 6.4 Post Hoc Results for Income and Hours Worked by Age Category.

Multiple Comparisons

Bonferroni

Dependent Variable	(I) 4 categories of age	(J) 4 categories of age	Mean Difference (I-J)	Std. Error	Sig.	95% Confidence Interval	
						Lower Bound	Upper Bound
RINCOM2	1 18-29	2 30-39	-2.1643*	.4486	.000	-3.3513	-.9774
		3 40-49	-3.4576*	.4603	.000	-4.6754	-2.2397
		4 50+	-3.0903*	.4935	.000	-4.3960	-1.7846
	2 30-39	1 18-29	2.1643*	.4486	.000	.9774	3.3513
		3 40-49	-1.2932*	.3972	.007	-2.3443	-.2421
		4 50+	-.9259	.4353	.203	-2.0776	.2258
	3 40-49	1 18-29	3.4576*	.4603	.000	2.2397	4.6754
		2 30-39	1.2932*	.3972	.007	.2421	2.3443
		4 50+	.3673	.4473	1.000	-.8163	1.5509
	4 50+	1 18-29	3.0903*	.4935	.000	1.7846	4.3960
		2 30-39	.9259	.4353	.203	-.2258	2.0776
		3 40-49	-.3673	.4473	1.000	-1.5509	.8163
HRS2	1 18-29	2 30-39	-.7112	1.2585	1.000	-4.0412	2.6187
		3 40-49	-.1694	1.2913	1.000	-3.5861	3.2473
		4 50+	-5.929E-03	1.3844	1.000	-3.6690	3.6572
	2 30-39	1 18-29	.7112	1.2585	1.000	-2.6187	4.0412
		3 40-49	.5418	1.1145	1.000	-2.4070	3.4907
		4 50+	.7053	1.2211	1.000	-2.5258	3.9364
	3 40-49	1 18-29	.1694	1.2913	1.000	-3.2473	3.5861
		2 30-39	-.5418	1.1145	1.000	-3.4907	2.4070
		4 50+	.1634	1.2549	1.000	-3.1570	3.4839
	4 50+	1 18-29	6.020E-03	1.3844	1.000	-3.6572	3.6690
		2 30-39	-.7053	1.2211	1.000	-3.9364	2.5258
		3 40-49	-.1634	1.2549	1.000	-3.4839	3.1570

Based on observed means.

* The mean difference is significant at the .05 level.

SECTION 6.4 MANOVA SAMPLE STUDY AND ANALYSIS

This section provides a complete example that applies the entire process of conducting MANOVA: development of research questions and hypotheses, data screening methods, test methods, interpretation of output, and presentation of results. The SPSS data set *gssft.sav* is utilized. Our previous example demonstrates a one-way MANOVA, while this example will present a two-way MANOVA.

Problem

This time, we are interested in determining the degree to which gender and job satisfaction affects income and years of education among employees. Since two IVs are tested in this analysis, questions must also take into account the possible interaction between factors. The following research questions and respective null hypotheses address the multivariate main effects for each IV and the possible interaction between factors.

Research Questions	*Null Hypotheses*
RQ1: Do income and years of education differ by gender among employees?	$H_0$1: Income and years of education will not differ by gender among employees.
RQ2: Do income and years of education differ by job satisfaction among employees?	$H_0$2: Income and years of education will not differ by job satisfaction among employees.
RQ3: Do gender and job satisfaction interact in the effect on income and years of education?	$H_0$3: Gender and job satisfaction will not interact in the effect on income and years of education.

Both IVs are categorical and include gender (*sex*) and job satisfaction (*satjob*). One should note that *satjob* represents four levels: very satisfied, moderately satisfied, a little dissatisfied, and very dissatisfied. The DVs are respondent's income (*rincom2*) and years of education (*educ*); both are quantitative. The variable, *rincom2*, is a transformation of *rincom91* from the previous example.

Method

Data should first be examined for missing data, outliers and fulfillment of test assumptions. The **Explore** procedure was conducted to identify outliers and evaluate normality. Boxplots (see Figure 6.5) indicate extreme values in *educ*. Consequently, *educ* was transformed to *educ2* in order to eliminate subjects with 6 years of education or less. **Explore** was conducted again to evaluate normality. Tests indicate significant non-normality for both *rincom2* and *educ2* in many categories (see Figure 6.6). Since MANOVA is fairly robust to non-normality, no further transformations will be performed. However, the significant non-normality coupled with the unequal group sample sizes, as in this example, may lead to violation of homogeneity of variance-covariance. The next step in examining test assumptions was to determine linearity between the DVs. A scatterplot was created; Pearson correlation coefficients were calculated (see Figure 6.7). Both indicate a linear relationship. Although the correlation coefficient is significant, it is still fairly weak (r=.337, p<.001). The final test assumption of homogeneity of variance-covariance will be tested with the MANOVA procedure. MANOVA was then conducted using **Multivariate**.

Figure 6.5 Boxplots for Years of Education by Gender and Job Satisfaction.

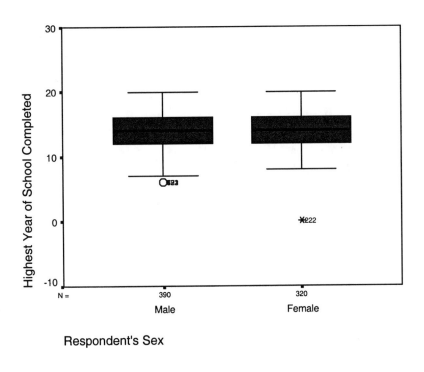

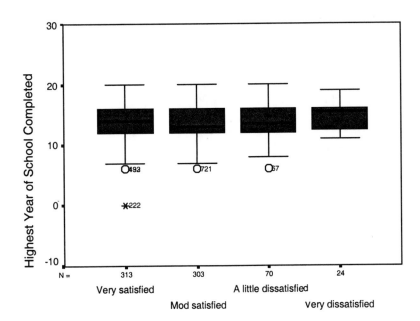

Figure 6.6 Tests of Normality of Income and Years of Education.

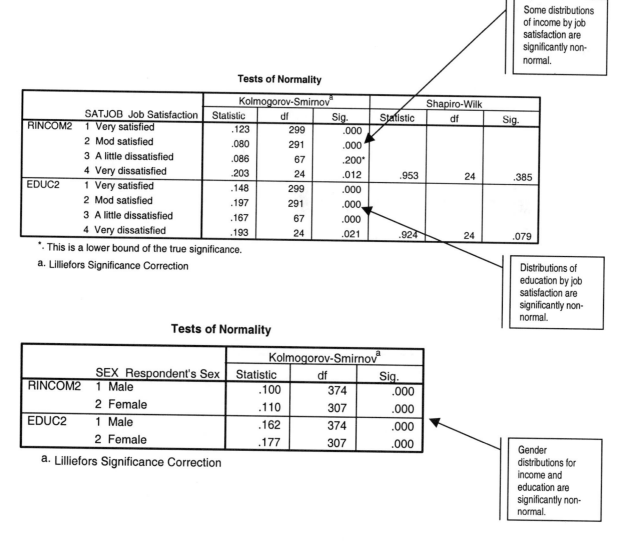

Tests of Normality

		Kolmogorov-Smirnov[a]			Shapiro-Wilk		
	SATJOB Job Satisfaction	Statistic	df	Sig.	Statistic	df	Sig.
RINCOM2	1 Very satisfied	.123	299	.000			
	2 Mod satisfied	.080	291	.000			
	3 A little dissatisfied	.086	67	.200*			
	4 Very dissatisfied	.203	24	.012	.953	24	.385
EDUC2	1 Very satisfied	.148	299	.000			
	2 Mod satisfied	.197	291	.000			
	3 A little dissatisfied	.167	67	.000			
	4 Very dissatisfied	.193	24	.021	.924	24	.079

*. This is a lower bound of the true significance.

a. Lilliefors Significance Correction

Some distributions of income by job satisfaction are significantly non-normal.

Distributions of education by job satisfaction are significantly non-normal.

Tests of Normality

		Kolmogorov-Smirnov[a]		
	SEX Respondent's Sex	Statistic	df	Sig.
RINCOM2	1 Male	.100	374	.000
	2 Female	.110	307	.000
EDUC2	1 Male	.162	374	.000
	2 Female	.177	307	.000

a. Lilliefors Significance Correction

Gender distributions for income and education are significantly non-normal.

Figure 6.7 Correlation Coefficients for Income and Years of Education.

Correlations

		EDUC2	RINCOM2
EDUC2	Pearson Correlation	1.000	.337**
	Sig. (2-tailed)		.000
	N	742	681
RINCOM2	Pearson Correlation	.337**	1.000
	Sig. (2-tailed)	.000	
	N	681	686

Correlation coefficient indicates low relationship.

**. Correlation is significant at the 0.01 level (2-tailed).

Output and Interpretation of Results

Figures 6.8 – 6.11 present some of the MANOVA output. The Box's Test (see Figure 6.8) is not significant and indicates that homogeneity of variance-covariance is fulfilled, $F(21, 20370)=1.245$, $p=.201$, so Wilks' Lambda test statistic will be used in interpreting the MANOVA results. The multivariate tests are presented in Figure 6.9. Factor interaction was then examined and revealed nonsignificance, $F(6, 1344)=.749$, $p=.610$, $\eta^2=.003$. The main effects of job satisfaction ($F(6, 1344)=3.98$, $p=.001$, $\eta^2=.017$) and gender ($F(2, 672)=3.98$, $p<.001$, $\eta^2=.024$) were both significant. However, multivariate effect sizes are very small. Prior to examining the univariate ANOVA results, the alpha level was adjusted to $\alpha=.025$ since two DVs were analyzed. Univariate ANOVA results (see Figure 6.10) indicate that income significantly differs for job satisfaction ($F(3, 673)=7.17$, $p<.001$, $\eta^2=.031$) and gender ($F(1, 673)=16.14$, $p<.001$, $\eta^2=.023$). Years of education do not significantly differ for job satisfaction ($F(3, 673)=2.18$, $p=.089$, $\eta^2=.010$) or gender ($F(1, 673)=1.03$, $p=.310$, $\eta^2=.002$). Scheffé post hoc results (see Figure 6.11) for income and job satisfaction indicate that individuals very satisfied significantly differ from those with only moderate satisfaction. Figures 6.12 and 6.13 present the unadjusted and adjusted group means for income and years of education.

Presentation of Results

The following narrative summarizes the results for the two-way MANOVA example.

A two-way MANOVA was conducted to determine the effect of job satisfaction and gender on the two dependent variable of respondent's income and years of education. Data were first transformed to eliminate outliers. Respondent's income was transformed to remove cases with income of zero or equal to or exceeding 22. Years of education was also transformed to eliminate cases with 6 or fewer years. MANOVA results indicate that job satisfaction (Wilks' $\Lambda=.965$, $F(6, 1344)=3.98$, $p=.001$, $\eta^2=.017$) and gender (Wilks' $\Lambda=.976$, $F(2, 672)=8.14$, $p<.001$, $\eta^2=.024$) significantly affect the combined DV of income and years of education. However, multivariate effect sizes are very small. Univariate ANOVA and Scheffé post hoc tests were conducted as follow-up tests. ANOVA results indicate that income significantly differs for job satisfaction ($F(3, 673)=7.17$, $p<.001$, $\eta^2=.031$) and gender ($F(1, 673)=16.14$, $p<.001$, $\eta^2=.023$). Years of education does not significantly differ for job satisfaction ($F(3, 673)=2.18$, $p=.089$, $\eta^2=.010$) or gender ($F(1, 673)=1.03$, $p=.310$, $\eta^2=.002$). Scheffé post hoc results for income and job satisfaction indicate that individuals very satisfied significantly differ from those with only moderate satisfaction. Table 1 presents the adjusted and unadjusted group means for income and years of education by job satisfaction and gender.

Table 1 Adjusted and Unadjusted Means for Income and Years of Education by Job Satisfaction and Gender

	Income		Years of Education	
	Adjusted M	**Unadjusted M**	**Adjusted M**	**Unadjusted M**
Gender				
Male	14.95	15.15	14.37	14.07
Female	12.89	13.07	14.04	14.12
Job Satisfaction				
Very Satisfied	14.93	15.02	14.33	14.32
Mod. Satisfied	13.42	13.52	13.84	13.83
Little Dissatisfied	13.74	13.81	13.99	14.00
Very Dissatisfied	13.61	13.71	14.68	14.79

Figure 6.8 Box's Test for Homogeneity of Variance-Covariance.

Box's Test of Equality of Covariance Matrices[a]

Box's M	26.935
F	1.245
df1	21
df2	20370
Sig.	.201

Box's Test is not significant. Use Wilks' Lambda criteria.

Tests the null hypothesis that the observed covariance matrices of the dependent variables are equal across groups.

a. Design: Intercept+SATJOB+SEX+SATJOB * SEX

SECTION 6.5   SPSS "HOW TO" FOR MANOVA

This section presents the steps for conducting multivariate analysis of variance (MANOVA) using the **Multivariate** procedure for the preceding example, which utilizes the *gssft.sav* data set. To open the Multivariate dialogue box as shown in Figure 6.14, select the following:

```
Analyze
        General Linear Model
                Multivariate
```

Multivariate Dialogue Box (see Figure 6.14)

Once in this box, click the DVs (*rincom2* and *educ2*) and move each to the Dependent Variables box. Click the IVs (*satjob* and *sex*) and move each to the Fixed Factor(s) box. Then click **Options**.

Multivariate Options Dialogue Box (see Figure 6.15)

Move each IV to the Display Means box. Select **Descriptive Statistics, Estimates of Effect Size,** and **Homogeneity Tests** under Display. These options are described in Chapter 4. Click **Continue**. Back in the **Multivariate** Dialogue Box, click **Post Hoc**.

Figure 6.9 MANOVA Summary Table.

Multivariate Tests[c]

Effect		Value	F	Hypothesis df	Error df	Sig.	Eta Squared
Intercept	Pillai's Trace	.923	4042.110[a]	2.000	672.000	.000	.923
	Wilks' Lambda	.077	4042.110[a]	2.000	672.000	.000	.923
	Hotelling's Trace	12.030	4042.110[a]	2.000	672.000	.000	.923
	Roy's Largest Root	12.030	4042.110[a]	2.000	672.000	.000	.923
SATJOB	Pillai's Trace	.035	3.967	6.000	1346.000	.001	.017
	Wilks' Lambda	.965	3.984[a]	6.000	1344.000	.001	.017
	Hotelling's Trace	.036	4.002	6.000	1342.000	.001	.018
	Roy's Largest Root	.033	7.291[b]	3.000	673.000	.000	.031
SEX	Pillai's Trace	.024	8.135[a]	2.000	672.000	.000	.024
	Wilks' Lambda	.976	8.135[a]	2.000	672.000	.000	.024
	Hotelling's Trace	.024	8.135[a]	2.000	672.000	.000	.024
	Roy's Largest Root	.024	8.135[a]	2.000	672.000	.000	.024
SATJOB * SEX	Pillai's Trace	.007	.750	6.000	1346.000	.609	.003
	Wilks' Lambda	.993	.749[a]	6.000	1344.000	.610	.003
	Hotelling's Trace	.007	.748	6.000	1342.000	.611	.003
	Roy's Largest Root	.005	1.051[b]	3.000	673.000	.370	.005

a. Exact statistic

b. The statistic is an upper bound on F that yields a lower bound on the significance level.

c. Design: Intercept+SATJOB+SEX+SATJOB * SEX

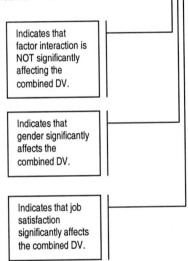

Indicates that factor interaction is NOT significantly affecting the combined DV.

Indicates that gender significantly affects the combined DV.

Indicates that job satisfaction significantly affects the combined DV.

Figure 6.10 Univariate ANOVA Summary Table.

> Indicates that job satisfaction significantly affects income but NOT years of education.

Tests of Between-Subjects Effects

Source	Dependent Variable	Type III Sum of Squares	df	Mean Square	F	Sig.	Eta Squared
Corrected Model	RINCOM2	1101.382[a]	7	157.340	9.676	.000	.091
	EDUC2	64.156[b]	7	9.165	1.358	.220	.014
Intercept	RINCOM2	47955.386	1	47955.386	2949.195	.000	.814
	EDUC2	49956.968	1	49956.968	7404.472	.000	.917
SATJOB	RINCOM2	349.728	3	116.576	7.169	.000	.031
	EDUC2	44.078	3	14.693	2.178	.089	.010
SEX	RINCOM2	262.463	1	262.463	16.141	.000	.023
	EDUC2	6.952	1	6.952	1.030	.310	.002
SATJOB * SEX	RINCOM2	26.942	3	8.981	.552	.647	.002
	EDUC2	15.304	3	5.101	.756	.519	.003
Error	RINCOM2	10943.317	673	16.261			
	EDUC2	4540.639	673	6.747			
Total	RINCOM2	149640.0	681				
	EDUC2	139907.0	681				
Corrected Total	RINCOM2	12044.699	680				
	EDUC2	4604.796	680				

a. R Squared = .091 (Adjusted R Squared = .082)

b. R Squared = .014 (Adjusted R Squared = .004)

> Indicates that gender significantly affects income but NOT years of education.

Multivariate Post Hoc Dialogue Box (see Figure 6.16)

Under Factors, select the IVs (*satjob*) and move to Post Hoc Tests box. For our example, only *satjob* was selected since gender has only two categories. Under Equal Variances Assumed, select the desired post hoc test. We selected Scheffé.

Figure 6.11 Post Hoc Tests for Income and Years of Education by Job Satisfaction.

Multiple Comparisons

Scheffe

Dependent Variable	(I) Job Satisfaction	(J) Job Satisfaction	Mean Difference (I-J)	Std. Error	Sig.	95% Confidence Interval Lower Bound	95% Confidence Interval Upper Bound
RINCOM2	1 Very satisfied	2 Mod satisfied	1.5045*	.3321	.000	.5739	2.4351
		3 A little dissatisfied	1.2174	.5450	.174	-.3101	2.7450
		4 Very dissatisfied	1.3151	.8555	.501	-1.0826	3.7127
	2 Mod satisfied	1 Very satisfied	-1.5045*	.3321	.000	-2.4351	-.5739
		3 A little dissatisfied	-.2871	.5464	.964	-1.8184	1.2443
		4 Very dissatisfied	-.1894	.8564	.997	-2.5895	2.2107
	3 A little dissatisfied	1 Very satisfied	-1.2174	.5450	.174	-2.7450	.3101
		2 Mod satisfied	.2871	.5464	.964	-1.2443	1.8184
		4 Very dissatisfied	9.764E-02	.9593	1.000	-2.5908	2.7861
	4 Very dissatisfied	1 Very satisfied	-1.3151	.8555	.501	-3.7127	1.0826
		2 Mod satisfied	.1894	.8564	.997	-2.2107	2.5895
		3 A little dissatisfied	-9.764E-02	.9593	1.000	-2.7861	2.5908
EDUC2	1 Very satisfied	2 Mod satisfied	.4929	.2139	.152	-.1066	1.0923
		3 A little dissatisfied	.3211	.3511	.841	-.6629	1.3050
		4 Very dissatisfied	-.4706	.5511	.866	-2.0150	1.0738
	2 Mod satisfied	1 Very satisfied	-.4929	.2139	.152	-1.0923	.1066
		3 A little dissatisfied	-.1718	.3520	.971	-1.1582	.8146
		4 Very dissatisfied	-.9635	.5516	.385	-2.5095	.5825
	3 A little dissatisfied	1 Very satisfied	-.3211	.3511	.841	-1.3050	.6629
		2 Mod satisfied	.1718	.3520	.971	-.8146	1.1582
		4 Very dissatisfied	-.7917	.6179	.650	-2.5234	.9401
	4 Very dissatisfied	1 Very satisfied	.4706	.5511	.866	-1.0738	2.0150
		2 Mod satisfied	.9635	.5516	.385	-.5825	2.5095
		3 A little dissatisfied	.7917	.6179	.650	-.9401	2.5234

Based on observed means.

* The mean difference is significant at the .05 level.

II. MANCOVA

As with univariate ANCOVA, researchers often wish to control for the effects of concomitant variables in a multivariate design. The appropriate analysis technique for this situation is a multivariate analysis of covariance, or MANCOVA. Multivariate analysis of covariance is essentially a combination of MANOVA and ANCOVA. MANCOVA asks if there are statistically significant mean differences among groups after adjusting the newly created DV (a linear combination of all original DVs) for differences on one or more covariates.

SECTION 6.6   PRACTICAL VIEW

Purpose

The main advantage of MANCOVA over MANOVA is the fact that the researcher can incorporate one or more covariates into the analysis. The effects of these covariates are then removed from the analysis, leaving the researcher with a clearer picture of the true effects of the IV(s) on the multiple DVs. There are two main reasons for including several (i.e., more than one) covariates in the analysis (Stevens, 1992). First, the inclusion of several covariates will result in a greater reduction in error variance than would result from incorporation of one covariate. Recall that in ANCOVA, the main reason for including a covariate is to remove from the error term unwanted sources of variability (variance within the groups), which could be attributed to the covariate. This ultimately results in a more sensitive

F-test, which increases the likelihood of rejecting the null hypothesis. By including more covariates in a MANCOVA analysis, we can reduce this unwanted error by an even greater amount, improving the chances of rejecting a null hypothesis that is really false.

Figure 6.12 Unadjusted Means for Income and Years of Education by Gender and Job Satisfaction.

Descriptive Statistics

	SATJOB Job Satisfaction	SEX Respondent's Sex	Mean	Std. Deviation	N
RINCOM2	1 Very satisfied	1 Male	15.8193	3.7389	166
		2 Female	14.0301	4.0093	133
		Total	15.0234	3.9565	299
	2 Mod satisfied	1 Male	14.5157	4.3237	159
		2 Female	12.3182	4.0367	132
		Total	13.5189	4.3298	291
	3 A little dissatisfied	1 Male	15.2571	4.1398	35
		2 Female	12.2187	3.7566	32
		Total	13.8060	4.2184	67
	4 Very dissatisfied	1 Male	14.2143	5.0563	14
		2 Female	13.0000	2.8674	10
		Total	13.7083	4.2475	24
	Total	1 Male	15.1524	4.1183	374
		2 Female	13.0717	4.0376	307
		Total	14.2144	4.2087	681
EDUC2	1 Very satisfied	1 Male	14.2590	2.8856	166
		2 Female	14.3985	2.2086	133
		Total	14.3211	2.6030	299
	2 Mod satisfied	1 Male	13.7484	2.6766	159
		2 Female	13.9242	2.4762	132
		Total	13.8282	2.5847	291
	3 A little dissatisfied	1 Male	14.1429	2.7880	35
		2 Female	13.8438	2.5541	32
		Total	14.0000	2.6629	67
	4 Very dissatisfied	1 Male	15.3571	2.4685	14
		2 Female	14.0000	2.1602	10
		Total	14.7917	2.3953	24
	Total	1 Male	14.0722	2.7860	374
		2 Female	14.1238	2.3635	307
		Total	14.0954	2.6023	681

A second reason for including more than one covariate is that it becomes possible to make better adjustments for initial differences in situations where the research design includes the use of intact groups (Stevens, 1992). The researcher has even more information upon which to base the statistical matching procedure. In this case, the means of the linear combination of DVs for each group are adjusted to what they would be if all groups had scored equally on the combination of covariates.

Again, the researcher needs to be cognizant of the choice of covariates in a multivariate analysis. There should exist a significant relationship between the set of DVs and the covariate or set of covariates (Stevens, 1992). Similar to ANCOVA, if more than one covariate is being used, there should be relatively low intercorrelations among all covariates (roughly < .40). In ANCOVA, the amount of error reduction was a result of the magnitude of the correlation between the DV and the covariate. In

MANCOVA, if several covariates are being used, the amount of error reduction is determined by the magnitude of the multiple correlation (R^2) between the newly created DV and the set of covariates (Stevens, 1992). A higher value for R^2 is directly associated with low intercorrelations among covariates, which means a greater degree of error reduction.

Figure 6.13 Adjusted Means for Income and Years of Education by Gender and Job Satisfaction.

1. Respondent's Sex

Dependent Variable	Respondent's Sex	Mean	Std. Error	95% Confidence Interval	
				Lower Bound	Upper Bound
RINCOM2	Male	14.952	.338	14.288	15.615
	Female	12.892	.386	12.135	13.649
EDUC2	Male	14.377	.218	13.950	14.804
	Female	14.042	.248	13.554	14.529

2. Job Satisfaction

Dependent Variable	Job Satisfaction	Mean	Std. Error	95% Confidence Interval	
				Lower Bound	Upper Bound
RINCOM2	Very satisfied	14.925	.235	14.464	15.385
	Mod satisfied	13.417	.237	12.951	13.883
	A little dissatisfied	13.738	.493	12.770	14.706
	Very dissatisfied	13.607	.835	11.968	15.246
EDUC2	Very satisfied	14.329	.151	14.032	14.626
	Mod satisfied	13.836	.153	13.536	14.137
	A little dissatisfied	13.993	.318	13.370	14.617
	Very dissatisfied	14.679	.538	13.623	15.734

Figure 6.14 Multivariate Dialogue Box.

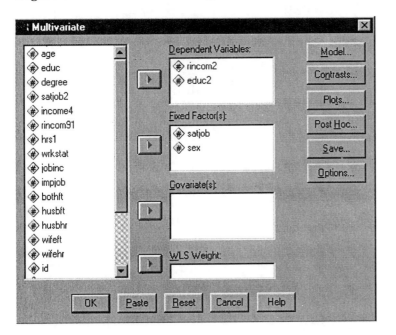

Figure 6.15 Multivariate Options Dialogue Box.

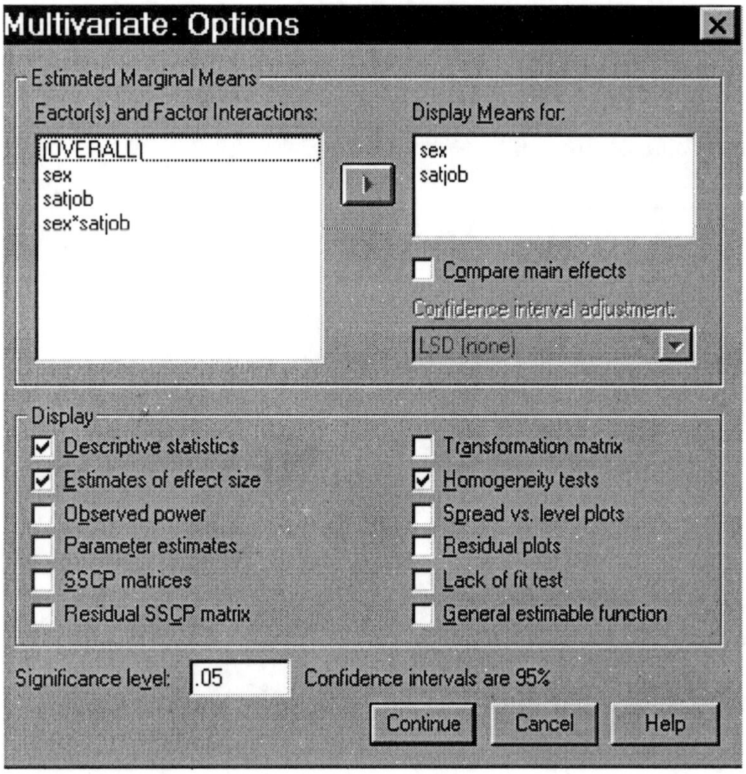

Figure 6.16 Multivariate Post Hoc Dialogue Box.

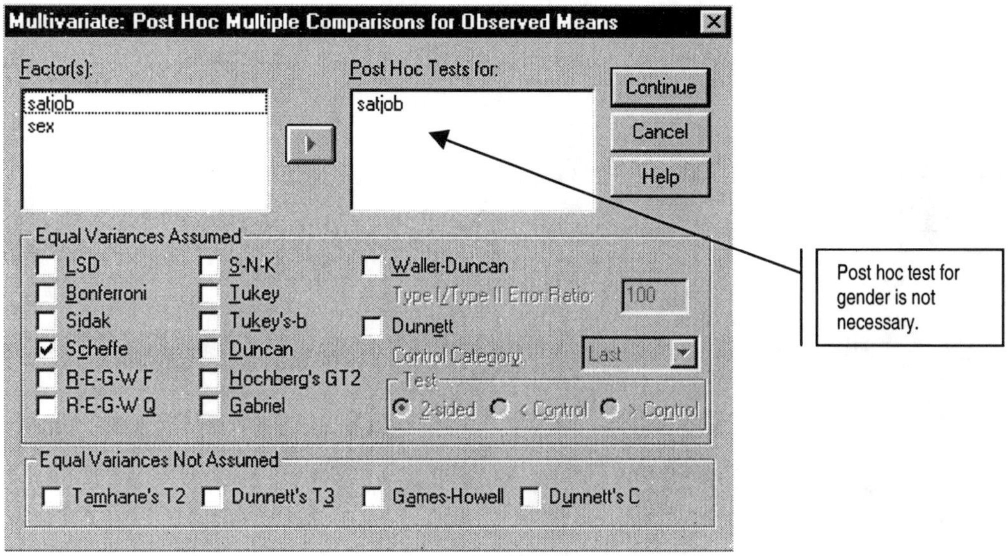

The null hypothesis being tested in MANCOVA is that the *adjusted* population mean vectors are equal:

$$H_o: \quad \mu_{1_{adj}} = \mu_{2_{adj}} = \mu_{3_{adj}} = \dots = \mu_{k_{adj}}$$

Wilks' Lambda (Λ) is again the most common test statistic used in MANCOVA. However, in this case, the sum-of-squares and cross-products (SSCP) matrices are first adjusted for the effects of the covariate(s).

The procedure to be used in conducting MANCOVA mirrors that used in conducting MANOVA. Following the statistical adjustment of newly created DV scores, the overall multivariate null hypothesis is evaluated using Wilks' Λ. If the null is retained, interpretation of the analysis ceases at this point. However, if the overall null hypothesis is rejected, the researcher then examines the results of univariate ANCOVAs in order to discover which DVs are being affected by the IV(s). A Bonferroni-type adjustment to protect from the potential of an inflated Type I error rate is again appropriate at this point.

The reader may recall from Chapter 5 explicit mention of a specific application of MANCOVA that is used to assess the contribution of each individual DV to any significant differences in the IVs. This procedure is accomplished by removing the effects of all other DVs by treating them as covariates in the analysis.

Sample Research Questions

In the sample study presented earlier in this chapter, we investigated the differences in worker productivity (measured by income level, DV_1, and hours worked, DV_2) for individuals in different age categories (IV). Assume that the variable of years of education has been shown to relate to both DVs and we want to remove its effect from the analysis. Consequently, we decide to include years of education as a covariate in our analysis. Therefore, the design we have is now a one-way MANCOVA. Accordingly, this study would address the following research questions:

(1) Are there significant mean differences in worker productivity (as measured by the combination of income and hours worked) for individuals of different ages, after removing the effect of years of education?

(2) Are there significant mean differences in income levels for individuals of different ages, after removing the effect of years of education?

   (2a) If so, which age categories differ?

(3) Are there significant mean differences in hours worked for individuals of different ages, after removing the effect of years of education?

   (3a) If so, which age categories differ?

For our second MANCOVA example, we will add a covariate to our two-factor design presented earlier. This two-way MANCOVA will investigate differences in the combined DV of income level (DV_1,) and years of education (DV_2) for individuals of different gender (IV_1) and of different levels of job satisfaction (IV_2), while controlling for age. Again, one should note that the following research questions address both the multivariate and univariate analyses within MANCOVA:

(1) a. Are there significant mean differences in the combined DV of income and years of education between males and females, after removing the effect of age?

b. Are there significant mean differences in the combined DV of income and years of education for different levels of job satisfaction, after removing the effect of age? If so, which job satisfaction categories differ?

c. Is there a significant interaction between gender and job satisfaction on the combined DV of income and years of education, after removing the effect of age?

(2) a. Are there significant mean differences on income between males and females, after removing the effect of age?

b. Are there significant mean differences on income among different levels of job satisfaction, after removing the effect of age? If so, which job satisfaction categories differ?

c. Is there a significant interaction between gender and job satisfaction on income, after removing the effect of age?

(3) a. Are there significant mean differences in years of education between males and females, after removing the effect of age?

b. Are there significant mean differences in years of education among different levels of job satisfaction, after removing the effect of age? If so, which job satisfaction categories differ?

c. Is there a significant interaction between gender and job satisfaction on years of education, after removing the effect of age?

SECTION 6.7 ASSUMPTIONS AND LIMITATIONS

Multivariate analysis of covariance rests on the same basic assumptions as univariate ANCOVA. However, the assumptions for MANCOVA must accommodate multiple DVs. The following list presents the assumptions for MANCOVA, with an asterisk indicating modification from the ANCOVA assumption.

(1) The observations within each sample must be randomly sampled and must be independent of each other.

(2*) The distributions of scores on the dependent variables must be normal in the populations from which the data were sampled.

(3*) The distributions of scores on the dependent variables must have equal variances.

(4*) Linear relationships must exist between all pairs of DVs, all pairs of covariates, and all DV-covariate pairs in each cell.

(5*) If two covariates are used, the regression planes for each group must be homogeneous or parallel. If more than two covariates are used, the regression hyperplanes must be homogeneous.

(6) The covariates are reliable and are measured without error.

The first and sixth assumptions essentially remain unchanged. Assumptions #2 and #3 are simply modified in order to include multiple DVs. Assumption #4 has a substantial modification in that we must now assume linear relationships not only between the DV and the covariate, but also among several other pairs of variables (Tabachnick & Fidell, 1996). There also exists an important modification to assumption number 5. Recall that if only one covariate is included in the analysis, there exists the assumption that covariate regression slopes for each group are homogeneous. However, if the MANCOVA analysis involves more than one covariate, the analogous assumption involves homogeneity of regression planes (for 2 covariates) and hyperplanes (for 3 or more covariates).

Our discussion of assessing MANCOVA assumptions will center on the two substantially modified assumptions (i.e., #4 and #5). Similar procedures, as have been discussed earlier, are used for testing the remaining assumptions.

Methods of Testing Assumptions

The assumption of normally distributed DVs is assessed in the usual manner. Initial assessment of normality is done through inspection of histograms, boxplots, and normal Q-Q plots. Statistical assessment of normality is accomplished by examining the values (and the associated significance tests) for skewness and kurtosis, and through the use of the Kolmogorov-Smirnov test. The assumption of homoscedasticity is assessed primarily with Box's Test or using one of three different statistical tests discussed in previous chapters (i.e., Chapters 3 and 5), namely Hartley's F-max test, Cochran's test, or Levene's test.

The assumption of linearity among all pairs of DVs and covariates is crudely assessed by inspecting the within-cells bivariate scatterplots between all pairs of DVs, all pairs of covariates, and all DV-covariate pairs. This process is feasible if the analysis includes only a small number of variables. However, the process becomes much more cumbersome (and potentially unmanageable!) with analyses involving the examination of numerous DVs and/or covariates—just imagine all of the possible bivariate pairings! If the researcher is involved in such an analysis, one recommendation is to engage in "spot checks" of random bivariate relationships or bivariate relationships in which nonlinearity may be likely (Tabachnick & Fidell, 1996).

Once again, if curvilinear relationships are indicated, they may be corrected by transforming some or all of the variables. Bear in mind that transforming the variables may create difficulty in interpretations. One possible solution might be to eliminate the covariate that appears to produce nonlinearity and replacing it with another appropriate covariate (Tabachnick & Fidell, 1996).

The reader will remember that a violation of the assumption of homogeneity of regression slopes (as well as regression planes and hyperplanes) is an indication that there is a covariate by treatment (IV) interaction, meaning that the relationship between the covariate and the newly created DV is different at different levels of the IV(s). A preliminary or custom MANCOVA can be conducted to test the assumption of homogeneity of regression planes (in the case of two covariates) or regression hyperplanes (in the case of three or more covariates). If the analysis contains more than one covariate, there is an interaction effect for each covariate. The effects are lumped together and tested as to whether the combined interactions are significant (Stevens, 1992).

The null hypothesis being tested in these cases is that all regression planes/hyperplanes are equal and parallel. Rejecting this hypothesis means that there is a significant interaction between covariates and IVs and that the planes/hyperplanes are not equal. If the researcher is to continue in the use of multivariate analysis of covariance, one would hope to fail to reject this particular null hypothesis. In SPSS, this is determined by examining the results of the F-test for the interaction of the IV(s) by the covariate(s).

SECTION 6.8 PROCESS AND LOGIC

The Logic Behind MANCOVA

The calculations for MANCOVA are nearly identical to those for MANOVA. The only substantial difference is that the sum-of-squares and cross-products (SSCP) matrices must first be adjusted for the effects of the covariate(s). The adjusted matrices are symbolized by T^* (adjusted total sum-of-squares and cross-products matrix), **W*** (adjusted within sum-of-squares and cross-products matrix), and **B*** (adjusted between sum-of-squares and cross-products matrix).

Wilks' Λ is again calculated by using the SSCP matrices (Stevens, 1992). We can compare the MANOVA and MANCOVA formulas for Λ:

MANOVA

$$\Lambda \;=\; \frac{|\mathbf{W}|}{|\mathbf{T}|} \;=\; \frac{|\mathbf{W}|}{|\mathbf{B} + \mathbf{W}|}$$

MANCOVA

$$\Lambda^* \;=\; \frac{|\mathbf{W}^*|}{|\mathbf{T}^*|} \;=\; \frac{|\mathbf{W}^*|}{|\mathbf{B}^* + \mathbf{W}^*|} \qquad \text{(Equation 6.4)}$$

The interpretation of Λ remains as it was in MANOVA. If there is no treatment effect or group differences, then **B*** = 0 and Λ^* = 1 indicating no differences between groups on the linear combination of DVs after removing the effects of the covariate(s); whereas, if **B*** were very large, then Λ^* would approach 0, indicating significant group differences on the combination of DVs, after controlling for the covariate(s).

As in MANOVA, eta squared for MANCOVA is obtained in the following manner:

$$\eta^2 = 1 - \Lambda$$

In the multivariate analysis of covariance situation, η^2 is interpreted as the variance accounted for in the best linear combination of DVs by the IV(s) and/or interactions of IV(s), after removing the effects of any covariate(s).

Interpretation of Results

Interpretation of MANCOVA results is quite similar to that of MANOVA; however, with the inclusion of covariates, interpretation of a preliminary MANCOVA is necessary in order to test the assumption of homogeneity of regression slopes. Essentially, this analysis tests for the interaction between the factors (IVs) and covariates. This preliminary or custom MANCOVA will also test homogeneity of variance-covariance (Box's Test), which is actually interpreted first since it helps in identifying the appropriate test statistic to be utilized in examining the homogeneity of regression and the final MANCOVA results. If the Box's Test is significant at $p<.001$ and group sample sizes are extremely unequal, then Pillai's Trace is utilized when interpreting the homogeneity of regression test and the MANOVA results. If equal variances are assumed, Wilks' Lambda should be used as the multivariate test statistic. Once the test statistic has been determined, then the homogeneity of regression slopes or planes results are interpreted by examining the F ratio and p value for the interaction. If factor-covariate interaction is significant, then MANCOVA is not an appropriate analysis technique. If interaction is not significant, then one can proceed with conducting the full MANCOVA analysis. Using the F ratio and p value for a test statistic that was identified in the preliminary analysis through the Box's Test, factor interaction should be examined if two or more IVs are utilized in the analysis. If factor interaction is significant, then main effects for each factor on the combined DV is not a valid indicator of effect. If factor interaction is not significant, the main effects for each IV can be accurately interpreted by examining the F ratio, p value, and effect size for the appropriate test statistic. When main effects are significant, univariate ANOVA results indicate group differences for each DV. Since MANCOVA does not provide post hoc analyses, examining group means (before and after covariate adjustment) for each DV can assist in determining how groups differed for each DV.

In summary, the first step in interpreting the MANCOVA results is to evaluate the preliminary MANCOVA results that include the Box's Test and the test for homogeneity of regression slopes. If Box's Test is not significant, utilize the Wilks' Lambda statistic when interpreting the homogeneity of regression slopes and the subsequent multivariate tests. If Box's Test is significant, use Pillai's Trace. Once the multivariate test statistic has been identified, examine the significance (F ratios and p values) of factor-covariate interaction (homogeneity of regression slopes). If factor-covariate interaction is not significant, then proceed with the full MANCOVA. To interpret the full MANCOVA results, examine the significance (F ratios and p values) of factor interaction. This is necessary only if two or more IVs are included. Next evaluate the F ratio, p value, and effect size for each factor's main effect. If multivariate significance is found, interpret the univariate ANOVA results to determine significant group differences for each DV.

For our example that investigates age category (*agecat4*) differences in respondent's income (*rincom91*) and hours worked per week (*hrs1*) when controlling for education level (*educ*), the previously transformed variables of *rincom2* and *hrs2* were utilized. These transformations are described in Section 6.3. Linearity of the two DVs and the covariate was then tested by creating a matrix scatterplot and calculating Pearson correlation coefficients. Results indicate linear relationships. Although the correlation coefficients are statistically significant, all are quite low. The last assumption, homogeneity of variance-covariance, was tested within a preliminary MANCOVA analysis utilizing **Multivariate**. The Box's Test (see Figure 6.17) reveals that equal variances can be assumed, $F(9, 2827520)=.634$, $p=.769$; therefore, Wilks' Lambda will be used as the multivariate statistic. Figure 6.18 presents the MANOVA results for the homogeneity of regression test. The interaction between *agecat4* and *educ2* is not significant, Wilks' $\Lambda=.993$, $F(6,1342)=.815$, $p=.558$. A full MANCOVA was then conducted using **Multivariate** (see Figure 6.19). Wilks' Lambda criteria indicates significant groups differences in age category with respect to income and hours worked per week, Wilks' $\Lambda=.898$, $F(6,1348)=12.36$, $p<.001$, multivariate $\eta^2=.052$. Univariate ANOVA results (see Figure 6.20) reveal that age category significantly differs for only income ($F(3, 675)=24.18$, $p<.001$, partial $\eta^2=.097$) and not hours worked per week ($F(3, 675)=.052$, $p=.984$, partial $\eta^2=.000$). A comparison of adjusted means shows that individuals 18-29 years of age have income that is more than 3 points lower than those 40-49 and older than 50 (see Figure 6.21).

Figure 6.17 Box's Test for Homogeneity of Variance-Covariance.

Box's Test of Equality of Covariance Matrices[a]

Box's M	5.740
F	.634
df1	9
df2	2827520
Sig.	.769

Box's Test is not significant. Use Wilks' Lambda criteria.

Tests the null hypothesis that the observed covariance matrices of the dependent variables are equal across groups.

a. Design: Intercept+AGECAT4+EDUC2+AGECAT4 * EDUC2

Figure 6.18 MANCOVA Summary Table: Test for Homogeneity of Regression Slopes.

Multivariate Tests[c]

Effect		Value	F	Hypothesis df	Error df	Sig.	Eta Squared
Intercept	Pillai's Trace	.284	133.096[a]	2.000	671.000	.000	.284
	Wilks' Lambda	.716	133.096[a]	2.000	671.000	.000	.284
	Hotelling's Trace	.397	133.096[a]	2.000	671.000	.000	.284
	Roy's Largest Root	.397	133.096[a]	2.000	671.000	.000	.284
AGECAT4	Pillai's Trace	.004	.504	6.000	1344.000	.805	.002
	Wilks' Lambda	.996	.504[a]	6.000	1342.000	.806	.002
	Hotelling's Trace	.005	.503	6.000	1340.000	.806	.002
	Roy's Largest Root	.004	.833[b]	3.000	672.000	.476	.004
EDUC2	Pillai's Trace	.106	39.974[a]	2.000	671.000	.000	.106
	Wilks' Lambda	.894	39.974[a]	2.000	671.000	.000	.106
	Hotelling's Trace	.119	39.974[a]	2.000	671.000	.000	.106
	Roy's Largest Root	.119	39.974[a]	2.000	671.000	.000	.106
AGECAT4 * EDUC2	Pillai's Trace	.007	.816	6.000	1344.000	.558	.004
	Wilks' Lambda	.993	.815[a]	6.000	1342.000	.558	.004
	Hotelling's Trace	.007	.814	6.000	1340.000	.559	.004
	Roy's Largest Root	.005	1.169[b]	3.000	672.000	.321	.005

a. Exact statistic

b. The statistic is an upper bound on F that yields a lower bound on the significance level.

c. Design: Intercept+AGECAT4+EDUC2+AGECAT4 * EDUC2

Indicates that factor-covariate interaction is NOT significant.

Figure 6.19 MANCOVA Summary Table.

Indicates that the covariate significantly influences the combined DV.

Multivariate Tests[c]

Effect		Value	F	Hypothesis df	Error df	Sig.	Eta Squared
Intercept	Pillai's Trace	.298	142.742[a]	2.000	674.000	.000	.298
	Wilks' Lambda	.702	142.742[a]	2.000	674.000	.000	.298
	Hotelling's Trace	.424	142.742[a]	2.000	674.000	.000	.298
	Roy's Largest Root	.424	142.742[a]	2.000	674.000	.000	.298
EDUC2	Pillai's Trace	.126	48.428[a]	2.000	674.000	.000	.126
	Wilks' Lambda	.874	48.428[a]	2.000	674.000	.000	.126
	Hotelling's Trace	.144	48.428[a]	2.000	674.000	.000	.126
	Roy's Largest Root	.144	48.428[a]	2.000	674.000	.000	.126
AGECAT4	Pillai's Trace	.102	12.037	6.000	1350.000	.000	.051
	Wilks' Lambda	.898	12.356[a]	6.000	1348.000	.000	.052
	Hotelling's Trace	.113	12.673	6.000	1346.000	.000	.053
	Roy's Largest Root	.113	25.371[b]	3.000	675.000	.000	.101

a. Exact statistic

b. The statistic is an upper bound on F that yields a lower bound on the significance level.

c. Design: Intercept+EDUC2+AGECAT4

Indicates that age category significantly affects the combined DV when years of education is controlled.

Figure 6.20 Univariate ANOVA Summary Table.

Tests of Between-Subjects Effects

Source	Dependent Variable	Type III Sum of Squares	df	Mean Square	F	Sig.	Eta Squared
Corrected Model	RINCOM2	2411.930[a]	4	602.983	42.456	.000	.201
	HRS2	1490.452[b]	4	372.613	2.948	.020	.017
Intercept	RINCOM2	922.346	1	922.346	64.943	.000	.088
	HRS2	33502.664	1	33502.664	265.018	.000	.282
EDUC2	RINCOM2	1355.655	1	1355.655	95.452	.000	.124
	HRS2	1439.392	1	1439.392	11.386	.001	.017
AGECAT4	RINCOM2	1030.099	3	343.366	24.177	.000	.097
	HRS2	19.820	3	6.607	.052	.984	.000
Error	RINCOM2	9586.657	675	14.202			
	HRS2	85331.025	675	126.416			
Total	RINCOM2	149199.0	680				
	HRS2	1567026	680				
Corrected Total	RINCOM2	11998.587	679				
	HRS2	86821.476	679				

a. R Squared = .201 (Adjusted R Squared = .196)

b. R Squared = .017 (Adjusted R Squared = .011)

Indicates that age category significantly affects income but NOT hours worked.

Figure 6.21 Unadjusted and Adjusted Means for Income and Hours Worked per Week by Age Category.

Descriptive Statistics

	AGECAT4 4 categories of age	Mean	Std. Deviation	N
RINCOM2	1 18-29	11.8672	4.1438	128
	2 30-39	14.0315	3.8810	222
	3 40-49	15.3247	3.8660	194
	4 50+	14.9574	4.4173	141
	Total	14.1839	4.2155	685
HRS2	1 18-29	46.3203	10.3200	128
	2 30-39	47.0315	11.4182	222
	3 40-49	46.4897	11.7545	194
	4 50+	46.3262	11.5149	141
	Total	46.6000	11.3189	685

Figure 6.21 is continued on the next page.

Figure 6.21 Unadjusted and Adjusted Means for Income and Hours Worked per Week by Age Category. (*Continued*)

4 categories of age

Dependent Variable	4 categories of age	Mean	Std. Error	95% Confidence Interval	
				Lower Bound	Upper Bound
RINCOM2	1 18-29	11.993[a]	.333	11.339	12.648
	2 30-39	13.887[a]	.253	13.389	14.384
	3 40-49	15.356[a]	.272	14.822	15.890
	4 50+	15.165[a]	.321	14.535	15.795
HRS2	1 18-29	46.450[a]	.995	44.497	48.403
	2 30-39	46.882[a]	.756	45.398	48.366
	3 40-49	46.528[a]	.811	44.935	48.122
	4 50+	46.660[a]	.957	44.780	48.540

[a]. Evaluated at covariates appeared in the model: EDUC2 = 14.0985.

Writing Up Results

The process of summarizing MANCOVA results is almost identical to MANOVA; however, MANCOVA results will obviously include a statement of how the covariate influenced the DVs. One should note that although the preliminary MANCOVA results are quite important in the analysis process, these results are not reported since it is understood that if a full MANCOVA has been conducted, such assumptions have been fulfilled. Consequently, the MANCOVA results narrative should address the following:

(1) Subject elimination and/or variable transformation;
(2) Full MANCOVA results (test statistic, *F* ratio, degrees of freedom, *p* value, and effect size);
    (a) Main effects for each IV and covariate on the combined DV;
    (b) Main effect for the interaction between IVs;
(3) Univariate ANOVA results (*F* ratio, degrees of freedom, *p* value, and effect size);
    (a) Main effect for each IV and DV; and
    (b) Comparison of means to indicate which groups differ on each DV.

Often a table is created that compares the unadjusted and adjusted group means for each DV. For our example, the results statement includes all of these components with the exception of factor interaction since only one IV is utilized. The following results narrative applies the results from Figures 6.17 – 6.21.

Multivariate analysis of covariance (MANCOVA) was conducted to determine the effect of age category on employee productivity as measured by income and hours worked per week while controlling for years of education. Prior to the test, variables were transformed to eliminate outliers. Cases with income equal to zero or equal to or exceeding 22 were eliminated. Hours worked per week was transformed; those less than or equal to 16 were recoded 17 and those greater than or equal to 80 were recoded 79. Years of education was also transformed to eliminate cases with 6 or fewer years. MANOVA results revealed significant differences among the age categories on the combined dependent variable, Wilks' Λ=.898, $F(6,1348)$=12.36, $p<.001$, multivariate η^2=.052. The covariate (years of education) significantly influenced the combined dependent variable, Wilks' $\Lambda=.874$, $F(2,674)=48.43$, $p<.001$, multivariate $\eta^2=.126$. Analysis of covariance (ANCOVA) was conducted on each dependent variable as a follow-up test to MANCOVA. Age category differences were significant for income, $(F(3, 675)=24.18, p<.001$, partial $\eta^2=.097$) but not hours worked per week $(F(3, 675)=.052, p=.984,$ partial $\eta^2=.000)$. A comparison of adjusted means revealed that income of those 18-29 years differs by more than 3 points from those 40-49 years and those 50 years and older. Table 1 presents adjusted and unadjusted means for income and hours worked per week by age category.

Table 1 Adjusted and Unadjusted Means for Income and Hours Worked per Week
          by Age Category

Age	Income		Hours Worked per Week	
	Adjusted M	**Unadjusted M**	**Adjusted M**	**Unadjusted M**
18-29 years	11.99	11.87	46.45	46.32
30-39 years	13.89	14.03	46.88	47.03
40-49 years	15.36	15.32	46.53	46.49
50+ years	15.17	14.96	46.66	46.33

SECTION 6.9 MANCOVA SAMPLE STUDY AND ANALYSIS

This section provides a complete example that applies the entire process of conducting MANCOVA: development of research questions and hypotheses, data screening methods, test methods, interpretation of output, and presentation of results. The SPSS data set *gssft.sav* is utilized. Our previous example demonstrates a one-way MANCOVA, while this example will present a two-way MANCOVA.

Problem

Utilizing the two-way MANOVA example previously presented, in which we examined the degree to which gender and job satisfaction affects income and years of education among employees, we are now interested in adding the covariate of age. Since two IVs are tested in this analysis, questions must also take into account the possible interaction between factors. The following research questions and respective null hypotheses address the multivariate main effects for each IV and the possible interaction between factors.

Research Questions	*Null Hypotheses*
RQ1: Do income and years of education differ by gender among employees when controlling for age? →	$H_0 1$: Income and years of education will not differ by gender among employees when controlling for age.
RQ2: Do income and years of education differ by job satisfaction among employees when controlling for age? →	$H_0 2$: Income and years of education will not differ by job satisfaction among employees when controlling for age.
RQ3: Do gender and job satisfaction interact in the effect on income and years of education when controlling for age? →	$H_0 3$: Gender and job satisfaction will not interact in the effect on income and years of education when controlling for age.

Both IVs are categorical and include gender (*sex*) and job satisfaction (*satjob*). The DVs are respondent's income (*rincom2*) and years of education (*educ2*); both are quantitative. The covariate is years of age (*age*) and is quantitative. One should note that the variables *rincom2* and *educ2* are transformed variables of *rincom91* and *educ*, respectively. Transformations of these variables are described in section 6.3 of this chapter.

Method

Since variables were previously transformed to eliminate outliers, data screening is complete. MANCOVA test assumptions should then be examined. Linearity between the DVs and covariate is first assessed by creating a scatterplot matrix and calculating Pearson correlation coefficients. Scatterplots and correlation coefficients indicate linear relationships. Although three of the four correlation coefficients are significant ($p<.001$), coefficients are still fairly weak. The final test assumptions of homogeneity of variance-covariance and homogeneity of regression slopes will be tested in a preliminary MANCOVA using **Multivariate**. For our example, Box's Test (see Figure 6.22) indicates homogeneity of variance-covariance, $F(21,20374)=1.24$, $p=.204$. Therefore, Wilks' Lambda will be utilized as the test statistic for all the multivariate tests. Figure 6.23 reveals that factor and covariate interaction is not significant, Wilks' $\Lambda=.976$, $F(14, 1332)=1.143$, $p=.315$. Full MANCOVA was then conducted using **Multivariate**.

Figure 6.22 Box's Test for Homogeneity of Variance-Covariance.

Box's Test of Equality of Covariance Matrices[a]

Box's M	26.868
F	1.242
df1	21
df2	20374
Sig.	.204

Box's Test is not significant. Use Wilks' Lambda.

Tests the null hypothesis that the observed covariance matrices of the dependent variables are equal across groups.

a. Design: Intercept+SEX+SATJOB+AGE+SEX * SATJOB * AGE

Output and Interpretation of Results

Figure 6.24 presents the unadjusted group means for each DV, while Figure 6.25 displays the adjusted means. MANCOVA results are presented in Figure 6.26 and indicate no significant interaction between the two factors of gender and job satisfaction, Wilks' $\Lambda=.993$, $F(6, 1340)=.839$, $p=.539$. The main effects of gender (Wilks' $\Lambda=.974$, $F(2, 670)=9.027$, $p<.001$, multivariate $\eta^2=.026$) and job satisfaction (Wilks' $\Lambda=.972$, $F(6, 1340)=3.242$, $p=.004$, multivariate $\eta^2=.014$) indicate significant effect on the combined DV. However, one should note the extremely small effect sizes for each IV. The covariate significantly influenced the combined DV, Wilks' $\Lambda=.908$, $F(2, 670)=33.912$, $p<.001$, multivariate $\eta^2=.092$. Univariate ANOVA results (see Figure 6.27) indicate that only the DV of income was significantly effected by the IVs and covariate.

Figure 6.23 MANCOVA Summary Table: Test for Homogeneity of Regression Slopes.

Multivariate Tests[c]

Effect		Value	F	Hypothesis df	Error df	Sig.	Eta Squared
Intercept	Pillai's Trace	.399	220.855[a]	2.000	666.000	.000	.399
	Wilks' Lambda	.601	220.855[a]	2.000	666.000	.000	.399
	Hotelling's Trace	.663	220.855[a]	2.000	666.000	.000	.399
	Roy's Largest Root	.663	220.855[a]	2.000	666.000	.000	.399
SEX	Pillai's Trace	.001	.173[a]	2.000	666.000	.841	.001
	Wilks' Lambda	.999	.173[a]	2.000	666.000	.841	.001
	Hotelling's Trace	.001	.173[a]	2.000	666.000	.841	.001
	Roy's Largest Root	.001	.173[a]	2.000	666.000	.841	.001
SATJOB	Pillai's Trace	.010	1.100	6.000	1334.000	.360	.005
	Wilks' Lambda	.990	1.100[a]	6.000	1332.000	.360	.005
	Hotelling's Trace	.010	1.100	6.000	1330.000	.360	.005
	Roy's Largest Root	.009	1.947[b]	3.000	667.000	.121	.009
AGE	Pillai's Trace	.029	9.912[a]	2.000	666.000	.000	.029
	Wilks' Lambda	.971	9.912[a]	2.000	666.000	.000	.029
	Hotelling's Trace	.030	9.912[a]	2.000	666.000	.000	.029
	Roy's Largest Root	.030	9.912[a]	2.000	666.000	.000	.029
SEX * SATJOB * AGE	Pillai's Trace	.024	1.143	14.000	1334.000	.315	.012
	Wilks' Lambda	.976	1.143[a]	14.000	1332.000	.315	.012
	Hotelling's Trace	.024	1.142	14.000	1330.000	.315	.012
	Roy's Largest Root	.018	1.702[b]	7.000	667.000	.105	.018

a. Exact statistic

b. The statistic is an upper bound on F that yields a lower bound on the significance level.

c. Design: Intercept+SEX+SATJOB+AGE+SEX * SATJOB * AGE

Factor-covariate interaction is NOT significant.

Presentation of Results

The following narrative summarizes the results from this two-way MANCOVA example.

A two-way MANCOVA was conducted to determine the effect of gender and job satisfaction on income and years of education while controlling for years of age. Data were first transformed to eliminate outliers. Respondent's income was transformed to eliminate cases with income of zero and equal to or exceeding 22. Years of education was also transformed to eliminate cases with 6 or fewer years. The main effects of gender (Wilks' Λ=.974, $F(2, 670)$=9.027, p<.001, multivariate η^2=.026) and job satisfaction (Wilks' Λ=.972, $F(6, 1340)$=3.24, p=.004, multivariate η^2=.014) indicate significant effect on the combined DV. The covariate significantly influenced the combined DV, Wilks' Λ=.908, $F(2, 670)$=33.91, p<.001, multivariate η^2=.092. Univariate ANOVA results (Figure 6.27) indicate that only the DV of income was significantly effected by gender ($F(1, 671)$=17.73, p<.001, partial η^2=.026), job satisfaction ($F(3, 671)$=5.64, p=.001, partial η^2=.025) and the covariate of age ($F(1, 671)$=54.16, p<.001, partial η^2=.075). Table 1 presents the adjusted and unadjusted group means for income and years of education. Comparison of adjusted income means indicates that those very satisfied have higher incomes than those less satisfied.

Table 1 Adjusted and Unadjusted Group Means for Income and Years of Education

	Income		Years of Education	
	Adjusted *M*	**Unadjusted *M***	**Adjusted *M***	**Unadjusted *M***
Gender				
Male	15.00	15.15	14.37	14.07
Female	12.92	13.05	14.04	14.13
Job Satisfaction				
Very Sat	14.80	15.00	14.35	14.33
Mod Sat	13.51	13.52	13.83	13.83
A Little Dis	13.73	13.81	13.99	14.00
Very Dis	13.80	13.71	14.67	14.79

Figure 6.24 Unadjusted Group Means for Years of Education and Income.

Descriptive Statistics

	SEX Respondent's Sex	SATJOB Job Satisfaction	Mean	Std. Deviation	N
EDUC2	1 Male	1 Very satisfied	14.2590	2.8856	166
		2 Mod satisfied	13.7484	2.6766	159
		3 A little dissatisfied	14.1429	2.7880	35
		4 Very dissatisfied	15.3571	2.4685	14
		Total	14.0722	2.7860	374
	2 Female	1 Very satisfied	14.4167	2.2070	132
		2 Mod satisfied	13.9242	2.4762	132
		3 A little dissatisfied	13.8438	2.5541	32
		4 Very dissatisfied	14.0000	2.1602	10
		Total	14.1307	2.3642	306
	Total	1 Very satisfied	14.3289	2.6039	298
		2 Mod satisfied	13.8282	2.5847	291
		3 A little dissatisfied	14.0000	2.6629	67
		4 Very dissatisfied	14.7917	2.3953	24
		Total	14.0985	2.6029	680
RINCOM2	1 Male	1 Very satisfied	15.8193	3.7389	166
		2 Mod satisfied	14.5157	4.3237	159
		3 A little dissatisfied	15.2571	4.1398	35
		4 Very dissatisfied	14.2143	5.0563	14
		Total	15.1524	4.1183	374
	2 Female	1 Very satisfied	13.9773	3.9779	132
		2 Mod satisfied	12.3182	4.0367	132
		3 A little dissatisfied	12.2187	3.7566	32
		4 Very dissatisfied	13.0000	2.8674	10
		Total	13.0458	4.0185	306
	Total	1 Very satisfied	15.0034	3.9479	298
		2 Mod satisfied	13.5189	4.3298	291
		3 A little dissatisfied	13.8060	4.2184	67
		4 Very dissatisfied	13.7083	4.2475	24
		Total	14.2044	4.2037	680

Figure 6.25 Adjusted Group Means for Years of Education and Income by Gender and Job Satisfaction.

1. Respondent's Sex

Dependent Variable	Respondent's Sex	Mean	Std. Error	95% Confidence Interval Lower Bound	Upper Bound
EDUC2	1 Male	14.374[a]	.218	13.946	14.802
	2 Female	14.043[a]	.249	13.555	14.532
RINCOM2	1 Male	14.996[a]	.325	14.358	15.633
	2 Female	12.921[a]	.371	12.193	13.649

a. Evaluated at covariates appeared in the model: AGE Age of Respondent = 40.34.

2. Job Satisfaction

Dependent Variable	Job Satisfaction	Mean	Std. Error	95% Confidence Interval Lower Bound	Upper Bound
EDUC2	1 Very satisfied	14.345[a]	.152	14.046	14.643
	2 Mod satisfied	13.830[a]	.153	13.530	14.131
	3 A little dissatisfied	13.994[a]	.318	13.370	14.618
	4 Very dissatisfied	14.666[a]	.538	13.609	15.723
RINCOM2	1 Very satisfied	14.798[a]	.226	14.353	15.242
	2 Mod satisfied	13.507[a]	.229	13.058	13.955
	3 A little dissatisfied	13.727[a]	.474	12.797	14.658
	4 Very dissatisfied	13.801[a]	.803	12.225	15.378

a. Evaluated at covariates appeared in the model: AGE Age of Respondent = 40.34.

Figure 6.26 MANCOVA Summary Table.

Multivariate Tests[c]

Effect		Value	F	Hypothesis df	Error df	Sig.	Eta Squared
Intercept	Pillai's Trace	.670	679.424[a]	2.000	670.000	.000	.670
	Wilks' Lambda	.330	679.424[a]	2.000	670.000	.000	.670
	Hotelling's Trace	2.028	679.424[a]	2.000	670.000	.000	.670
	Roy's Largest Root	2.028	679.424[a]	2.000	670.000	.000	.670
AGE	Pillai's Trace	.092	33.912[a]	2.000	670.000	.000	.092
	Wilks' Lambda	.908	33.912[a]	2.000	670.000	.000	.092
	Hotelling's Trace	.101	33.912[a]	2.000	670.000	.000	.092
	Roy's Largest Root	.101	33.912[a]	2.000	670.000	.000	.092
SEX	Pillai's Trace	.026	9.027[a]	2.000	670.000	.000	.026
	Wilks' Lambda	.974	9.027[a]	2.000	670.000	.000	.026
	Hotelling's Trace	.027	9.027[a]	2.000	670.000	.000	.026
	Roy's Largest Root	.027	9.027[a]	2.000	670.000	.000	.026
SATJOB	Pillai's Trace	.028	3.231	6.000	1342.000	.004	.014
	Wilks' Lambda	.972	3.242[a]	6.000	1340.000	.004	.014
	Hotelling's Trace	.029	3.252	6.000	1338.000	.004	.014
	Roy's Largest Root	.026	5.894[b]	3.000	671.000	.001	.026
SEX * SATJOB	Pillai's Trace	.007	.840	6.000	1342.000	.539	.004
	Wilks' Lambda	.993	.839[a]	6.000	1340.000	.539	.004
	Hotelling's Trace	.008	.838	6.000	1338.000	.540	.004
	Roy's Largest Root	.005	1.189[b]	3.000	671.000	.313	.005

The covariate of age significantly influences the combined DV.

Gender significantly influences the combined DV.

Job satisfaction significantly influences the combined DV.

Factor interaction is NOT significant.

a. Exact statistic

b. The statistic is an upper bound on F that yields a lower bound on the significance level.

c. Design: Intercept+AGE+SEX+SATJOB+SEX * SATJOB

Figure 6.27 Univariate ANOVA Summary Table.

Tests of Between-Subjects Effects

Source	Dependent Variable	Type III Sum of Squares	df	Mean Square	F	Sig.	Eta Squared
Corrected Model	EDUC2	69.142^a	8	8.643	1.280	.251	.015
	RINCOM2	1917.923^b	8	239.740	15.958	.000	.160
Intercept	EDUC2	9098.497	1	9098.497	1347.329	.000	.668
	RINCOM2	4331.310	1	4331.310	288.305	.000	.301
AGE	EDUC2	3.586	1	3.586	.531	.466	.001
	RINCOM2	813.705	1	813.705	54.163	.000	.075
SEX	EDUC2	6.758	1	6.758	1.001	.317	.001
	RINCOM2	266.291	1	266.291	17.725	.000	.026
SATJOB	EDUC2	46.615	3	15.538	2.301	.076	.010
	RINCOM2	254.331	3	84.777	5.643	.001	.025
SEX * SATJOB	EDUC2	15.551	3	5.184	.768	.512	.003
	RINCOM2	29.773	3	9.924	.661	.577	.003
Error	EDUC2	4531.257	671	6.753			
	RINCOM2	10080.664	671	15.023			
Total	EDUC2	139763.0	680				
	RINCOM2	149199.0	680				
Corrected Total	EDUC2	4600.399	679				
	RINCOM2	11998.587	679				

a. R Squared = .015 (Adjusted R Squared = .003)

b. R Squared = .160 (Adjusted R Squared = .150)

Gender significantly affects income but NOT years of education.

Job satisfaction significantly affects income but NOT years of education.

SECTION 6.10 SPSS "HOW TO" FOR MANCOVA

This section describes the steps for conducting both the preliminary MANCOVA and the full MANCOVA using the **Multivariate** procedure. Again, the preceding example from the *gssft.sav* data set is utilized in these steps. The first series of steps describes the preliminary MANCOVA process for testing homogeneity of variance-covariance and homogeneity of regression slopes. To open the Multivariate dialogue box (see Figure 6.28), select the following:

```
Analyze
     General Linear Model
          Multivariate
```

Multivariate Dialogue Box (see Figure 6.28)

Once in this dialogue box, click each DV (*rincom2* and *educ2*) and move to the Dependent Variables box. Click each IV (*sex* and *satjob*) and move to the Fixed Factor(s) box. Then click each covariate (*age*) and move to the Covariate box. Then click **Model**.

Multivariate Model Dialogue Box (see Figure 6.29)

Under Specify Model, click **Custom**. Move each IV and covariate to the Model box. Then hold down the Ctrl key and highlight all IVs and covariate(s). Once highlighted, continue to hold down the shift key and move to the Model box. This should create the interaction between all IVs and covariate(s) (e.g., age*satjob*sex). Also check to make sure that Interaction is specified in the Build Terms box. Click **Continue**. Back in the **Multivariate** Dialogue Box, click **Options**.

Figure 6.28 Multivariate Dialogue Box.

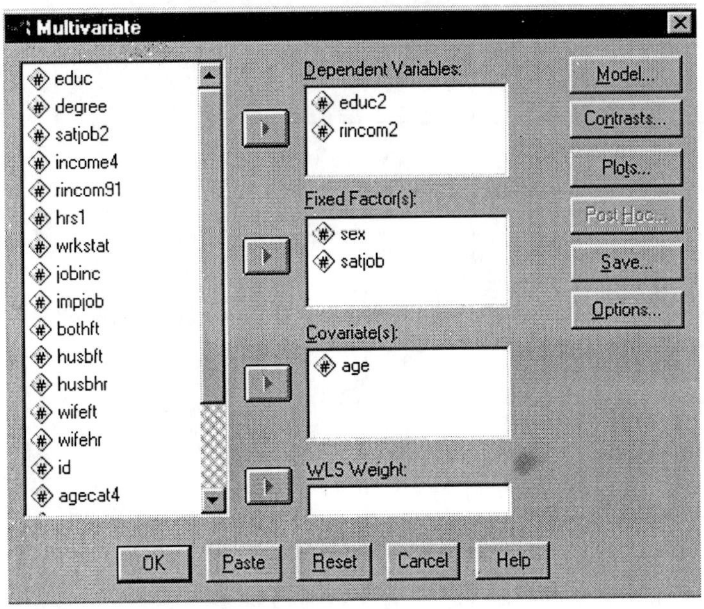

Figure 6.29 Multivariate Model Dialogue Box.

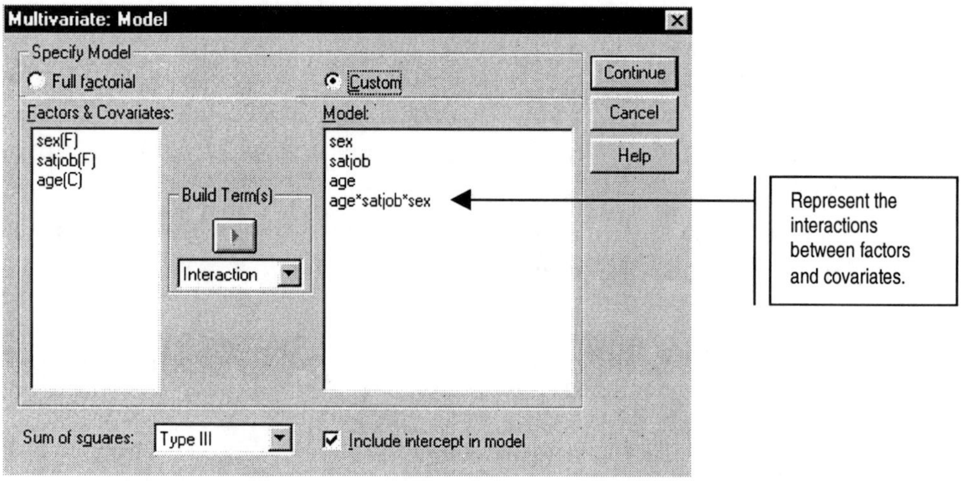

Multivariate Options Dialogue Box (see Figure 6.30)

Under Display, click Homogeneity Tests. Click **Continue**. Back in the **Multivariate** Dialogue Box, click **OK.**

These steps will create the output to evaluate homogeneity of variance-covariance and homogeneity of regression slopes. If interaction between the factors and covariates is not significant, then proceed with the following steps for conducting the full MANCOVA. The same dialogue boxes are opened, but different commands will be used. Open the Multivariate Dialogue Box by selecting the following:

Analyze
> **General Linear Model**
>> **Multivariate**

Figure 6.30 Multivariate Options Dialogue Box.

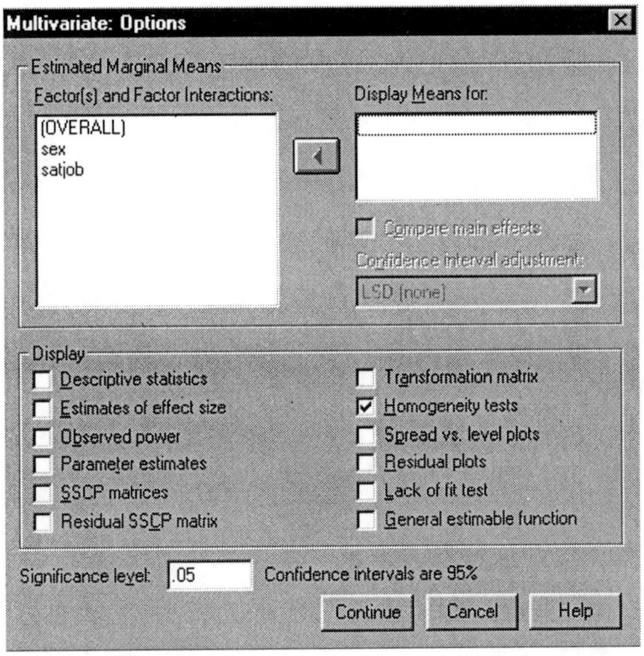

Multivariate Dialogue Box (see Figure 6.28)

If you have conducted the preliminary MANCOVA, variables should already be identified. If not, proceed with the following. Click each DV (*rincom2* and *educ2*) and move to the Dependent Variables box. Click each IV (*sex* and *satjob*) and move to the Fixed Factor(s) box. Then click each covariate (*age*) and move to the Covariate box. Then click **Model**.

Multivariate Model Dialogue Box (see Figure 6.31)

Under specify model, click Full. Click **Continue**. Back in the **Multivariate** Dialogue Box, click **Options**.

Figure 6.31 Multivariate Model Dialogue Box.

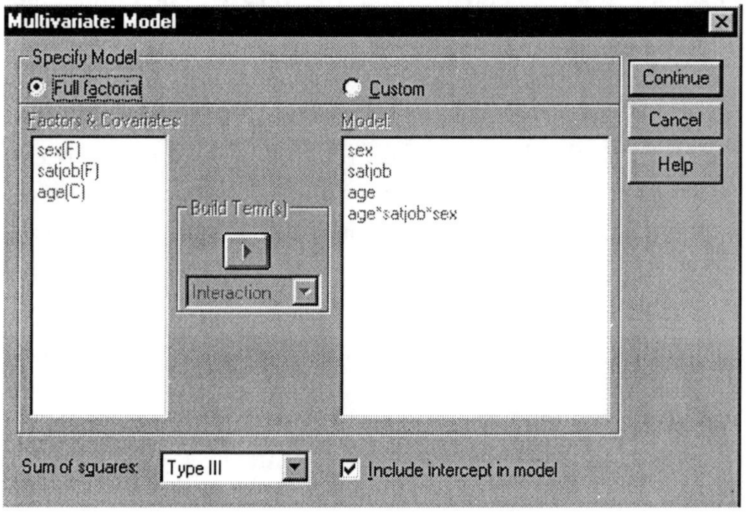

Multivariate Options Dialogue Box (see Figure 6.32)

Under Factor(s) and Factor Interaction, click each IV and move to the Display Means box. Under Display, click **Descriptive Statistics** and **Estimates of Effect Size**. Click **Continue**. Back in the **Multivariate** Dialogue Box, click **OK**.

Figure 6.32 Multivariate Options Dialogue Box.

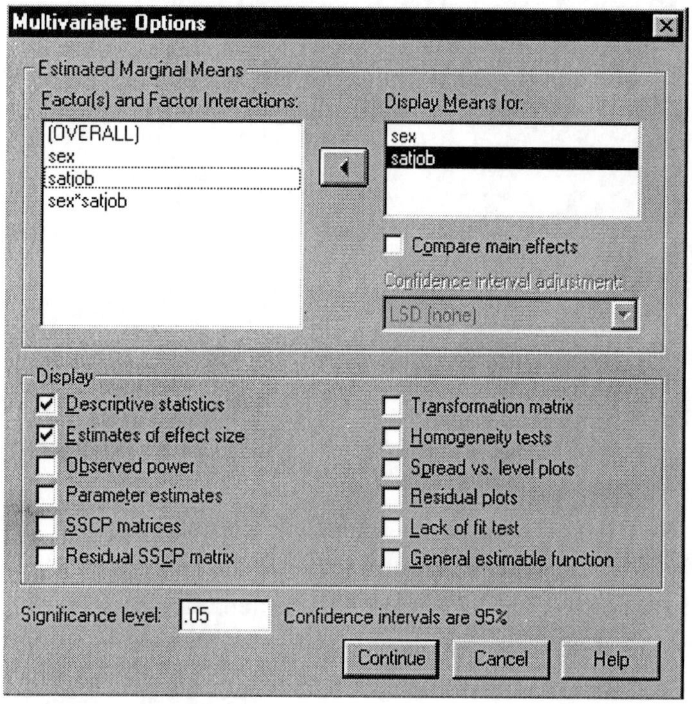

Summary

Multivariate analysis of variance (MANOVA) allows the researcher to examine group differences within a set of dependent variables. Factorial MANOVA will test the main effect for each factor on the combined DV as well as the interaction among factors on the combined DV. Usually follow-up tests, such as Univariate ANOVA and post hoc tests, are conducted within MANOVA to determine the specificity of group differences. Prior to conducting MANOVA, data should be screened for missing data and outliers. Data should also be examined for fulfillment of test assumptions: normality, homogeneity of variance-covariance, and linearity of DVs. Box's Test for homogeneity of variance-covariance will help determine which test statistic (e.g., Wilks' Lambda, Pillai's Trace) to utilize when interpreting the multivariate tests. The SPSS MANOVA table provides four different test statistics (Wilks' Lambda, Pillai's Trace, Hotelling's Trace, and Roy's Largest Root) with the F ratio, p value, and effect size that indicate the significance of factor main effects and interaction. Wilks' Lambda is the most commonly used criterion. If factor interaction is significant, then conclusions about main effects are limited. Univariate ANOVA and post hoc results determine group differences for each DV. Figure 6.33 provides a checklist for conducting MANOVA.

Figure 6.33 Checklist for Conducting MANOVA.

I. Screen Data
    a. Missing Data?
    b. Outliers?
       ❑ Run Outliers and review stem-and-leaf plots and boxplots within **Explore.**
       ❑ Eliminate or transform outliers if necessary.
    c. Normality?
       ❑ Run Normality Plots with Tests within **Explore.**
       ❑ Review boxplots and histograms.
       ❑ Transform data if necessary.
    d. Linearity of DVs?
       ❑ Create Scatterplots.
       ❑ Calculate Pearson correlation coefficients.
       ❑ Transform data if necessary.
    e. Homogeneity of Variance-Covariance?
       ❑ Run Box's Test within **Multivariate.**

II. Conduct MANOVA
    a. Run MANOVA with post hoc test.
       1. ⌐ **Analyze...**⌐ **General Linear Model...**⌐ **Multivariate.**
       2. Move DVs to Dependent Variable box.
       3. Move IVs to Fixed Factor box.
       4. ⌐ **Model.**
       5. ⌐ **Full.**
       6. ⌐ **Continue.**
       7. ⌐ **Options.**
       8. Move each IV to the Display Means box.
       9. Check **Descriptive Statistics, Estimates of Effect Size** and **Homogeneity Tests.**
       10. ⌐ **Continue.**
       11. ⌐ **Post hoc.**
       12. Move each IV to the Post Hoc Test box.
       13. Select post hoc method.
       14. ⌐ **Continue.**
       15. ⌐ **OK.**
    b. Homogeneity of Variance-Covariance?
       ❑ Examine F-ratio and p-value for Box's Test.
       ❑ If significant at $p<.001$ with extremely unequal group sample sizes, use Pillai's Trace for test statistic.
       ❑ If NOT significant at $p<.001$ with fairly equal group sample sizes, use Wilks' Lambda for test statistic.
    c. Interpret factor interaction.
       ❑ If factor interaction is significant, main effects are erroneous.
       ❑ If factor interaction is NOT significant, interpret main effects.
    d. Interpret main effects for each IV on the combined DV.
    e. Interpret Univariate ANCOVA results.
    f. Interpret post hoc results.

III. Summarize Results
    a. Describe any data elimination or transformation.
    b. Narrate Full MANOVA results.
       ❑ Main effects for each IV on the combined DV (test statistic, F-ratio, p-value, effect size).
       ❑ Main effect for factor interaction (test statistic, F-ratio, p-value, effect size).
    c. Narrate Univariate ANOVA results.
       ❑ Main effects for each IV and DV (F-ratio, p-value, effect size).
    d. Narrate post hoc results.
    e. Draw conclusions.

Multivariate analysis of covariance (MANCOVA) allows the researcher to examine group differences within a set of dependent variables while controlling for covariate(s). Essentially, the influence that the covariate(s) has on the combined DV is partitioned out before groups are compared, such that group means of the combined DV are adjusted to eliminate the effect of the covariate(s). One-way MANCOVA will test the main effects for the factor on the combined DV while controlling for the covariate(s). Factorial MANCOVA will do the same but will also test the interaction among factors on the combined DV while controlling for the covariate(s). Usually univariate ANCOVA is conducted within MANCOVA to determine the specificity of group differences. Prior to conducting MANCOVA, data should be screened for missing data and outliers. Data should also be examined for fulfillment of test assumptions: normality, homogeneity of variance-covariance, homogeneity of regression slopes, and linearity of DVs and covariates. A preliminary or custom MANCOVA must be conducted to test the assumptions of homogeneity of variance-covariance and homogeneity of regression slopes. Box's Test for homogeneity of variance-covariance will help determine which test statistic (e.g., Wilks' Lambda, Pillai's Trace) to utilize when interpreting the test for homogeneity of regression slopes and the full MANCOVA analyses. The test for homogeneity of regression slopes will indicate the degree to which the factors and covariate(s) interact to effect the combined DV. If interaction is significant, as indicated by the F ratio and p value for the appropriate test statistic, then the full MANCOVA should NOT be conducted. If interaction is not significant, then the full MANCOVA can be conducted. Once the full MANCOVA has been completed, factor interaction should be examined when two or more IVs are utilized. If factor interaction is significant, then conclusions about main effects are limited. Interpretation of the multivariate main effects and interaction is similar to MANOVA. Univariate ANOVA results determine the significance of group differences for each DV. Figure 6.34 provides a checklist for conducting MANCOVA.

Chapter 6   Multivariate Analysis of Variance and Covariance

Figure 6.34 Checklist for Conducting MANCOVA.

I.   Screen Data
  a.   Missing Data?
  b.   Outliers?
  - Run Outliers and review stem-and-leaf plots and boxplots within **Explore.**
  - Eliminate or transform outliers if necessary.
  c.   Normality?
  - Run Normality Plots with Tests within **Explore.**
  - Review boxplots and histograms.
  - Transform data if necessary.
  e.   Linearity of DVs and covariate(s)?
  - Create Scatterplots.
  - Calculate Pearson correlation coefficients.
  - Transform data if necessary.
  f.   Test remaining assumptions by conducting preliminary MANCOVA.

II.   Conduct Preliminary (Custom) MANCOVA.
  a.   Run Custom MANCOVA.
   1. ⟨⟩ **Analyze...**⟨⟩ **General Linear Model...**⟨⟩ **Multivariate.**
   2. Move DVs to Dependent Variable box.
   3. Move IVs to Fixed Factor box.
   4. Move covariate(s) to Covariate box.
   5. ⟨⟩ **Model.**
   6. ⟨⟩ **Custom.**
   7. Move each IV and covariate to the Model box.
   8. Hold down Ctrl key and highlight all IVs and covariate(s) ⟨⟩ ▶ while still holding down the Ctrl key in order to move interaction to Model box.
   9. ⟨⟩ **Continue.**
   10. ⟨⟩ **Options**.
   11. Check **Homogeneity Tests**.
   12. ⟨⟩ **Continue.**
   13. ⟨⟩ **OK.**
  b.   Homogeneity of Variance-Covariance?
  - Examine F-ratio and p-value for Box's Test.
  - If significant at $p<.001$ with extremely unequal group sample sizes, use Pillai's Trace for test statistic.
  - If NOT significant at $p<.001$ with fairly equal group sample sizes, use Wilks' Lambda for test statistic.
  c.   Homogeneity of Regression Slopes?
  - Using appropriate test statistic, examine F-ratio and p-value for interaction among IVs and covariates.
  - If interaction is significant, do not proceed with Full MANCOVA.
  - If interaction is NOT significant, proceed with Full MANCOVA.

161

Figure 6.34 Checklist for Conducting MANCOVA. (*Continued*)

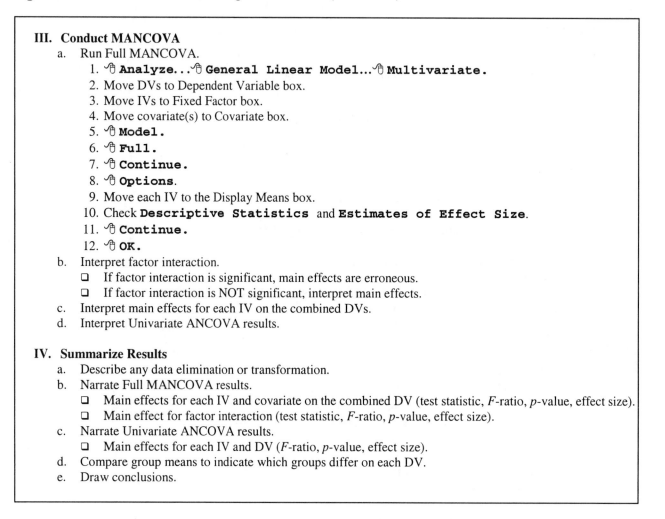

III. **Conduct MANCOVA**
  a. Run Full MANCOVA.
      1.   **Analyze...** **General Linear Model...** **Multivariate.**
      2. Move DVs to Dependent Variable box.
      3. Move IVs to Fixed Factor box.
      4. Move covariate(s) to Covariate box.
      5.   **Model.**
      6.   **Full.**
      7.   **Continue.**
      8.   **Options**.
      9. Move each IV to the Display Means box.
      10. Check **Descriptive Statistics** and **Estimates of Effect Size**.
      11. **Continue.**
      12. **OK.**
  b. Interpret factor interaction.
      ❏ If factor interaction is significant, main effects are erroneous.
      ❏ If factor interaction is NOT significant, interpret main effects.
  c. Interpret main effects for each IV on the combined DVs.
  d. Interpret Univariate ANCOVA results.

IV. **Summarize Results**
  a. Describe any data elimination or transformation.
  b. Narrate Full MANCOVA results.
      ❏ Main effects for each IV and covariate on the combined DV (test statistic, *F*-ratio, *p*-value, effect size).
      ❏ Main effect for factor interaction (test statistic, *F*-ratio, *p*-value, effect size).
  c. Narrate Univariate ANCOVA results.
      ❏ Main effects for each IV and DV (*F*-ratio, *p*-value, effect size).
  d. Compare group means to indicate which groups differ on each DV.
  e. Draw conclusions.

Exercises for Chapter 6

The two exercises below utilize the data set *gssft.sav*, which can be downloaded at the SPSS Web site. Open the URL: **www.spss.com/tech/DataSets.html** in your Web browser. Scroll down until you see "Data Used in SPSS Guide to Data Analysis—8.0 and 9.0" and click on the link "dataset.exe." When the "Save As" dialogue appears, select the appropriate folder and save the file. Preferably, this should be a folder created in the SPSS folder of your hard drive for this purpose. Once the file is saved, double-click the "dataset.exe" file to extract the data sets to the folder.

1. You are interested in evaluating the effect of job satisfaction (*satjob2*) and age category (*agecat4*) on the combined DV of hours worked per week (*hrs1*) and years of education (*educ*).

    a. Develop the appropriate research questions and/or hypotheses for main effects and interaction.

b. Screen data for missing data and outliers. What steps, if any, are necessary for reducing missing data and outliers?

c. Test the assumptions of normality and linearity of DVs.

      i. What steps, if any, are necessary for increasing normality?

      ii. Are DVs linearly related?

d. Conduct MANOVA with post hoc (be sure to test for homogeneity of variance-covariance).

      i.   Can you conclude homogeneity of variance-covariance? Which test statistic is most appropriate for interpretation of multivariate results?

      ii.   Is factor interaction significant? Explain.

      iii.   Are main effects significant? Explain.

      iv.   What can you conclude from univariate ANOVA and post hoc results?

e. Write a results statement.

2. Building on the previous problem in which you investigated the effects of job satisfaction (*satjob2*) and age category (*agecat4*) on the combined dependent variable of hours worked per week (*hrs1*) and years of education (*educ*), you are now interested in controlling for respondent's income such that *rincom91* will be used as a covariate. Complete the following.

a. Develop the appropriate research questions and/or hypotheses for main effects and interaction.

b. Screen data for missing data and outliers. What steps, if any, are necessary for reducing missing data and outliers?

c. Test the assumptions of normality and linearity of DVs and covariate.

      i. What steps, if any, are necessary for increasing normality?

      ii.   Are DVs and covariate linearly related?

d.   Conduct a preliminary MANCOVA to test the assumptions of homogeneity of variance-covariance and homogeneity of regression slopes/planes.

      i.   Can you conclude homogeneity of variance-covariance? Which test statistic is most appropriate for interpretation of multivariate results?

      ii.   Do factors and covariate significantly interact? Explain.

d.   Conduct MANCOVA.

      i.   Is factor interaction significant? Explain.

      ii.   Are main effects significant? Explain.

      iii. What can you conclude from univariate ANOVA results?

f.   Write a results statement.

3. Compare the results from problems number 1 and number 2. Explain the differences in main effects.

CHAPTER 7

MULTIPLE REGRESSION

Up to this point, we have focused our attention on statistical analysis techniques that investigate the existence of differences between groups. In this chapter, we begin to redirect the focus of our attention to a second grouping of advanced/multivariate techniques—those that describe and test the existence of predictable relationships among a set of variables. Our discussion will include a brief review of simple linear regression, followed by an in-depth examination of multiple regression.

SECTION 7.1 PRACTICAL VIEW

Purpose

Regression analysis procedures have as their primary purpose the development of an equation that can be used for *predicting* values on some DV for all members of a population. (A secondary purpose is to use regression analysis as a means of *explaining* causal relationships among variables, the focus of Chapter 8.) The most basic application of regression analysis is the bivariate situation, to which the reader was undoubtedly exposed in an introductory statistics course. This is often referred to as *simple linear regression*, or just *simple regression*. Simple regression involves a single IV and a single DV. The goal of simple regression is to obtain a linear equation so that we can predict the value of the DV if we have the value of the IV. Simple regression capitalizes on the correlation between the DV and IV in order to make specific predictions about the DV (Sprinthall, 2000). The correlation tells us how much information about the DV is contained in the IV. If the correlation is perfect (i.e., $r = \pm 1.00$), the IV contains everything we need to know about the DV, and we will be able to perfectly predict one from the other. However, this is seldom, if ever, the case.

The idea behind simple regression is that we want to obtain the equation for the best-fitting line through a series of points. If we were to view a bivariate scatterplot for our fictitious IV (X) and DV (Y), we could then envision a line drawn through those points. Theoretically, an infinite number of lines could be drawn through those points (see Figure 7.1). However, only one of these lines would be the best-fitting line. Regression analysis is the means by which we determine the best-fitting line, called the *regression line*.

The regression line is the single straight line that lies closest to all points in a given scatterplot—this line is sometimes said to pass through the *centroid* of the scatterplot (Sprinthall, 2000). In order to make predictions, three important facts about the regression line must be known:

(1) the extent to which points are scattered around the line,
(2) the slope of the regression line, and
(3) the point at which the line crosses the *Y*-axis (Sprinthall, 2000).

These three facts are so important to regression that they serve as the basis for the calculation of the regression equation itself. The extent to which the points are scattered around the line is typically indicated by the degree of relationship between the IV (X) and the DV (Y). This relationship is measured by a correlation coefficient (e.g., the Pearson correlation, symbolized by r)—the stronger the relationship, the higher the degree of predictability between X and Y. (You will see in Section 7.3 just how important r is to the regression equation calculation.) The slope of the regression line can greatly affect prediction (Sprinthall, 2000). The degree of slope is determined by the amount of change in Y that accompanies a unit change (i.e., one point, one inch, one degree, etc.) in X. It is the slope that largely determines the predicted values of Y from known values for X. Finally, it is important to determine exactly where the regression line crosses the Y-axis (this value is known as the Y-intercept). Said another way, it is crucial to know what value is expected for Y when $X = 0$.

Figure 7.1 Bivariate Scatterplot Showing Several Possible Regression Lines.

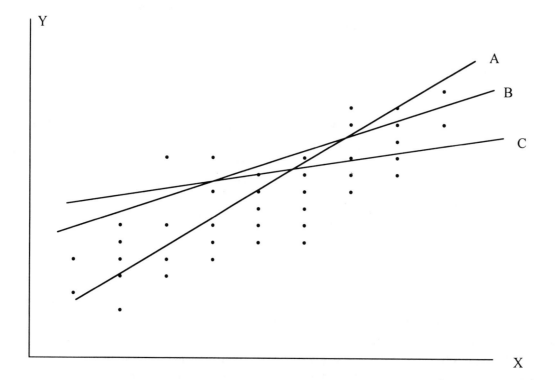

These three facts we have just discussed actually define the regression line. The regression line is essentially an equation that expresses Y as a function of X (Tate, 1992). The basic equation for simple regression is:

$$Y = bX + a \qquad\qquad \text{(Equation 7.1)}$$

where Y is the predicted value for the DV, X is the known raw score value on the IV, b is the slope of the regression line, and a is the Y-intercept. The significant role that both the slope and Y-intercept play in the regression equation should now be apparent to the reader. Oftentimes, you will see the above equation presented in the following analogous, although more precise, form:

$$\hat{Y} = B_0 + B_1 X_1 + \hat{e} \qquad\qquad \text{(Equation 7.2)}$$

where $\hat{y}$ is the predicted value for the DV, X is the raw score value on the IV, B_1 is the slope of the regression line, and B_0 is the Y-intercept. We have added one important term, $\hat{e}_i$, in Equation 7.2. This is the symbol for the errors of prediction, also referred to as the **residuals**. As previously mentioned, unless we have a perfect correlation between the IV and DV, the predicted values obtained by our regression equation will also be less than perfect—that is, there will be some errors. The residuals constitute an important measure of those errors and are essentially calculated as the difference between the actual value and predicted value for the DV (i.e., $\hat{e}_i = y_i - \hat{y}$).

Let us return momentarily to the concept of the best-fitting line (see Figure 7.2). The reason that we obtain the best-fitting line as our regression equation is because we mathematically calculate the line with the smallest amount of total squared error. This is commonly referred to as the **least squares solution** (Stevens, 1992; Tate, 1992) and actually provides us with values for the constants in the regression equation, B_1 and B_0 (also known as the **regression coefficients (B)**, **beta coefficients** or **beta weights (β)**) that minimize the sum of squared residuals—that is, $\Sigma(y_i - \hat{y})^2$ is minimized. In other words, the *total* amount of prediction error, both positive and negative, is as small as possible, giving us the best mathematically achievable line through the set of points in a scatterplot.

Multiple regression is merely an extension of simple linear regression involving more than one IV, or predictor variable. This technique is used to predict the value of a single DV from a weighted, linear combination of IVs (Harris, 1998). A multiple regression equation appears similar to its simple regression counterpart except that there are more coefficients, one for the Y-intercept and one for each of the IVs:

$$\hat{Y} = B_0 + B_1X_1 + B_2X_2 + ... + B_kX_k + \hat{e}_i \qquad \text{(Equation 7.3)}$$

where there is a corresponding B coefficient for each IV (X_k) in the equation and the best linear combination of weights and raw score X values will again minimize the total squared error in our regression equation.

Let us consider a concrete example: Suppose we wanted to determine the extent to which we could predict female life expectancy from a set of predictor variables for a selected group of countries throughout the world. The predictor variables we have selected include percent urban population; gross domestic product per capita; birthrate per 1,000; number of hospital beds per 10,000; number of doctors per 10,000; number of radios per 100; and number of telephones per 100. In our analysis, we would be looking to obtain the regression coefficients for each IV that would provide us with the best linear combination of IVs—and their associated weights—in order to predict, as accurately as possible, female life expectancy. The regression equation predicting female life expectancy is as follows:

Female life exp. $= B_{urban}\,X_{urban} + B_{GDP}X_{GDP} + B_{birthrate}X_{birthrate} + B_{beds}X_{beds} + B_{docs}X_{docs} + B_{radios}X_{radios} + B_{phones}X_{phones} + \hat{e}_i$

We will return to this example in greater detail a bit later in the chapter, but first there are several important issues related to multiple regression that warrant our attention.

A first issue of interest is a set of measures unique to multiple regression. Another way of looking at the previously mentioned concept of the minimization of total error is to consider multiple regression as a means of seeking the linear combination of IVs that *maximally* correlate with the DV (Stevens, 1992). This maximized correlation is called the **multiple correlation** and is symbolized by R. The multiple correlation is essentially equivalent to the Pearson correlation between the actual, or observed, values and the predicted values on the DV (i.e., $R = r_{yi\hat{y}i}$). Analogous to our earlier interpretation of the Pearson correlation, the multiple correlation tells us how much information about a DV (e.g., female life expectancy) is contained in the combination of IVs (e.g., percent urban population, gross domestic product, birthrate, number of hospital beds, number of doctors, number of radios, and number of telephones). In multiple regression, there is a test of significance (F-test) to determine whether the relationship between the set of IVs and the DV is large enough to be meaningful.

Figure 7.2 Graphical Representation of a Linear Regression Model and the Least Squares Criterion.

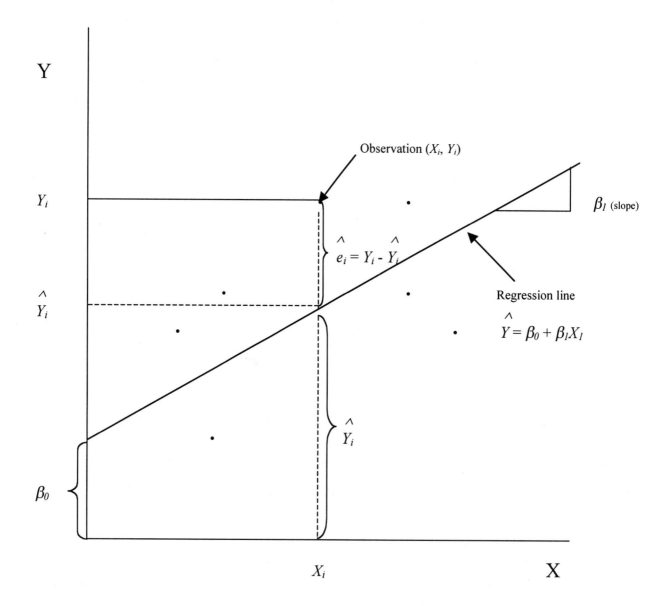

You may recall from an earlier course in statistics a term called the ***coefficient of determination***, or r^2. For the Pearson r, this value was interpreted as the proportion of one variable in the pair that can be explained (or accounted for) by the other variable. In multiple regression, R^2 is also called the coefficient of determination and has a similar interpretation. The coefficient of determination is the proportion of DV variance that can be explained by the combination of the IVs (Levin & Fox, 2000; Sprinthall, 2000). In our example, an obtained value for R^2 would be interpreted as the proportion of variability in female life expectancy that could be accounted for by the combination of the seven predictor variables. If one multiplies this value by 100, R^2 becomes the percentage of explained variance (Sprinthall, 200).

A second issue is one that has an associated word of caution, which we will address momentarily. The issue at hand is that of ***multicollinearity***. Multicollinearity is a problem that arises when there exists moderate to high intercorrelations among predictor variables (IVs) to be used in a regression analysis. (Recall from Chapter 1 that the opposite of multicollinearity is *orthogonality*, or complete independence among variables.) The underlying problem of multicollinearity is that if two variables are highly correlated, they are essentially containing the same—or at least much of the same—information and are therefore measuring the same thing (Sprinthall, 2000). Not only does one gain little by adding to a regression analysis variables that are measuring the same thing, but multicollinearity can cause real problems for the analysis itself. Stevens (1992) points out three reasons why multicollinearity can be problematic for researchers:

(1) Multicollinearity severely limits the size of R since the IVs are "going after" much of the same variability on the DV.
(2) When trying to determine the importance of individual IVs, multicollinearity causes difficulty because individual effects are confounded due to the overlapping information.
(3) Multicollinearity tends to increase the variances of the regression coefficients, which ultimately results in a more unstable prediction equation.

Multicollinearity should be addressed by the researcher prior to the execution of the regression analysis. The simplest method for diagnosing multicollinearity is to examine the correlation matrix for the predictor variables, looking for moderate to high intercorrelations. However, it is preferable to use one of two statistical methods to assess multicollinearity. First, tolerance statistics can be obtained for each IV. ***Tolerance*** is a measure of collinearity among IVs, where possible values range from 0 to 1. A value for tolerance close to zero is an indication of multicollinearity. Typically, a value of 0.1 serves as the cutoff point—if the tolerance value for a given IV is less than 0.1, multicollinearity is a distinct problem (Norusis, 1998). A second method is to examine values for the ***variance inflation factor*** for each predictor. The variance inflation factor (VIF) for a given predictor "indicates whether there exists a strong linear association between it and all remaining predictors" (Stevens, 1992). The VIF is defined by the quantity $1/(1-R_j^2)$ and is obtainable on most computer programs. Although there is no steadfast rule of thumb, values of VIF that are greater than 10 are generally cause for concern (Stevens, 1992).

There are several methods for combating multicollinearity in a regression analysis; two of the most straightforward are presented here. The simplest method is to delete the problematic variable from the analysis (Sprinthall, 2000). If the information in one variable is being "captured" by another, no real information is being lost by deleting one of them. A second approach is to combine the variables involved so as to create a single measure that addresses a single construct, thus deleting the repetition (Stevens, 1992). One might consider this approach for variables with intercorrelations of .80 or higher. Several other approaches to dealing with multicollinear relationships exist, but they are beyond the scope of this text. If interested, the reader is advised to pursue the discussion in Stevens (1992).

A third issue of great importance in multiple regression is the method of specifying the regression model; in other words, determining or selecting a good set of predictor variables. Keeping in mind that the goal of any analysis should be to achieve a *parsimonious* solution, we want to select IVs that will give us an efficient regression equation without including "everything under the sun." Initially, one of the most efficient methods of selecting a group of predictors is to rely on the researcher's substantive knowledge (Stevens, 1992). Being familiar with and knowledgeable about your population, sample, and data will provide you with meaningful information about the relationships among variables and the likely predictive power of a set of predictors. Furthermore, for reasons we will discuss later, a recommended ratio of subjects to IVs (i.e., n/k) of at least 15 to 1 will provide a reliable regression equation (Stevens, 1992). Keeping the number of predictor variables low tends to improve this ratio, since most researchers do not have the luxury of increasing their sample size at will, which would be necessary if one were to continue to add predictors to the equation.

Once a set of predictors has been selected, there are several methods by which they may be incorporated into the regression analysis and subsequent equation. Tabachnick and Fidell (1996) identify three such strategies: standard multiple regression, sequential multiple regression, and stepwise multiple regression. (The reader should recall—and possibly *revisit*—the discussion of standard and sequential analyses as presented in Chapter 1.) It should be noted that decisions about model specification can and do affect the nature of the research questions being investigated. In *standard multiple regression*, all IVs are entered into the analysis simultaneously. The effect of each IV on the DV is assessed as if it had been entered into the equation after all other IVs had been entered. Each IV is then evaluated in terms of what it adds to the prediction of the DV, as specified by the regression equation (Tabachnick & Fidell, 1996).

In *sequential multiple regression*, sometimes referred to as *hierarchical multiple regression*, a researcher may want to examine the influence of several predictor IVs in a specific order. Using this approach, the researcher specifies the order in which variables are entered into the analysis. Substantive knowledge, as previously mentioned, may lead the researcher to believe that one variable may be more influential than others in the set of predictors and that variable is entered into the analysis first. Subsequent variables are then added in order to determine the specific amount of variance they can account for, above and beyond, what has been explained by any variables entered prior (Aron & Aron, 1999). Individual effects are assessed at the point of entry of a given variable (Tabachnick & Fidell, 1996).

Finally, *stepwise multiple regression*, also sometimes referred to as *statistical multiple regression*, is often used in studies that are exploratory in nature (Aron & Aron, 1999). The researcher may have a large set of predictors and may want to determine which specific IVs make meaningful contributions to the overall prediction. There are essentially three variations of stepwise regression, listed and described below:

(1) *Forward selection* — The bivariate correlations among all IVs and the DV are calculated. The IV that has the highest correlation with the DV is entered into the analysis first. It is assessed in terms of its contribution (in terms of R^2) to the DV. The next variable to be entered into the analysis is the IV that contributes most to the prediction of the DV, after partialing out the effects of the first variable. This effect is measured by the increase in R^2 (ΔR^2) due to the second variable. This process continues until, at some point, predictor variables stop making significant contributions to the prediction of the DV. It is important to remember that once a variable has been entered into the analysis, it remains there (Stevens, 1992; Pedhazur, 1982).

(2) *Stepwise selection* — Stepwise selection is a variation of forward selection. It is an improvement over the previous method in that, at each step, tests are performed to determine the significance of each IV already in the equation as if it were to enter last. In other words, if a variable entered into the analysis is measuring much of the same construct as another, this reassessment may determine that the first variable to enter may no longer contribute anything to the overall analysis. In this procedure, that variable would then be dropped out of the analysis.

Even though it was at one time a "good" predictor, in conjunction with others, it may no longer serve as a substantial contributor (Pedhazur, 1982).

(3) **Backward deletion** — The initial step here is to compute an equation with all predictors included. Then, a significance test (a partial F-test) is conducted for every predictor, as if each were entered last, in order to determine the level of contribution to overall prediction. The smallest partial F is compared to a preselected "F to remove" value. If the value is less than the "F to remove" value (not significant), that predictor is removed from the analysis and a new equation with the remaining variables is computed, followed by another test of the resulting smallest partial F. This process continues until only significant predictors remain in the equation (Stevens, 1992).

It is important to note that both sequential and stepwise approaches to regression contain a distinct advantage over standard multiple regression—one variable is added at a time and each is continually checked for significant improvement to prediction. However, the important difference between these two is that sequential regression orders and adds variables based on some *theory or plan by the researcher*; whereas, in stepwise regression, those decisions are being made by a computer based *solely on statistical analyses* (Aron & Aron, 1999). Sequential regression should be used in research based on theory or some previous knowledge; stepwise regression should be used where exploration is the purpose of the analysis.

A fourth issue of consequence in multiple regression is that of model validation, sometimes called *model cross-validation*. A regression equation is developed in order to be able to predict DV values for individuals in a population, but remember that the equation itself was developed based only on a sample from that population. The multiple correlation, R, will be at its maximum value for the sample from which the equation was derived. If the predictive power drops off drastically when applied to an independent sample from the same population, the regression equation is of little use since it has little or no generalizability (Stevens, 1992). If the equation is not predicting well for other samples, it is not fulfilling its designed and intended purpose.

In order to obtain a reliable equation, substantial consideration must be given to the sample size (n) and the number of predictors (k). As mentioned earlier, a recommended ratio of these two factors is about 15 subjects for every predictor (Stevens, 1992). This results in a equation that will cross-validate with relatively little loss in its ability to predict. Another recommendation for this ratio is identified by Tabachnick and Fidell (1996). The simplest rule of thumb they offer is that $n \geq 50 + 8k$, for testing multiple correlations, and $n \geq 104 + k$, for testing individual predictors. They suggest calculating n both ways and using the larger value.

Cross-validation can be accomplished in several ways. Ideally, one should wait a period of time, select an independent sample from the same population, and test the previously obtained regression equation (Tatsuoka, 1988). This is not always feasible, so an alternative would be to split the original sample into two "subsamples." Then one subsample can be used to develop the equation, while the other is used to cross-validate it (Stevens, 1992). Of course, this would only be feasible if one had a large enough sample, based on the criteria set forth above.

A final issue of importance in regression is the effect that outliers can have on a regression solution. Recall that regression is essentially a maximization procedure (i.e., we are trying to maximize the correlation between observed and predicted DV scores). Because of this fact, multiple regression can be very sensitive to extreme cases. One or two outliers have been shown to adversely affect the interpretation of regression analysis results (Stevens, 1992). It is therefore recommended that outliers be identified and dealt with appropriately prior to running the regression analysis. This is typically accomplished by initial screenings of boxplots, but more precisely with the statistical procedure known as *Mahalanobis distance* (as described in Chapter 3).

One final note regarding multiple regression. There does exist a multivariate version of multiple regression (i.e., multivariate multiple regression), but it is so similar in its approach and conduct that it will not be discussed in detail in this text. Basically, multivariate multiple regression involves the prediction of several DVs from a set of predictor IVs. This procedure is a variation of multiple regression in that the regression equations realized are those that would be obtained if each DV were regressed *separately* on the set of IVs. The actual correlations among DVs in the analysis are ignored (Stevens, 1992).

Sample Research Questions

Building on the example we began discussing in the previous section, we can now specify the research questions to be addressed by our multiple regression analysis. The methods by which the regression model is developed often dictates the type of research question(s) to be addressed. For example, if we were entering all seven IVs from our data set into the model, the appropriate research questions would be:

(1)   Which of the seven predictor variables (i.e., percent urban population, GDP, birthrate, number of hospital beds, number of doctors, number of radios, and number of telephones) are most influential in predicting female life expectancy? Are there any predictor variables that do not contribute significantly to the prediction model?

(2)   Does the obtained regression equation resulting from a set of seven predictor variables allow us to reliably predict female life expectancy?

However, if we were using a stepwise method of specifying the model, the revised questions would be:

(1)   Which of the possible seven predictor variables (i.e., percent urban population, GDP, birthrate, number of hospital beds, number of doctors, number of radios, and number of telephones) are included in an equation for predicting female life expectancy?

(2)   Does the obtained regression equation resulting from a subset of the seven predictor variables allow us to reliably predict female life expectancy?

SECTION 7.2 ASSUMPTIONS AND LIMITATIONS

In multiple regression, there are actually two sets of assumptions—assumptions about the raw scale variables and assumptions about the residuals (Pedhazur, 1982). With respect to the raw scale variables, the following conditions are assumed:

(1)   The independent variables are fixed (i.e., the same values of the IVs would have to be used if the study were to be replicated).

(2)   The independent variables are measured without error.

(3)   The relationship between the independent variables and the dependent variable is linear (in other words, the regression of the DV on the combination of IVs is linear).

The remaining assumptions concern the residuals. Recall again from Chapter 3 that *residuals*, or *prediction errors*, are the portions of scores not accounted for by the multivariate analyses. Meeting these assumptions is necessary in order to achieve the best linear estimations (Pedhazur, 1982). These assumptions are:

(4)   The mean of the residuals for each observation on the dependent variable over many replications is zero.

(5)   Errors associated with any single observation on the dependent variable are independent of (i.e., not correlated with) errors associated with any other observation on the dependent variable.

(6)   The errors are not correlated with the independent variables.

(7)   The variance of the residuals across all values of the independent variables is constant (i.e., homoscedasticity of the variance of the residuals).

(8)   The errors are normally distributed.

Assumptions 1, 2, and 4 are largely research design issues. We will focus our attention on assumptions 3, 5, and 6—which address the issue of linearity—and assumptions 7 and 8—which address homoscedasticity and normality, respectively.

Methods of Testing Assumptions

There are essentially two approaches to testing the assumptions in multiple regression (Tabachnick & Fidell, 1996). The first approach involves the routine pre-analysis data screening procedures that have been discussed in the preceding several chapters. As a reminder, linearity can be assessed through examination of the various bivariate scatterplots. Normality is evaluated in similar fashion, as well as through the assessment of the values for skewness, kurtosis, and Kolmogorov-Smirnov statistics. Finally, homoscedasticity is assessed by interpreting the results of Box's M Test.

The alternative approach to the routine procedure is to examine the residuals scatterplots. These scatterplots resemble bivariate scatterplots in that they are plots of values on the combination of two "variables"—in this case, these are the predicted values of the DV ($\hat{y}$) and the standardized residuals or prediction errors ($\hat{e}_i$). Examination of these residual scatterplots provides a test of *all three* of these crucial assumptions (Tabachnick & Fidell, 1996). If the assumptions of linearity, normality, and homoscedasticity are tenable, we would expect to see the points cluster along the horizontal line defined by $\Lambda_i = 0$, in a somewhat rectangular pattern (see Figure 7.3).

Any systematic, differential patterns or clusters of points are an indication of possible model violations (Tabachnick & Fidell, 1996; Stevens, 1992). Examples of residuals plots depicting violations of the three assumptions are shown in Figure 7.4. (*It is important to note that the plots shown in this figure are idealized and have been constructed to show clear violations of assumptions. A word of caution—with real data, the patterns are seldom this obvious.*) If the assumption of linearity is tenable, we would expect to see a relatively straight line relationship among the points in the plot. This typically appears as a rectangle (Tabachnick & Fidell, 1996), as depicted in Figure 7.3. However, as shown in Figure 7.4(a), the points obviously appear in a nonlinear pattern. In fact, this example is so extreme as to depict a clearly curvilinear pattern. This is an unmistakable violation of the assumption of linearity.

If the assumption of normality is defensible, we would expect to see an even distribution of points both above and below the line defined by $\hat{e}_i = 0$. In Figure 7.4(b), there appears to be a clustering of points the farther we move both above and below that reference line, indicating a non-normal (in this case, bimodal) distribution of residuals (Tate, 1992).

Finally, Figure 7.4(c) shows a violation of the assumption of homoscedasticity. If this assumption is tenable, we would expect to see the points dispersed evenly about the reference line—again, defined by $\hat{e}_i = 0$—across all predicted values for the DV. In Figure 7.4(c), notice that the width is very narrow at small predicted values for the DV; however, the width increases rapidly as the predicted DV value increases. This is a clear indication of heteroscedasticity, or a lack of constant variance.

Figure 7.3 Residuals Plot of Standardized Residuals ($\hat{e}_i$) Versus Predicted Values ($\hat{y}_i$) When Assumptions Are Met.

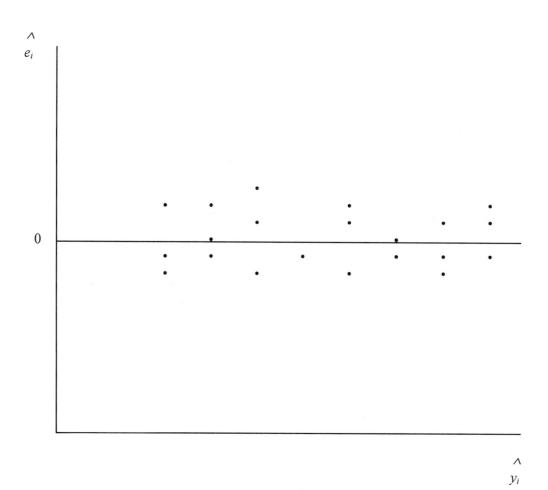

Residuals scatterplots may be examined *in place of* the routine pre-analysis data screening or *following* those procedures (Tabachnick & Fidell, 1996). If examination of the residuals scatterplots is conducted instead of the routine procedures—and if no violations are evident, no outliers exist, there are sufficient number of cases, and there is no evidence of multicollinearity—then one can be safe in interpreting that single regression run on the computer. However, if the initial residuals scatterplots do not look "clean," then further data screening using the routine procedures is warranted (Tabachnick & Fidell, 1996). In many cases, this may involve the transformation of one or more variables in order to meet the assumptions. If a curvilinear pattern appears, one possible remedy is to use a polynomial (i.e., nonlinear) model (Stevens, 1992), which is beyond the scope of this book.

In cases that involve moderate violations of linearity and homoscedasticity, one should be aware that these violations merely weaken the regression analysis, but do not invalidate it (Tabachnick & Fidell, 1996). Furthermore, moderate violations of the normality assumption may often be ignored—especially with larger sample sizes—since there are no adverse effects on the analysis (Tate, 1992). It may still be possible to proceed with the analysis, depending on the subjective judgments of the researcher. Unfortunately, however, there are no rules to explicitly define that which constitutes a "moderate" violation. In reality, we would probably be justified in expecting some slight departures from the "ideal" situation, as depicted in Figure 7.3, due to sampling fluctuations (Tate, 1992).

Figure 7.4   Residuals Plots Showing Violations of (a) Linearity, (b) Normality, and (c) Homoscedasticity.

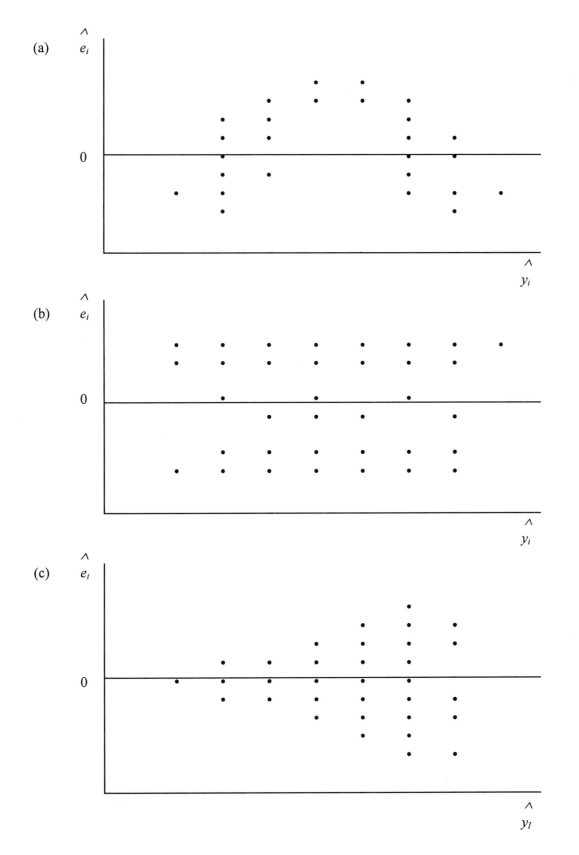

SECTION 7.3 PROCESS AND LOGIC

The Logic Behind Multiple Regression

You will recall from your previous exposure to simple regression that the statistical calculations basically involve the determination of the constants a and b. The slope of the line (i.e., b) is first calculated by multiplying the correlation coefficient between X and Y—recall we discussed earlier in this chapter the important role played by the correlation between X and Y—by the standard deviation of Y and dividing that term by the standard deviation of X:

$$b = \frac{(r)(SD_Y)}{(SD_X)} \qquad \text{(Equation 7.4)}$$

The constant a (the Y-intercept) is then calculated in the following manner:

$$a = \bar{Y} - b\bar{X} \qquad \text{(Equation 7.5)}$$

There are analogous equations for the multivariate regression situation, although they appear slightly more ominous and, therefore, will not be shown here. Recall from Equation 7.3 that in multiple regression there are *at least* two regression coefficients (specifically, the slope coefficients B_1 and B_2) that must be calculated. The calculations mirror Equation 7.4; the only substantial difference is that they incorporate a concept known as partial correlation. *Partial correlation* is a measure of the relationship between an IV and DV, holding all other IVs constant. For example, the calculated value for B_1 tells us how much of a change in Y can be expected for a given change in X_1 when the effects of X_2 are held constant (Sprinthall, 2000).

The other main calculation in multiple regression is the determination of the value for R^2 and its associated significance test. Recall that R^2 is a measure of variance accounted for in the DV by the predictors. One can think of this as being similar to analysis of variance, in that we must partition the sum of squares variability. In regression analysis, we separate the total variability into variability due to regression (see Equation 7.6) and variability about the regression, also known as the sum of squares residual (see Equation 7.7).

$$SS_{reg} = \Sigma(\hat{y}_i - \bar{y})^2 \qquad \text{(Equation 7.6)}$$

$$SS_{res} = \Sigma(y_i - \hat{y})^2 \qquad \text{(Equation 7.7)}$$

The total sum of squares is simply the sum of these two terms and is symbolized by $\Sigma(y_i - \bar{y})^2$. The squared multiple correlation is then calculated by dividing the sum of squares due to regression (SS_{reg}) by the sum of squares total (SS_{tot}):

$$R^2 = \frac{SS_{reg}}{SS_{tot}} \qquad \text{(Equation 7.8)}$$

The standard F-test from analysis of variance can be written making some simple algebraic substitutions:

$$F = \frac{R^2 / k}{(1 - R^2) / (n - k - 1)} \qquad \text{(Equation 7.9)}$$

where k and $n - k - 1$ are the appropriate degrees of freedom for the numerator and denominator, respectively (Stevens, 1992). From this point, the significance of the obtained value for R^2 can be tested using the standard F-test criteria, or by simply examining the associated p value from the computer printout. This then tells the researcher whether the set of IV predictor variables is accounting for, or explaining, a statistically significant amount of variance in the DV.

Interpretation of Results

Interpretation of multiple regression focuses on determining the adequacy of the regression model(s) that has been developed. Conducting multiple regression typically generates output that can be divided into three parts: model summary, ANOVA, and coefficients. Our discussion on how to interpret regression results will address these three parts. The first part of the regression output, model summary, displays several multiple correlation indices—multiple correlation (R), squared multiple correlation (R^2), and adjusted squared multiple correlation (R^2_{adj})—all of which indicate how well an IV or combination of IVs predicts the criterion variable (DV). The multiple correlation (R) is a Pearson correlation coefficient between the predicted and actual scores of the DV. The squared multiple correlation (R^2) represents the degree of variance accounted for by the IV or combination of IVs. Unfortunately, R and R^2 typically overestimate their corresponding population values especially with small samples; thus R^2_{adj} is calculated to account for such bias. Change in R^2 (ΔR^2) is also calculated for each step and represents the change in variance that is accounted for by the set of predictors once a new variable has been added to the model. Change in R^2 is important since it is used to determine which variables significantly contribute to the model, or in the case of a stepping method, which variables are added or removed from the model. If a stepping method is used, the model summary will present these statistics for each model or step that is generated.

The ANOVA table presents the F-test and corresponding level of significance for each step or model generated. This test examines the degree to which the relationship between the DV and IVs is linear. If the F-test is significant, then the relationship is linear and therefore the model significantly predicts the DV.

The final part of the output is the coefficients table that reports the following: unstandardized regression coefficient (B), the standardized regression coefficient (beta or β), t and p values, and three correlation indices. The unstandardized regression coefficient (B), also known as the *partial regression coefficient*, represents the slope weight for each variable in the model and is used to create the regression equation. B weights also indicate how much the value of the DV changes when the IV increases by 1 and the other IVs remain the same. A positive B specifies a positive change in the DV when the IV increases, whereas a negative B indicates a negative change in the DV when the IV increases. Since it is difficult to interpret the relative importance of the predictors when the slope weights are not standardized, beta weights (β) or standardized regression coefficients are often utilized to create a prediction equation for the standardized variables. Beta weights are based upon z-scores with a mean of 0 and standard deviation of 1. The coefficients table also presents t and p values, which indicate the significance of the B weights, beta weights, and the subsequent part and partial correlation coefficients. Actually, three correlation coefficients are displayed in the coefficients table. The zero-order correlation represents the bivariate correlation between the IV and DV. The partial correlation coefficient indicates the relationship between the IV and DV after partialing out all other IVs. The part correlation, rarely used when interpreting the output, represents the correlation between the DV and IVs after partialing only one of the IVs.

The final important statistic in the coefficient table is tolerance, which is a measure of multicollinearity among the IVs. Since the inclusion of IVs that are highly dependent upon each other can create an erroneous regression model, determining which variables account for a high degree of common variance in the DV is critical. Tolerance is reported for all the IVs included and excluded in the generated model. This statistic represents the proportion of variance in a particular IV that is not explained by its linear relationship with the other IVs. Tolerance ranges from 0 to 1, with 0 indicating multicollinearity. Typically, if tolerance of an IV is less than .1, the regression procedure should be repeated without the violating IV.

As one can see, there is a lot to interpret when conducting multiple regression. Since tolerance is an indicator of the appropriateness of IVs utilized in the regression, this statistic should be interpreted first. If some IVs violate the tolerance criteria, regression should be conducted again without the violating variables. If the value for tolerance is acceptable, one should proceed with interpreting the model summary, ANOVA table, and table of coefficients.

Let us now apply this process to our example. Since we will utilize the Forward stepping method, our research question is more exploratory in nature: Which IVs (% urban population [*urban*]; gross domestic product per capita [*gdp*]; birthrate per 1,000 [*birthrat*]; hospital beds per 10,000 [*beds*]; doctors per 10,000 [*docs*]; radios per 100 [*radios*]; and phones per 100 [*phones*]) are predictors of female life expectancy? Data were first screened for missing data and outliers and then examined for test assumptions. Outliers were identified by calculating Mahalanobis distance in a preliminary **Regression** procedure (see Chapter 3 for SPSS "How To"). **Explore** was then conducted on the newly generated Mahalanobis variable (*mah_1*) to determine which cases exceeded the chi square (χ^2) criteria (See Figure 7.5). Using a chi square table, we found the critical value of chi square at $p<.001$ with $df=8$ to be 26.125. Case #83 exceeds this critical value and so was deleted from our analysis. Linearity was then analyzed by creating a scatterplot matrix (see Figure 7.6). Scatterplots display nonlinearity for the following variables: *gdp, beds, docs, radio,* and *phone.* These variables were transformed by taking the natural log of each. The reader should note that the data set already includes these transformations as *lngdp, lnbeds, lndocs, lnradio,* and *lnphone.* A scatterplot matrix (see Figure 7.7) with the transformed variables displays elliptical shapes that indicate linearity and normality. Univariate normality was also assessed by conducting **Explore**. Histograms and normality tests (see Figure 7.8) indicate some non-normal distributions; however, the distributions are not extreme. Multivariate normality and homoscedasticity were examined through the generation of a residuals plot within another preliminary **Regression** (see Chapter 3 for SPSS "How To"). The residuals plot is somewhat scattered but again is not extreme (see Figure 7.9). Thus, multivariate normality and homoscedasticity will be assumed.

Regression was then conducted using the Forward method. The three major parts of the output—model summary, ANOVA table, and coefficient table—are presented in Figures 7.10 –7.12, respectively. Tolerance among the IVs is adequate since coefficients for all IVs included and excluded are above .1 (see Figure 7.12). Since the Forward method was utilized, only some of the IVs were entered into the model. The model summary (see Figure 7.10) indicates that three of the eight IVs were entered into the model. For the first step, *lnphone* was entered as it accounted for the most unique variance in the DV ($R^2=.800$). The variables of *birthrat* and *lndocs* were entered in the next two steps, respectively, creating a model that accounted for 86.9% of the variance in female life expectancy. The ANOVA table (see Figure 7.11) presents the F-test for each step/model. The final model significantly predicts the DV, $F(3, 102)=226.50$, $p<.001$. The table of coefficients (see Figure 7.12) is then utilized to create a prediction equation for the DV. The following equation is generated using the B weights.

$$\textit{Female life expectancy} = 2.245X_{lnphone} - .241X_{birthrat} + 2.172X_{lndocs} + 68.159$$

If we utilize the beta weights, we develop the following equation for predicting the standardized DV.

$$Z_{\textit{Female life expectancy}} = .394\ Z_{lnphone} - .288\ Z_{birthrat} + .306\ Z_{lndocs}$$

Bivariate and partial correlation coefficients should also be noted in the coefficients table.

Figure 7.5 Outliers for Mahalanobis Distance.

Extreme Values

			Case Number	Value	
MAH_1	Highest	1	83	36.89903	← Case #83 exceeds χ2 critical value.
		2	72	20.98981	
		3	19	18.35770	
		4	99	18.26913	
		5	81	17.51827	
	Lowest	1	26	2.41654	
		2	107	2.96129	
		3	102	3.26488	
		4	108	3.37951	
		5	6	3.38166	

Writing Up Results

The summary of multiple regression results should always include a description of how variables have been transformed or cases deleted. Typically, descriptive statistics (e.g., correlation matrix, means and standard deviations for each variable) are presented in tables unless only a few variables are analyzed. The reader should note that our example of a results summary will not include these descriptive statistics due to space limitations. The overall regression results are summarized in the narrative by identifying the variables in the model: R^2, R^2_{adj}, F and p values with degrees of freedom. If a step approach has been utilized, you may want to report each step (R^2, R^2_{adj}, R^2 change, and level of significance for change) within a table. Finally, you may want to report the B weight, beta weight, bivariate correlation coefficients, and partial correlation coefficients of the predictors with the DV in a table. If you do not present these coefficients in a table, you may want to report the prediction equation, either standardized or unstandardized. The following results statement applies the results presented in Figures 7.10 – 7.12.

Forward multiple regression was conducted to determine which independent variables (% urban population [urban]; gross domestic product per capita [gdp]; birthrate per 1,000 [birthrate]; hospital beds per 10,000 [beds]; doctors per 10,000 [docs]; radios per 100 [radios]; and phones per 100 [phones]) were the predictors of female life expectancy. Data screening led to the elimination of one case. Evaluation of linearity led to the natural log transformation of gdp, beds, docs, radios, and phones. Regression results indicate an overall model of three predictors (phone, birthrate, and docs) that significantly predict female life expectancy, $R^2=.869$, $R^2_{adj}=.866$, $F(3, 102,)=226.50$, $p<.001$. This model accounted for 86.9% of variance in female life expectancy.

A summary of the regression model is presented in Table 1. In addition, bivariate and partial correlation coefficients between each predictor and the dependent variable are presented in Table 2.

Table 1 Model Summary

Step	R	R^2	R^2_{adj}	ΔR^2	F_{chg}	p	df_1	df_2
1. Phones	.894	.800	.796	.800	416.03	<.001	1	104
2. Birthrate	.921	.849	.846	.049	33.52	<.001	1	103
3. Doctors	.932	.869	.866	.020	15.92	<.001	1	102

Table 2 Coefficients for Final Model

	B	β	t	Bivariate r	Partial r
Phones per 100	2.245	.394	5.078*	.894	.449
Birthrate per 1,000	−.241	−.288	−4.263*	−.861	−.389
Doctors per 10,000	2.172	.306	3.990*	.881	.367

Note: * Indicates significance at p<.001.

Figure 7.6 Scatterplot Matrix for Original IVs and DV.

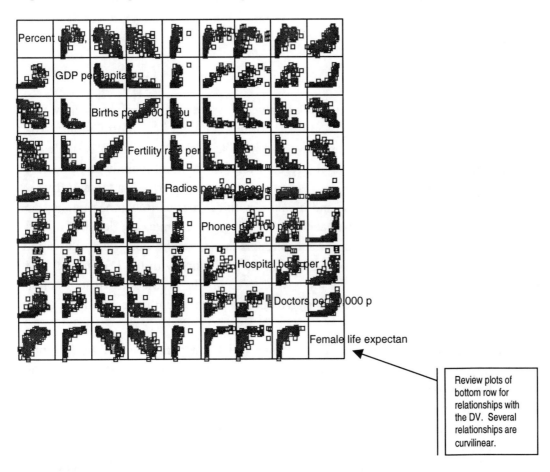

Review plots of bottom row for relationships with the DV. Several relationships are curvilinear.

Figure 7.7 Scatterplot Matrix of Transformed IVs with DV.

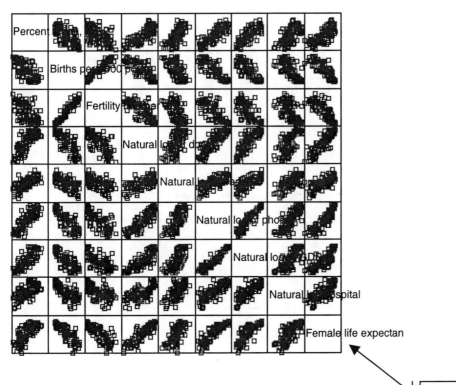

Plots display more
elliptical shapes—
representative of
linearity and
normality.

Figure 7.8 Test of Normality.

Tests of Normality

	Kolmogorov-Smirnov[a]		
	Statistic	df	Sig.
URBAN	.088	104	.045
LIFEEXPF	.125	104	.000
BIRTHRAT	.139	104	.000
FERTRATE	.125	104	.000
LNDOCS	.140	104	.000
LNRADIO	.086	104	.054
LNPHONE	.088	104	.044
LNGDP	.095	104	.021
LNBEDS	.060	104	.200*

Indicates that most
distributions are
non-normal.

* This is a lower bound of the true significance.

a. Lilliefors Significance Correction

Figure 7.9 Residuals Plot.

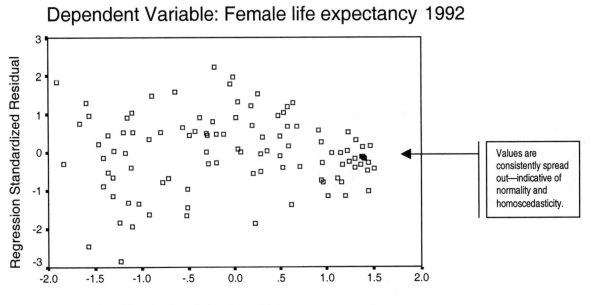

Scatterplot

Dependent Variable: Female life expectancy 1992

Values are
consistently spread
out—indicative of
normality and
homoscedasticity.

Figure 7.10 Model Summary Table for Female Life Expectancy.

Model Summary[d]

Model	R	R Square	Adjusted R Square	Std. Error of the Estimate	R Square Change	F Change	df1	df2	Sig. F Change
					Change Statistics				
1	.894[a]	.800	.798	4.98	.800	416.027	1	104	.000
2	.921[b]	.849	.846	4.34	.049	33.516	1	103	.000
3	.932[c]	.869	.866	4.06	.020	15.919	1	102	.000

a. Predictors: (Constant), LNPHONE

b. Predictors: (Constant), LNPHONE, BIRTHRAT

c. Predictors: (Constant), LNPHONE, BIRTHRAT, LNDOCS

d. Dependent Variable: LIFEEXPF

Represents each
step in the model
building.

Figure 7.11 ANOVA Summary Table.

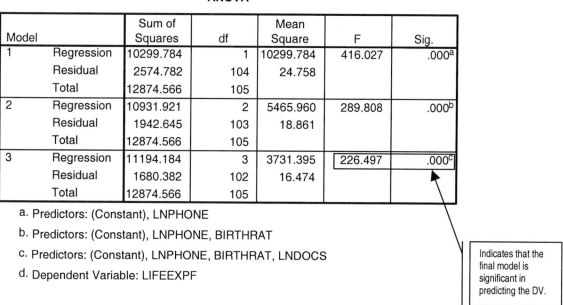

ANOVA[d]

Model		Sum of Squares	df	Mean Square	F	Sig.
1	Regression	10299.784	1	10299.784	416.027	.000[a]
	Residual	2574.782	104	24.758		
	Total	12874.566	105			
2	Regression	10931.921	2	5465.960	289.808	.000[b]
	Residual	1942.645	103	18.861		
	Total	12874.566	105			
3	Regression	11194.184	3	3731.395	226.497	.000[c]
	Residual	1680.382	102	16.474		
	Total	12874.566	105			

a. Predictors: (Constant), LNPHONE

b. Predictors: (Constant), LNPHONE, BIRTHRAT

c. Predictors: (Constant), LNPHONE, BIRTHRAT, LNDOCS

d. Dependent Variable: LIFEEXPF

Indicates that the final model is significant in predicting the DV.

SECTION 7.4 SAMPLE STUDY AND ANALYSIS

This section provides a complete example of the process of conducting multiple regression. This process includes the development of research questions and hypotheses, data screening methods, test methods, interpretation of output, and presentation of results. The example utilizes the data set, *country.sav* from the SPSS Web site.

Problem

In the previous example, we identified predictors of female life expectancy. For this example, we will utilize the same IVs but change the DV to male life expectancy. In addition, the Enter method will be used, such that all IVs will be entered into the model. The following research question is gener-ated to address this scenario:

How accurately do the IVs [% urban population (*urban*); gross domestic product per capita (*gdp*); birthrate per 1,000 (*birthrat*); hospital beds per 10,000 (*beds*); doctors per 10,000 (*docs*); radios per 100 (*radios*); and phones per 100 (*phones*)] predict male life expectancy?

Figure 7.12 Coefficients Tables for Variables Included and Excluded from Model.

Coefficients[a]

Model		Unstandardized Coefficients		Standardized Coefficients	t	Sig.	Correlations			Collinearity Statistics	
		B	Std. Error	Beta			Zero-order	Partial	Part	Tolerance	VIF
1	(Constant)	60.513	.585		103.493	.000					
	LNPHONE	5.102	.250	.894	20.397	.000	.894	.894	.894	1.000	1.000
2	(Constant)	72.700	2.166		33.564	.000					
	LNPHONE	3.284	.383	.576	8.583	.000	.894	.646	.329	.326	3.070
	BIRTHRAT	-.325	.056	-.388	-5.789	.000	-.861	-.495	-.222	.326	3.070
3	(Constant)	68.159	2.322		29.349	.000					
	LNPHONE	2.245	.442	.394	5.078	.000	.894	.449	.182	.213	4.696
	BIRTHRAT	-.241	.056	-.288	-4.263	.000	-.861	-.389	-.152	.281	3.565
	LNDOCS	2.172	.544	.306	3.990	.000	.881	.367	.143	.218	4.592

a. Dependent Variable: LIFEEXPF

Coefficients used to develop regression equation.

Coefficients used to develop a regression equation for standardized variables.

Tolerance statistics should be greater than .1.

Excluded Variables[d]

Model		Beta In	t	Sig.	Partial Correlation	Tolerance	VIF	Minimum Tolerance
1	LNBEDS	.098[a]	1.431	.155	.140	.409	2.444	.409
	LNGDP	-.022[a]	-.166	.868	-.016	.112	8.927	.112
	LNRADIO	.067[a]	1.043	.299	.102	.466	2.144	.466
	LNDOCS	.428[a]	5.564	.000	.481	.253	3.956	.253
	BIRTHRAT	-.388[a]	-5.789	.000	-.495	.326	3.070	.326
	URBAN	.096[a]	1.267	.208	.124	.335	2.984	.335
2	LNBEDS	.017[b]	.274	.785	.027	.386	2.587	.259
	LNGDP	-.159[b]	-1.372	.173	-.135	.108	9.293	.103
	LNRADIO	.043[b]	.759	.450	.075	.464	2.156	.250
	LNDOCS	.306[b]	3.990	.000	.367	.218	4.592	.213
	URBAN	.103[b]	1.566	.121	.153	.335	2.985	.195
3	LNBEDS	.013[c]	.229	.819	.023	.386	2.588	.184
	LNGDP	-.153[c]	-1.412	.161	-.139	.108	9.295	8.728E-02
	LNRADIO	.047[c]	.900	.370	.089	.464	2.157	.176
	URBAN	.003[c]	.049	.961	.005	.280	3.573	.175

a. Predictors in the Model: (Constant), LNPHONE
b. Predictors in the Model: (Constant), LNPHONE, BIRTHRAT
c. Predictors in the Model: (Constant), LNPHONE, BIRTHRAT, LNDOCS
d. Dependent Variable: LIFEEXPF

Method

Data are screened to identify missing data and outliers and to evaluate the fulfillment of test assumptions. Outliers were identified by calculating Mahalanobis distance in a preliminary **Regression** procedure. **Explore** was then conducted on the newly generated Mahalanobis variable (*mah_1*) to determine which cases exceeded the chi square (χ^2) criteria (see Figure 7.13). Using a chi square table, we found the critical value of chi square at $p<.001$ with $df=8$ to be 26.125. Cases #69, #72, and #67 exceed this critical value and so were deleted from the analysis. Linearity was then analyzed by creating a scatterplot matrix (see Figure 7.14). Scatterplots display nonlinearity for the following variables: *gdp, beds, docs, radios,* and *phones.* These variables were transformed by taking the natural log of each. The data set already includes these transformations as *lngdp, lnbeds, lndocs, lnradio,* and *lnphone.* A scatterplot of the transformed variables indicates linearity and normality (see Figure 7.15). Univariate normality was also assessed by conducting **Explore**. Histograms and normality tests (see Figure 7.16) indicate some non-normal distributions; however, the distributions are not too extreme. Multivariate normality and homoscedasticity were examined through the generation of a residuals plot within another preliminary **Regression**. The residuals plot is somewhat scattered but again is not extreme (see Figure 7.17). Thus, multivariate normality and homoscedasticity will be assumed. Multiple **Regression** was then conducted using the Enter method. See the section on SPSS "How To" for more details on how to generate the following output.

Figure 7.13 Outliers for Mahalanobis Distance.

Extreme Values

			Case Number	Value
MAH_1	Highest	1	67	50.81753
		2	72	27.23509
		3	69	26.69299
		4	108	25.67930
		5	83	21.00443
	Lowest	1	21	1.19032
		2	40	1.29138
		3	87	1.59614
		4	53	1.73529
		5	25	1.81816

Outliers exceed χ2 critical value.

Output and Interpretation of Results

Figures 7.18 – 7.20 present the three primary parts of regression output: model summary, ANOVA table, coefficients table. Review of the tolerance statistics presented in the coefficients table (see Figure 7.20) indicate that all IVs were tolerated in the model. The model summary (see Figure 7.18) and the ANOVA summary (see Figure 7.19) indicate that the overall model of the seven IVs significantly predicts male life expectancy, $R^2=.845$, $R^2_{adj}=.834$, $F(7, 96,)=74.69$, $p<.001$. However, review of the beta weights in Figure 7.20 specify that only three variables, *birthrate* $\beta=-.241$, $t(96)=-3.02$, $p=.003$; *doctors* $\beta=.412$, $t(96)=4.26$, $p<.001$; and *phones* $\beta=-.548$, $t(96)=3.88$, $p<.001$, significantly contributed to the model. The reader should note that although the same three variables created the model for predicting female life expectancy despite a different method being utilized, the significance of the model predicting male life expectancy is much lower since all seven variables were entered into the model.

Figure 7.14 Scatterplot Matrix of Original IVs with DV.

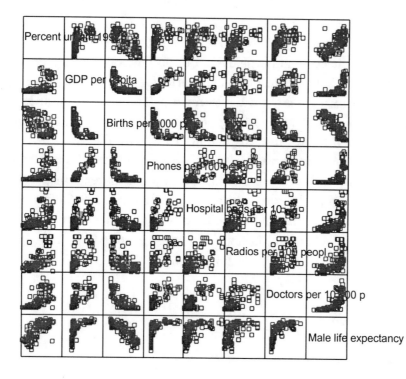

Figure 7.15 Scatterplot Matrix of Transformed IVs with DV.

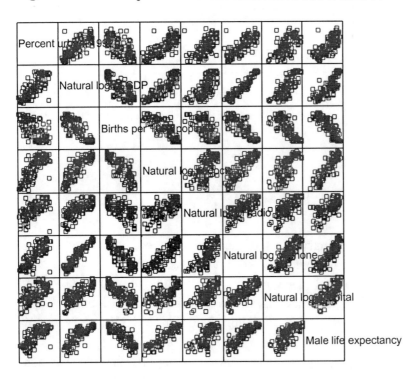

Figure 7.16 Tests of Normality.

Tests of Normality

	Kolmogorov-Smirnov[a]		
	Statistic	df	Sig.
URBAN	.091	104	.034
LIFEEXPM	.136	104	.000
BIRTHRAT	.136	104	.000
LNDOCS	.134	104	.000
LNRADIO	.102	104	.009
LNPHONE	.088	104	.044
LNGDP	.097	104	.017
LNBEDS	.062	104	.200*

*. This is a lower bound of the true significance.

a. Lilliefors Significance Correction

Figure 7.17 Residuals Plot.

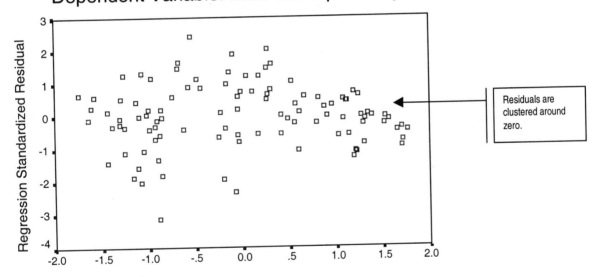

Scatterplot

Dependent Variable: Male life expectancy 1992

Figure 7.18 Model Summary Predicting Male Life Expectancy.

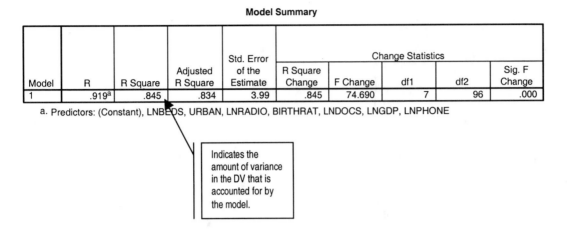

Model Summary

Model	R	R Square	Adjusted R Square	Std. Error of the Estimate	R Square Change	F Change	df1	df2	Sig. F Change
					Change Statistics				
1	.919ᵃ	.845	.834	3.99	.845	74.690	7	96	.000

a. Predictors: (Constant), LNBEDS, URBAN, LNRADIO, BIRTHRAT, LNDOCS, LNGDP, LNPHONE

Indicates the amount of variance in the DV that is accounted for by the model.

Presentation of Results

Standard multiple regression was conducted to determine the accuracy of the independent variables (% urban population [urban]; gross domestic product per capita [gdp]; birthrate per 1,000 [birthrate]; hospital beds per 10,000 [beds]; doctors per 10,000 [docs]; radios per 100 [radios]; and phones per 100 [phone]) predicting male life expectancy. Data screening led to the elimination of three cases. Evaluation of linearity led to the natural log transformation of gdp, beds, docs, radios, and phones. Regression results indicate that the overall model significantly predicts male life expectancy, R^2=.845, R^2_{adj}=.834, $F(7, 96,)$=74.69, p<.001. This model accounts for 84.5% of variance in male life expectancy. A summary of regression coefficients is presented in Table 1 and indicates that only three (birthrate, doctors, and phones) of the seven variables significantly contributed to the model.

Table 1 Coefficients for Model Variables

	B	*β*	*t*	*p*	Bivariate *r*	Partial *r*
Urban	−.007	−.017	−.243	.808	.701	−.025
Birthrate	−.179	−.241	−3.021	.003	−.829	−.295
Doctors	2.586	.412	4.258	<.001	.883	.399
Radios	.0651	.006	.104	.918	.639	.011
Phones	2.745	.548	3.881	<.001	.874	.368
GDP	−1.182	−.185	−1.462	.147	.816	−.148
Beds	.627	−.058	−.850	.398	.681	−.086

Figure 7.19 ANOVA Summary Table.

ANOVA[b]

Model		Sum of Squares	df	Mean Square	F	Sig.
1	Regression	8311.675	7	1187.382	74.690	.000[a]
	Residual	1526.162	96	15.898		
	Total	9837.837	103			

a. Predictors: (Constant), LNBEDS, URBAN, LNRADIO, BIRTHRAT, LNDOCS, LNGDP, LNPHONE

b. Dependent Variable: LIFEEXPM

F-ratio and level of significance indicate the degree to which the model predicts the DV.

Figure 7.20 Coefficients Table.

Coefficients[a]

Model		Unstandardized Coefficients		Standardized Coefficients			Correlations			Collinearity Statistics	
		B	Std. Error	Beta	t	Sig.	Zero-order	Partial	Part	Tolerance	VIF
1	(Constant)	71.271	6.205		11.486	.000					
	URBAN	-6.55E-03	.027	-.017	-.243	.808	.701	-.025	-.010	.342	2.927
	BIRTHRAT	-.179	.059	-.241	-3.021	.003	-.829	-.295	-.121	.253	3.949
	LNDOCS	2.586	.607	.412	4.258	.000	.883	.399	.171	.173	5.791
	LNRADIO	6.505E-02	.628	.006	.104	.918	.639	.011	.004	.448	2.232
	LNPHONE	2.745	.707	.548	3.881	.000	.874	.368	.156	.081	12.343
	LNGDP	-1.182	.808	-.185	-1.462	.147	.816	-.148	-.059	.101	9.898
	LNBEDS	-.627	.738	-.058	-.850	.398	.681	-.086	-.034	.341	2.929

a. Dependent Variable: LIFEEXPM

Tolerance statistics exceed .1.

SECTION 7.5 SPSS "HOW TO"

This section demonstrates the steps for conducting multiple regression using the **Regression** procedure with the previous example. To conduct regression, select the following menus:

Analyze
      Regression
            Linear

Multiple Linear Regression Dialogue Box (see Figure 7.21)

Once in this dialogue box, identify the DV (*lifeexpm*) and move it to the Dependent Variable Box. Identify each of the IVs and move each to the Independent(s) Box. Next, select the appropriate regression method. SPSS provides five different methods:

Enter—Enters all IVs, one at a time, into the model regardless of significant contribution.

Backward—Enters all IVs one at a time and then removes them one at a time based upon a level of significance for removal (default is $p \geq .10$). The process ends when no more variables meet the removal requirement.

Forward—Only enters IVs that significantly contribute to the model (account for a significant amount of unique variance in the DV). Variables are entered one variable at a time. When no more variables account for a significant amount of variance, the process ends.

Stepwise—Combines Forward and Backward methods. Two criteria are utilized—one for entering and one for removing. Basically at each step, an IV is entered that meets the criteria for entering. Then the model is analyzed to determine if any IVs should be removed. If an entered variable meets the removal criteria, it is removed. The process continues until no more IVs meet the enter or removal criteria. Although Stepwise is quite common, it has recently come under great criticism for not creating a model of the best combination of predictors (Thompson, 1998).

Remove—This method first utilizes the **Enter** method, after which specified variable(s) is removed from the model and **Enter** is conducted again.

For this example, we selected **Enter**. Once the method is selected, click **Statistics**.

Multiple Linear Regression: Statistics Dialogue Box (see Figure 7.22)

This box provides several statistical options. Under Regression Coefficients, three options are provided:

Estimates—The default. This option produces B and beta weights with associated standard error, t and p values.

Confidence Intervals—Calculates confidence intervals for B weights at 95%.

Covariance Matrix—Creates a covariance-variance-correlation matrix that can help assess collinearity. The matrix is organized with covariances below the diagonal, variances on the diagonal, and correlations above the diagonal.

For our example, we used the default of **Estimates**. Just to the right of **Regression Coefficients** is a list of more statistical options. These are described below.

Model Fit—Produces Multiple R, R^2, an ANOVA table, and corresponding F and p values.

R Squared Change—Reports the change in R^2 as a new variable is entered into the model. This is appropriate when using a stepping method.

Descriptives—Calculates variable means, standard deviations, and correlation matrix.

Part and Partial Correlations—Calculates part and partial correlation coefficients.

Collinearity Diagnostics—Calculates tolerance of each IV.

Figure 7.21 Multiple Linear Regression Dialogue Box.

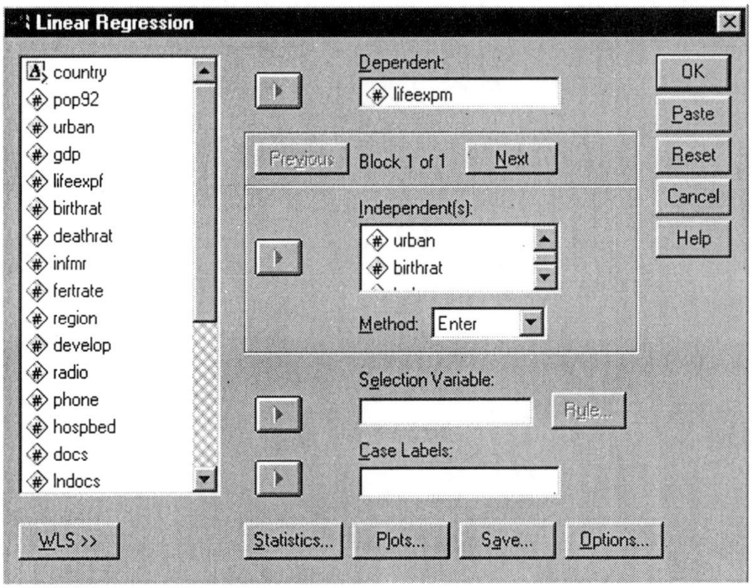

Figure 7.22 Multiple Linear Regression Statistics Dialogue Box.

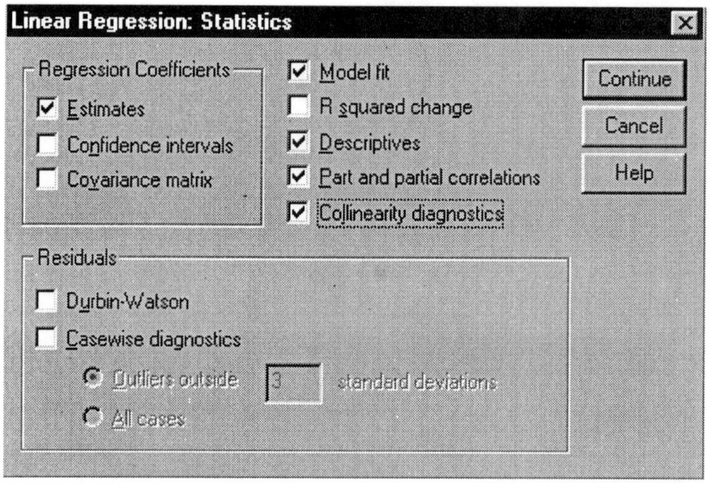

For this example, the following were checked: **Model Fit, Descriptives, Part and Partial Correlations,** and **Collinearity Diagnostics.** These selections are described below. **R Square Change** was not selected since a stepping method was not utilized. The last set of statistical options is under **Residuals** and is not utilized a great deal. Click **Continue.** Back in the **Regression** Box, click **Options.**

Multiple Linear Regression: Options Dialogue Box (see Figure 7.23)

This box provides criteria options for entering or removing variables from the model. Two criteria options are provided: use of probability of F or use of F-value. The default method is use of probability of F with .05 criteria for entry and .01 for removal. Since this example utilizes the **Enter** method, selection of a stepping criteria is unnecessary. Another option in this box is removing the constant from the regression equation. Including the constant is the default. Click **Continue**. Click **OK**. Although this completes the steps for our example problem, two dialogue boxes provide additional options that will be described for future use. Actually we have already used both in evaluating test assumptions. The first is **Save**.

Figure 7.23 Multiple Linear Regression: Options Dialogue Box.

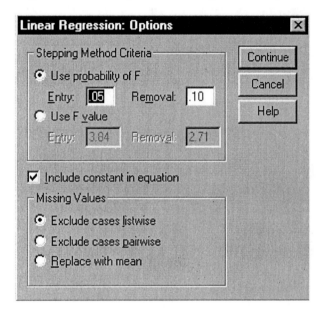

Multiple Linear Regression: Save Dialogue Box (see Figure 7.24)

Several types of residuals can be saved in the data file with new variable names. Saving these residuals is necessary if you seek to create plots with these residuals. Types of residuals include **Unstandardized** and **Standardized**. Three other types are provided and are described as follows:

> **Studentized**—residuals standardized on a case-by-case basis, depending on how far the case is from the mean.
> **Deleted**—residual for a case if the case was deleted from the analysis.
> **Studentized Deleted**—combines the concepts of studentized and deleted residuals.

You can also save predicted values and distances, such as Mahalanobis distance that was utilized to screen for outliers. The next dialogue box is **Plots**.

Figure 7.24 Multiple Linear Regression: Save Dialogue Box.

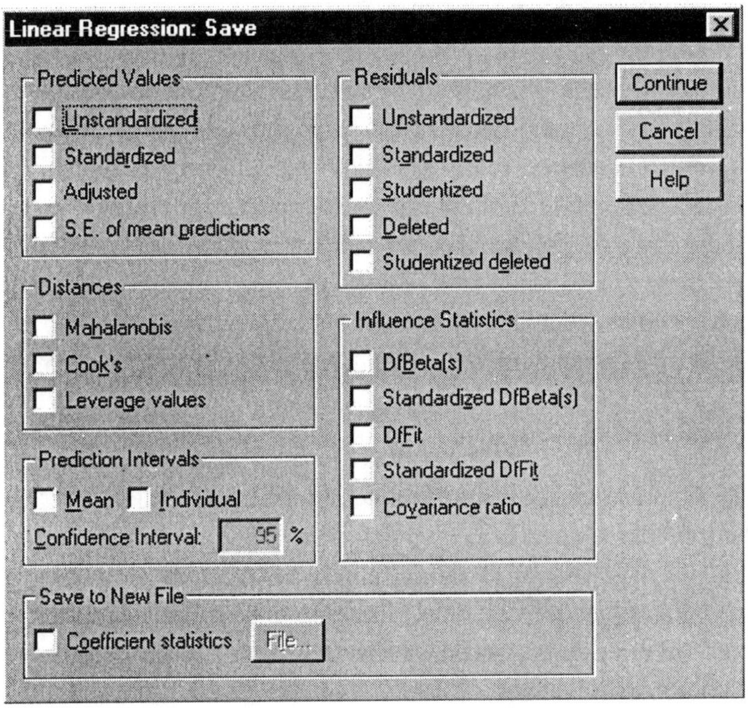

Figure 7.25 Multiple Linear Regression Plots Dialogue Box.

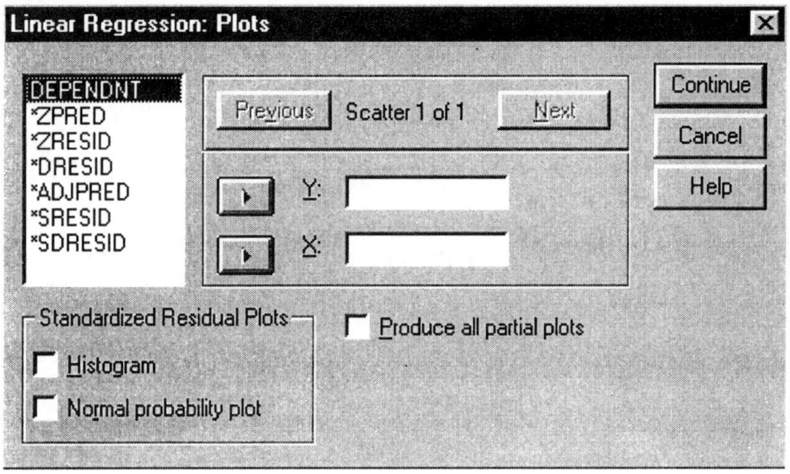

Multiple Linear Regression Plots: Dialogue Box (see Figure 7.25)

The **Plots** function provides several options for graphically analyzing the residuals or in other words for evaluating how well the generated model predicts the DV. Scatterplots can be created for any combination of the following: the DV, standardized predicted values (ZPRED), standardized residuals (ZRESID), deleted residuals (DRESID), studentized residuals (SRESID), or studentized deleted residuals (SDRESID). Earlier, we plotted the standardized predicted values (ZPRED) against the standardized residuals (ZRESID) to test linearity and homoscedasticity. Histograms of standardized residuals and normal probability plots comparing the distribution of the standardized residuals to a normal distribution can also be generated. The final option is Produce All Partial Plots. Checking this option will create scatterplots of residuals for each IV with the residuals of the DV when both variables are separately regressed on the remaining IVs. A minimum of two IVs must be in the equation to generate a partial plot.

Summary

The purpose of multiple regression is to model or group variables that best predict a criterion variable (DV). The procedure examines the significance of each IV to predict the DV as well as the significance of the entire model to predict the DV. A variety of methods (enter, forward, backward, remove, and stepwise) can be used to develop and test different models. Prior to conducting the regression, data should be screened for missing data and outliers as well as evaluated for test assumptions— linearity, normality, and homoscedasticity. Regression output typically includes three parts: model summary, ANOVA table, and coefficients table. The model summary table displays several multiple correlation indices—multiple correlation (R), squared multiple correlation (R^2), adjusted squared multiple correlation (R^2_{adj}), and change in R^2 (ΔR^2)—all of which indicate how well an IV or combination of IVs predicts the criterion variable (DV). The ANOVA table presents the F-test and corresponding level of significance for each step or model generated. This test examines the degree to which the relationship between the IVs and DV is linear. The coefficients table reports the following: unstandardized regression coefficient (B), the standardized regression coefficient (beta or β), t and p values, three correlation indices (bivariate r, partial r, and part r), and tolerance coefficient. When interpreting the output, tolerance should be examined first since this measures the degree to which IVs account for unique variance in the DV. If tolerance for an IV is less than .1, the regression analysis should be conducted again without the violating IV. If tolerance is acceptable, proceed with interpreting the model summary, ANOVA table, and table of coefficients. Figure 7.26 provides a checklist for conducting multiple regression.

Figure 7.26 Checklist for Conducting Multiple Regression.

I. Screen Data
  a. Missing Data?
  b. Multivariate Outliers?
      ❑ Run preliminary Regression to calculate Mahalanobis' Distance.
          1. 🖱 **Analyze…Regression…Linear.**
          2. Identify a variable that serves as a case number and move to Dependent Variable box.
          3. Identify all appropriate quantitative variables and move to Independent(s) box.
          4. 🖱 **Save.**
          5. Check **Mahalanobis'.**
          6. 🖱 **Continue.**
          7. 🖱 **OK.**
          8. Determine chi square χ^2 critical value at $p<.001$.
      ❑ Conduct **Explore** to test outliers for Mahalanobis chi square χ^2.
          1. 🖱 **Analyze…Descriptive Statistics…Explore**
          2. Move *mah_1* to Dependent Variable box.
          3. Leave Factor box empty.
          4. 🖱 **Statistics.**
          5. Check **Outliers.**
          6. 🖱 **Continue.**
          7. 🖱 **OK.**
      ❑ Delete outliers for subjects when χ^2 exceeds critical χ^2 at $p<.001$.
  c. Linearity, Normality, Homoscedasticity?
      ❑ Create Scatterplot Matrix of all IVs and DV.
          ❑ Scatterplot shapes are not close to elliptical shapes→reevaluate univariate normality and consider transformations.
      ❑ Run Normality Plots with Tests within **Explore.**
      ❑ Run preliminary Regression to create residual plot.
          1. 🖱 **Analyze…Regression…Linear**
          2. Move DV to Dependent Variable box.
          3. Move IVs to Independent(s) Variable box.
          4. 🖱 **Plot.**
          5. Select ZRESID for y-axis.
          6. Select ZPRED for x-axis.
          7. 🖱 **Continue.**
          8. 🖱 **OK.**
      ❑ If residuals are clustered at the top, bottom, left, or right area in plot→ reevaluate univariate normality and consider transformations.

II. Conduct Multiple Regression
  a. Run Regression using **Linear Regression**
          1. 🖱 **Analyze…**🖱 **Regression…**🖱 **Linear**
          2. Move DV to Dependent Variable box.
          3. Move IVs to Independent(s) box.
          4. Select appropriate method.
          5. 🖱 **Statistics.**
          6. Check **Estimates, Model Fit, R Squared Change, Descriptives, Part and Partial Correlations,** and **Collinearity Diagnostics.**
          7. 🖱 **Continue**
          8. 🖱 **Options.**
          9. Select appropriate criteria.
          10. 🖱 **Continue**
          11. 🖱 **OK.**
  b. Interpret tolerance.
  c. If tolerance for each IV is greater than .1, interpret model summary, ANOVA table, and coefficients table.

Figure 7.26 Checklist for Conducting Multiple Regression. (*Continued*)

III.   Summarize Results
  a.   Describe any data elimination or transformation.
  b.   Present descriptive statistics in tables (correlation matrix, means, and standard deviations).
  c.   Narrate the significance of the overall regression (R^2, R^2_{adj}, F and p-values with degrees of freedom).
  d.   If stepping method was used, summarize steps in a table (R^2, R^2_{adj}, R^2 change, and level of significance for change).
  e.   Create a tables that reports the B weights, β weights, bivariate r, and partial r for each IV in the model.
  f.   Draw conclusions.

Exercises for Chapter 7

1.   The following output was generated from conducting a forward multiple regression to identify which IVs (*urban, birthrat, lnphone,* and *lnradio*) predict *lngdp*. The data that were analyzed were from the SPSS *country.sav* data file.

Variables Entered/Removed[a]

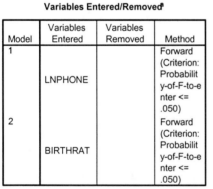

Model	Variables Entered	Variables Removed	Method
1	LNPHONE		Forward (Criterion: Probability-of-F-to-enter <= .050)
2	BIRTHRAT		Forward (Criterion: Probability-of-F-to-enter <= .050)

a. Dependent Variable: LNGDP

Model Summary

Model	R	R Square	Adjusted R Square	Std. Error of the Estimate	Change Statistics R Square Change	F Change	df1	df2	Sig. F Change
1	.941[a]	.886	.885	.5180	.886	862.968	1	111	.000
2	.943[b]	.890	.888	.5109	.004	4.095	1	110	.045

a. Predictors: (Constant), LNPHONE
b. Predictors: (Constant), LNPHONE, BIRTHRAT

ANOVA^c

Model		Sum of Squares	df	Mean Square	F	Sig.
1	Regression	231.539	1	231.539	862.968	.000^a
	Residual	29.782	111	.268		
	Total	261.321	112			
2	Regression	232.608	2	116.304	445.561	.000^b
	Residual	28.713	110	.261		
	Total	261.321	112			

a. Predictors: (Constant), LNPHONE

b. Predictors: (Constant), LNPHONE, BIRTHRAT

c. Dependent Variable: LNGDP

Coefficients^a

Model		Unstandardized Coefficients		Standardized Coefficients	t	Sig.	Correlations			Collinearity Statistics	
		B	Std. Error	Beta			Zero-order	Partial	Part	Tolerance	VIF
1	(Constant)	6.389	.058		110.662	.000					
	LNPHONE	.736	.025	.941	29.376	.000	.941	.941	.941	1.000	1.000
2	(Constant)	6.878	.248		27.744	.000					
	LNPHONE	.663	.044	.849	15.238	.000	.941	.824	.482	.322	3.104
	BIRTHRAT	-1.29E-02	.006	-.113	-2.024	.045	-.811	-.189	-.064	.322	3.104

a. Dependent Variable: LNGDP

Excluded Variables^c

Model		Beta In	t	Sig.	Partial Correlation	Collinearity Statistics		
						Tolerance	VIF	Minimum Tolerance
1	URBAN	.095^a	1.901	.060	.178	.404	2.475	.404
	BIRTHRAT	-.113^a	-2.024	.045	-.189	.322	3.104	.322
	LNRADIO	.026^a	.557	.579	.053	.461	2.171	.461
2	URBAN	.091^b	1.848	.067	.174	.403	2.479	.225
	LNRADIO	.021^b	.455	.650	.044	.459	2.178	.243

a. Predictors in the Model: (Constant), LNPHONE

b. Predictors in the Model: (Constant), LNPHONE, BIRTHRAT

c. Dependent Variable: LNGDP

a.   Evaluate the tolerance statistics. Is multicollinearity a problem?

b.   What variables create the model to predict *lngdp*? What statistics support your response?

c.   Is the model significant in predicting *lngdp*? Explain.

d.   What percentage of variance in *lngdp* is explained by the model?

e. Write the regression equation for *lngdp*.

2. This question utilizes the data set *gss.sav*, which can be downloaded at the SPSS Web site. Open the URL: **www.spss.com/tech/DataSets.html** in your Web browser. Scroll down until you see "Data Used in SPSS Guide to Data Analysis—8.0 and 9.0" and click on the link "dataset.exe." When the "Save As" dialog appears, select the appropriate folder and save the file. Preferably, this should be a folder created in the SPSS folder of your hard drive for this purpose. Once the file is saved, double-click the "dataset.exe" file to extract the data sets to the folder.
    You are interested in examining whether the variables shown here in brackets [years of age (*age*), hours worked per week(*hrs1*), years of education (*educ*), years of education for mother (*maeduc*), years of education for father (*paeduc*)] are predictors of individual income (*rincmdol*). Complete the following steps to conduct this analysis.

a. Conduct a preliminary regression to calculate Mahalanobis distance. Identify the critical value for chi square. Conduct Explore to identify outliers. Which cases should be removed from further analysis?

b. Create a scatterplot matrix. Can you assume linearity and normality?

c. Conduct a preliminary regression to create a residual plot. Can you assume normality and homoscedasticity?

d. Conduct multiple regression using the Enter method. Evaluate the tolerance statistics. Is multicollinearity a problem?

e. Does the model significantly predict *rincmdol*? Explain.

f. Which variables significantly predict *rincmdol*? Which variable is the best predictor of the DV?

g. What percentage of variance in *rincmdol* is explained by the model?

h. Write the regression equation for the standardized variables.

i. Explain why the variables of mother's and father's education are not significant predictors of *rincmdol*?

CHAPTER 8

PATH ANALYSIS

In the previous chapter, we discussed in detail one of the main purposes of multiple regression—that being *prediction*. In this chapter, we present a discussion of another use of multiple regression—regression as a technique for providing *explanations* of possible causal relationships among a set of variables. Path analysis is actually one of two techniques that are classified under the broad heading of causal modeling. Following a brief introduction to causal modeling, and the distinctions between the two major types of causal modeling, we present a detailed discussion of path analysis, focusing on appropriate uses and proper interpretations of the technique.

SECTION 8.1 PRACTICAL VIEW

Purpose

Regression can be used to establish the possibility of cause-and-effect relationships among a set of variables (Sprinthall, 2000). Using regression analysis in this manner constitutes a specific set of statistical analysis techniques known as causal modeling. *Causal modeling* techniques examine whether a pattern of intercorrelations among variables "fits" the researcher's underlying theory of which variables are causing other variables (Aron & Aron, 1997). It is important to remember, however, that in causal modeling we are attempting to draw causal inferences from correlational data—the degree of confidence in the validity of causal inference from correlational data is typically much weaker than inference drawn from data resulting from a well-designed experimental study where the important concept of random assignment to treatments has been incorporated (Tate, 1992). Conclusions drawn from causal modeling with correlational data must be confined to the following limitation: The results of causal modeling are valid and unbiased *only if* the assumed model adequately represents the *real* causal processes (Tate, 1992).

In causal modeling, the causal interrelationships are examined among a set of variables that have been logically ordered on the basis of time (Sprinthall, 2000). Logically, a causal variable must precede any variable that it supposedly affects—this establishes the causal ordering of the variables (Sprinthall, 2000). There are two types of causal modeling techniques: path analysis and structural equation modeling (the latter will be described at the end of this section). *Path analysis* begins with the researcher developing a diagram with arrows connecting variables and depicting the *causal flow*, or the direction of cause-and-effect. The precursor to path analysis is a simpler version of causal modeling in which the only effects represented are direct causal effects. Path analysis has a substantial advantage over the simpler model in that both *direct* and *indirect* causal effects can be estimated. We will first examine the simplest form of causal modeling, followed by a presentation of the more involved form—path analysis.

As we have mentioned, the simplest version of the causal modeling technique is one in which only direct causal effects are represented (Tate, 1992). This version is quite similar to multiple regression, as discussed in the previous chapter. The direct causal effect of an IV (X) on a DV (Y) is defined as the amount of change in Y resulting from a unit change in X, holding constant all other causal determinants of Y. The causal model is represented by a single regression equation in which the IVs are the causal determinants of the DV. For example, we might want to determine the direct causal paths of three IVs on a single DV. Assume we wanted to investigate the effects of a country's location in the world (*region*), its status as a developing nation (*develop*), and the number of doctors per 10,000 people (*docs*) on male life expectancy (*lifexpm*). Since we are assuming only direct causal paths, our single equation would simply attempt to explain the direct causes of each of the three IVs (*region, develop, docs*) on the DV (*lifexpm*).

The development of a causal model is probably the most difficult aspect of conducting any causal modeling study (Tate, 1992). The specification of the model is a formal declaration of the researcher's beliefs regarding the causal links among the variables. What was the basis of our decision to order the four variables as described above? These beliefs are typically influenced by several sources of information, including the research literature, formal and informal theories, personal observations and experiences with the phenomenon of interest, expert opinions, and last but certainly not least, common sense and logic (Tate, 1992). Specification of a hypothesized model is often complicated by several sources of difficulty:

- the vagueness of many theories in social science research;
- the potentially infinite number of possible causal determinants which are often posited in the related research literature; and
- the complexity of nearly all phenomena of interest in social science research (Tate, 1992)—which has, of course, been discussed on several occasions in this text.

The specified causal model can be represented in two ways: as an equation or in diagrammatic form. The assumed causal model, when stated as an equation, is often referred to as a ***structural equation***, and is typically stated in its standardized form. If we were to "define" our variables using z-score coefficients—i.e., the standardized form where $z_1 = region$, $z_2 = develop$, $z_3 = docs$, and $z_4 = lifexpm$—the structural equation for our working example would be:

$$z_4 = p_{41}z_1 + p_{42}z_2 + p_{43}z_3 + e_4 \qquad \text{(Equation 8.1)}$$

In this structural equation, the direct causal effects are represented by the p coefficients, often called ***path coefficients*** or ***structural coefficients***. These coefficients are analogous to standardized regression coefficients, β, resulting from a multiple regression analysis (Agresti & Finlay, 1997) and their interpretation is similar (Aron & Aron, 1997 & 1999; Tate, 1992; Asher, 1983). In other words, they are interpreted as the estimated change in the DV, expressed in standard deviation units, associated with a one standard deviation change in each IV, holding the other IVs constant. The subscripts that accompany the path coefficients indicate the direction of causation with the first subscript indicating the variable being determined and the second indicating the direct cause (Tate, 1992). The zs indicate the standardized raw score value on each variable. The final component in the structural equation is the residual or e_i. This residual term, called the ***disturbance term*** in causal modeling parlance, represents the composite effect of any other direct determinants of z_4, which have not been included in the causal model, plus any measurement error in z_4 (Tate, 1992; Tatsuoka, 1988).

Although we are really dealing with three IVs and one DV in our working example, it is not accurate to refer to them as such when conducting a causal modeling study. In the specific language of causal modeling, the variable that is being explained by the model (i.e., the DV, the effect, or, in our example, z_4) is referred to as the ***endogenous variable***, while all variables not explained by the model (i.e., the IVs, the causes, or z_1, z_2, and z_3) are referred to as ***exogenous variables*** (Tate, 1992; Tatsuoka, 1988). Endogenous variables are assumed to have their variance explained by the exogenous variables included in the model; whereas, the variability of exogenous variables is assumed to be explained by other variables outside the causal model under consideration (Pedhazur, 1982).

The second way that a specified causal model can be portrayed is with a path diagram. A ***path diagram*** is a pictorial representation of the theoretical explanations of cause-and-effect relationships among a set of variables (Agresti & Finlay, 1997). The path diagram for our simple working example is shown in Figure 8.1. It is important to note that a path diagram is not necessary for causal modeling analysis, but is helpful is presenting the results of the analysis (Pedhazur, 1982).

Figure 8.1 Sample Path Diagram for a Single Equation Causal Model.

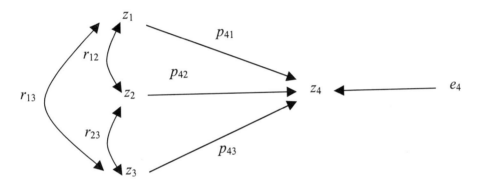

The direct causal effects of the exogenous variables z_1, z_2, and z_3 on the endogenous variable z_4 are shown with straight arrows, with the arrowheads indicating the assumed direction of causation (Tate, 1992). These arrows are often referred to as *causal paths* and are labeled with the associated path coefficients. Notice that the effect of the disturbance term is also included. Finally, the curved, double-headed arrows simply represent the bivariate correlations between exogenous variables in the model.

As we've previously mentioned, path analysis builds upon this simpler version by modeling both direct *and* indirect causal effects among the variables. An *indirect effect* occurs when a variable affects an endogenous variable through its effect on some other variable, known as an *intervening variable* (Agresti & Finlay, 1997). As in any causal modeling analysis, the first step is to specify the model of direct causal links among variables. This model will then imply indirect and total causal effects—this is a critical element that is missing in the simpler, single equation model previously discussed (Tate, 1992). Another distinct advantage of path analysis over the single equation model is the fact that it is now possible to test the overall fit of the model to the data in order to ascertain if the model (theory) is consistent with the observed correlations (actual data). The method by which we assess model fit will be described momentarily and elaborated upon in Section 8.3. If serious inconsistencies between the model and the data exist, it is recommended that the model be revised prior to describing any of the causal effects (Tate, 1992). It is important to note that consistency between the model and the observed correlations does not prove the validity of the model, but does represent support of the model (Tate, 1992).

We can now expand our single equation model into a path model. To do so, we decide to add another variable to our model—the number of deaths per 1,000 people (*deathrat*). Our initial path model is presented in Figure 8.2. The arrows indicate that *deathrat* and *docs* (although, we will actually use the natural log of *docs*, or *lndocs*) are the only important direct causal determinants of *lifexpm*. Furthermore, we hypothesize that *region* and *develop* have a direct causal effect on *deathrat*, and that *develop* has a direct causal effect on *lndocs*. Notice that in our path model, *lndocs* has changed from an exogenous (unexplained) variable to an endogenous (explained) variable.

Figure 8.2 Path Diagram for the Initial Model (Male Life Expectancy).

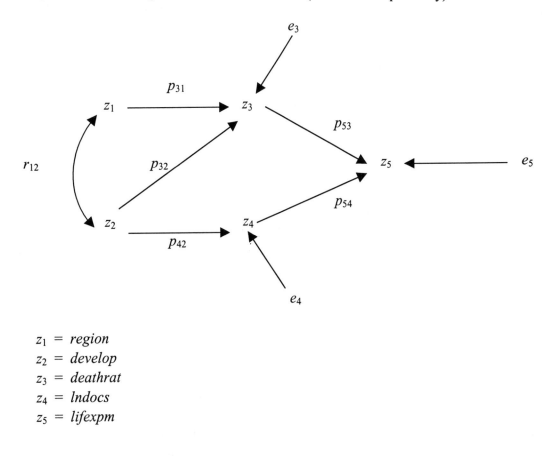

$z_1 = region$
$z_2 = develop$
$z_3 = deathrat$
$z_4 = lndocs$
$z_5 = lifexpm$

Model specification in path analysis becomes a much more convoluted process than in the single equation model (Tate, 1992). Correct specification in the single equation model necessitates that we have identified all causal effects of the lone endogenous variable. In our path analysis model, we have likely investigated the possible causes of our ultimate variable of interest—male life expectancy. After all, it is this variable in which we are most interested, in terms of explaining or describing its causal determinants. However, correct specification of the overall model in path analysis requires that *each* endogenous variable in the model is correctly specified. In other words, we have assumed that the model for *lifexpm* is correctly specified, but what about the models for *deathrat* and *lndocs*? We must ensure that the models for these additional endogenous variables are also correctly specified in order for our overall model to be valid and accurate (Tate, 1992).

Notice that we have also made informed decisions about excluding certain paths from the model. Specifically, our model has four missing paths—those from z_1 to z_4, z_1 to z_5, z_2 to z_5, and z_3 to z_4. It is important to note that these missing paths should also be consistent with theory and the associated literature (Tate, 1992). For example, excluding the path from *region* (z_1) to *lifexpm* (z_5) indicates the belief that *region* only affects *lifexpm* through its effect on *deathrat* (i.e., an indirect effect). Tate (1992) notes that a final theoretical path model should be "represented as much by the excluded paths as by the included paths."

Finally, when we are reasonably comfortable with our theoretical model, we can formally represent that model with a system of structural equations. This system of structural equations must include one for *each* endogenous variable in the model. For our current example, as depicted in Figure 8.2, these equations would be:

$$z_3 = p_{31}z_1 + p_{32}z_2 + e_3 \qquad \text{(Equation 8.2)}$$

$$z_4 = p_{42}z_2 + e_4 \qquad \text{(Equation 8.3)}$$

$$z_5 = p_{53}z_3 + p_{54}z_4 + e_5 \qquad \text{(Equation 8.4)}$$

One should notice that in a path diagram, indirect effects are identified by a chain of two or more straight arrows all going in the same direction (Tate, 1992). The value of an indirect path coefficient is determined by finding the product of all path coefficients in the chain. In Figure 8.2 for example, the paths from *region* to *deathrat* (p_{31}) and from *deathrat* to *lifexpm* (p_{53}) combine together to produce an indirect effect of *region* on *lifexpm* (equal to $p_{31}p_{53}$).

Multiple regression analysis provides the values for the unbiased estimates of the path coefficients. In order to obtain the coefficients, a separate regression run must be completed for each structural equation, each including only the direct causal effects for its associated endogenous variable. Using Equation 8.2 as an example, in order to obtain p_{31} and p_{32}, one must regress z_3 (*deathrat*) on z_1 (*region*) and z_2 (*develop*). In other words, a multiple regression analysis is conducted with *deathrat* as the DV and *region* and *develop* as the IVs. Similar procedures are then conducted for the two remaining structural equations.

Probably the most crucial part of the analysis in a causal modeling study is the assessment of model fit. Before the obtained estimates of path coefficients can be used to describe the causal effects among the variables, one should determine whether or not the model is consistent with the observed, empirical correlations among the variables. This is typically accomplished by obtaining the ***reproduced correlations***—those logically implied by the hypothetical or theoretical model—and comparing them to the empirical correlations (Agresti & Finlay, 1997; Tate, 1992). The reproduced correlations, therefore, are the bivariate correlations that *would* be produced *if* the causal model were correctly specified. If the observed and the reproduced correlations are reasonably close (say, within roughly .05 of each other), it can be assumed that the model is consistent with the empirical data (Tate, 1992). Larger discrepancies indicate that the model is not consistent with the data and model revisions should be considered. Unfortunately, the reproduced correlations, and subsequent comparisons to observed correlations, cannot be obtained via computer analysis and must be computed by hand. The procedures for doing so are described in detail in Section 8.3.

Earlier in this chapter, we alluded to a second type of causal modeling strategy, and we briefly introduce it here. This second type of causal modeling offers several advantages over path analysis. ***Structural equation modeling***, sometimes referred to as *latent variable modeling*, also involves diagrams with arrows showing causal flows among variables. However, one major advantage is that the computer analysis procedure provides an overall indication of the fit between the model and the theory. We have briefly mentioned, and will see later in some detail, how this assessment of model fit must be done by hand in path analysis. A second major advantage of structural equation modeling over path analysis is that it can incorporate latent variables. A *latent variable* is a variable that cannot actually be measured but can only be *approximated* with actual measures (Aron & Aron, 1997). For example, "intelligence" is a latent variable. We would be hard pressed to find a single measure for intelligence, but we can approximate measures for intelligence by obtaining values on several observable variables such as IQ, performance on academic achievement tests, etc. In structural equation modeling, a diagram is set up such that latent variables are combinations of observable, measurable variables. Path diagrams in structural equation modeling are much more involved, incorporating several additional components over and above those included in a path analysis. A disadvantage of structural equation modeling is that standard statistical analysis software packages (such as SPSS) are not able to conduct the required procedures. Special statistics programs are required in order to conduct this type of analysis. One such program, LISREL, takes its name from the purpose of the technique—that is, to uncover <u>li</u>near <u>s</u>tructural <u>rel</u>ations. Discussions of structural equation modeling and the LISREL program are beyond the scope and purpose of this text. If interested in reading further about structural equation modeling, brief descriptions and examples are included in Aron and Aron (1997 & 1999) and Johnson and Wichern (1998). If detailed information is required, the reader is directed to Tabachnick and Fidell (1996), Long (1983), and Pedhazur (1982).

Sample Research Questions

Returning to our working example for path analysis, we can specify our research questions for the study as follows:

(1)   Is our model—which describes the causal effects among the variables "region of the world," "status as a developing nation," "number of deaths," "number of doctors," and "male life expectancy"—consistent with our observed correlations among these variables?

(2)   If our model is consistent, what are the estimated direct, indirect, and total causal effects among the variables?

SECTION 8.2 ASSUMPTIONS AND LIMITATIONS

Since path analysis is essentially an extension and specific application of multiple regression, the assumptions that were discussed in the previous chapter are also appropriate here. As a reminder of those assumptions, we simply list them here:

(1)   The independent variables are fixed (i.e., the same values of the IVs would have to be used if the study were to be replicated).

(2)   The independent variables are measured without error.

(3)   The relationship between the independent variables and the dependent variable is linear (in other words, the regression of the DV on the combination of IVs is linear).

(4)   The mean of the residuals for each observation on the dependent variable over many replications is zero.

(5)   Errors associated with any single observation on the dependent variable are independent of (i.e., not correlated with) errors associated with any other observation on the dependent variable.

(6)   The errors are not correlated with the independent variables.

(7)   The variance of the residuals across all values of the independent variables is constant (i.e., homoscedasticity of the variance of the residuals).

(8)   The errors are normally distributed.

If additional specific information regarding the assumptions associated with multiple regression is required, the reader is advised to revisit Chapter 7.

As we have previously mentioned, valid causal inference requires the correct specification of the structural equation(s) in a path analysis. If, *and only if*, the model is correctly specified, the estimates of the various causal effects will be accurate and unbiased (Tate, 1992). In contrast, any specification errors that exist will result in the estimates of causal effects to be biased to some unknown degree. In order to use multiple regression in a manner to estimate the path coefficients, the following assumptions regarding correct model specification must be met:

(1)   The model must accurately reflect the actual causal sequence.

(2)   The structural equation for each endogenous variable includes all variables that are direct causes of that particular endogenous variable (i.e., variables that are not included in the model, and whose effects are therefore assumed to be "captured" by the residuals, are also assumed not to be correlated with any of the determinant variables).

(3)   There is a one-way causal flow in the model (i.e., there can be no reciprocal causation between variables).

(4)   The relationships among variables are assumed to be linear, additive, and causal in nature; any curvilinear relations, etc., are to be excluded.

(5)   All exogenous variables are measured without error (Tate, 1992; Pedhazur, 1982).

The reader should note that assumptions #1 through #4 for path analysis deal directly with the specification of the model which, as we have previously mentioned, can be based on a combination of factors (theory, experience, research literature, opinion, etc.). As we have seen with previous techniques, assumption #5 is largely an issue of research design and data collection.

We would be remiss if we did not discuss several limitations of path analysis. Earlier in the chapter, we referred to the fact that with path analysis we are attempting to estimate and describe causal relationships through the use of correlational data. Because of this fact, the degree of confidence we can have in the causal inferences drawn from the results of the analysis is bound to be much less than the confidence in inferences drawn from an experimental study.

Furthermore, if it is concluded that a model is not consistent with the empirical data, the model has been *misspecified*, which is a matter of degree. This degree of misspecification is subjective, to say the least, and must be evaluated by the researcher. Tate (1992) describes this limitation in the following manner:

> "A model which omits several relatively unimportant causes, ignores a real but weak causal feedback, and is based on measures with some modest measurement error may still produce estimates which are technically biased but still reasonable (and valuable) approximations to the true causal effects. On the other hand, completely misleading conclusions may result from a model which is perfect in every way except for the omission of a single important variable" (p. 319).

There is no statistical test that will definitively indicate whether or not the misspecification is within reasonable limits—those decisions are left to the researcher.

Due to the above limitations, it has been suggested (Tate, 1992) that the use of "conditional" statements in reporting the results of a path analysis study is warranted. For example, one might state obtained results in the following manner: "If this model accurately reflects reality, the estimated causal effects are ..." This serves as an appropriate reminder to ourselves—and to the readers of our research reports—of the limitations associated with drawing causal inferences from correlational data.

Methods of Testing Assumptions

With respect to the initial eight assumptions associated with the use of multiple regression analysis, a thorough discussion of the methods of assessing the tenability of those assumptions was presented in Chapter 7. As a reminder to the reader, these assumptions may be assessed through the use of routine data screening procedures (see Chapter 3), but are most appropriately tested through inspection of bivariate scatterplots and more accurately through inspection of the residuals plots (see Chapter 7). Recall that residuals plots may be used to assess assumptions of linearity, normality, and constant variance (homoscedasticity).

The method of assessing the validity of the assumptions specific to path analysis differs greatly from the assessment of assumptions for statistical inference (Tate, 1992). No statistical procedures exist for evaluating these assumptions since they deal specifically with the degree to which the causal model has been correctly specified. There is no empirical test that can tell us the extent to which we have selected and described the correct model. In order to evaluate these five assumptions, Tate (1992) suggests that we focus our attention on the credibility, reasonableness, and utility of a proposed model. In other words,

- a model should be plausible to those who are expert in the particular field of inquiry;
- the results should be reasonable within the context of the current research literature; and
- a model should be useful in predicting future events.

The responsibility for assessing the assumptions in this manner ultimately rests with the researcher and his/her subjective judgments.

SECTION 8.3 PROCESS AND LOGIC

The Logic Behind Path Analysis

You will recall that in Chapter 7 we provided a brief overview of the calculations involved in conducting a multiple regression analysis. The same logic and associated calculations hold true here, since we are again applying a regression analysis, albeit within the analysis of a causal model as opposed to a straightforward multiple regression. Therefore, we will again be calculating the β coefficients (i.e., the standardized versions in order to represent the path coefficients) for each causal determinant, the squared multiple correlation (R^2) for each structural equation, and the associated significance tests. This will be done in the same manner as described in the previous chapter.

Typically, we reserve this section of each chapter to explain the logic behind the calculations of each technique, without overwhelming the reader with mathematical equations and hand calculations, since the calculations are obtained via computer analysis. However, you will recall that we mentioned earlier that the assessment of model fit in a path analysis can only be accomplished through the use of hand calculations. The assessment of model fit is conducted by obtaining the reproduced correlations and comparing them to the empirical correlations, then evaluating them against the difference criterion of .05. Again, if all reproduced and observed correlations are relatively close to each other, the model is consistent with the empirical data; in other words, the model "fits" the data.

One commonly used approach to determining the reproduced correlations between two variables (and, therefore, among all variables in the set) involves the identification of all legitimate paths between the variables in the model in a process referred to as *path tracing* (Tate, 1992) or *path decomposition* (Pedhazur, 1982). **Path tracing** is a process that results in a correlation coefficient for each path, which is equal to the product of all coefficients in the path. A key is that one may only use legitimate paths, which are those paths that do not violate any of the following three rules:

(1)   no path may pass through the same variable more than once,

(2)   no path may go backward on an arrow after going forward on another arrow (although it is acceptable to go forward on an arrow after *first* going backward), and

(3)   no path may include more than one double-headed curved arrow (Tate, 1992).

To illustrate this process, refer to Figure 8.3, which represents the same model as in Figure 8.2 but which now includes the path coefficients resulting from our regression analysis. If we wanted to obtain the reproduced correlation between z_1 and z_3, the legitimate paths are:

Path	Component
z_1 to z_3	p_{31}
z_1 to z_2 to z_3	$r_{12}p_{32}$

Therefore, the resulting equation for the reproduced correlation (symbolized by $\hat{r}$) between z_1 and z_3 is represented by the following equation:

$$\hat{r}_{13} = p_{31} + r_{12}p_{32}$$

Making the appropriate substitutions of path coefficients, we now have:

$$\hat{r}_{13} = (-.395) + (-.621)(-.123) = -.319$$

As another example, let us consider the reproduced correlation between z_1 and z_5. The legitimate paths are:

Path	Component
z_1 to z_3 to z_5	$p_{31}p_{53}$
z_1 to z_2 to z_3 to z_5	$r_{12}p_{32}p_{53}$
z_1 to z_2 to z_4 to z_5	$r_{12}p_{42}p_{54}$

The resulting equation is obtained for $\hat{r}$ between z_1 and z_5:

$$\hat{r}_{15} = p_{31}p_{53} + r_{12}p_{32}p_{53} + r_{12}p_{42}p_{54}$$

Again, making the appropriate substitutions gives us:

$$\hat{r}_{15} = (-.395)(-.454) + (-.621)(-.123)(-.454) + (-.621)(-.596)(.573) = .356$$

Correlation decompositions such as these would be determined for all possible bivariate correlations in the model, with the exception of those between exogenous variables. The complete set of path decompositions and reproduced correlations for the model shown in Figure 8.3 is presented in Table 8.1. It is recommended that the reader reproduce these results in order to practice the identification of legitimate paths in a path model, while adhering to the path tracing rules.

The reader will notice that each path component in Table 8.1 includes an abbreviated "label" (i.e., D, I, S, or U). It is important to note the conceptual differences among the various types of path components—this is ultimately important when attempting to describe the direct, indirect, and total causal effects in a model (Tate, 1992). Causal effects are represented by paths consisting only of direct causal links—in other words, only straight arrows—that flow in only one direction. These causal effects may be *direct* (a causal path consisting of only one link; denoted "D" in Table 8.1) or *indirect* (consisting of two or more links; denoted "I" in Table 8.1). For example, the $\hat{r}_{15}$ decomposition shown above includes the indirect effect of z_1 on z_5, mediated through z_3 ($p_{31}p_{53}$).

Figure 8.3 Path Diagram for the Initial Model (Male Life Expectancy), Including Path Coefficients.

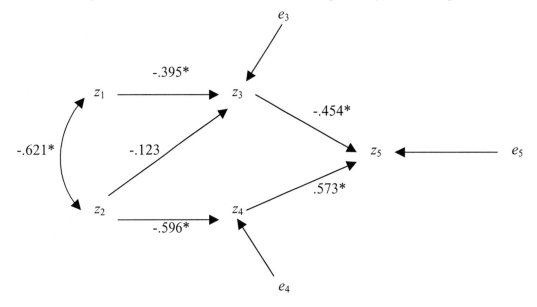

* Significant at the .05 level.

z_1 = *region*
z_2 = *develop*
z_3 = *deathrat*
z_4 = *lndocs*
z_5 = *lifexpm*

Any path components resulting from paths that have reversed causal direction at some point are called **spurious effects** (denoted "S" in Table 8.1), indicating that the relationship is caused by a common third factor (Tate, 1992). The paths may or may not include a double-headed curved arrow. For example, in the decomposition of $\hat{r}_{35}$, the component $p_{31}r_{12}p_{42}p_{54}$ represents a spurious effect—in other words, portions of r_{35} are <u>not</u> due to *either* direct or indirect causal effects of z_3 on z_5. Note that any path between two endogenous variables, which includes a curved arrow, will always represent a spurious effect (Tate, 1992).

Finally, in any model that contains more than one exogenous variable, as does Figure 8.3, the associated unexplained correlations among them will result in a degree of "undeterminability" with respect to the resolution of the direct and indirect effects of exogenous variables on endogenous variables (Tate, 1992). Since a model such as this does not explain the relationship among exogenous variables, we must recognize that this unanalyzed portion (denoted "U" in Table 8.1) may represent some degree of causal effect that has not been included in the model. In this situation, the total causal effect on an endogenous variable must be accompanied by a note specifying that there exists some uncertainty due to the unanalyzed component.

Table 8.1 Path Decompositions for the Initial Model (Male Life Expectancy) Shown in Figure 8.3.

Reproduced Correlation	Path Decomposition
$\hat{r}_{13}$	$p_{31} + r_{12}p_{32}$ (D) (U)
$\hat{r}_{14}$	$r_{12}p_{42}$ (U)
$\hat{r}_{15}$	$p_{31}p_{53} + r_{12}p_{32}p_{53} + r_{12}p_{42}p_{54}$ (I) (U) (U)
$\hat{r}_{23}$	$p_{32} + r_{12}p_{31}$ (D) (U)
$\hat{r}_{24}$	p_{42} (D)
$\hat{r}_{25}$	$p_{32}p_{53} + p_{42}p_{54} + r_{12}p_{31}p_{53}$ (I) (I) (U)
$\hat{r}_{34}$	$p_{32}p_{42} + p_{31}r_{12}p_{42}$ (S) (S)
$\hat{r}_{35}$	$p_{53} + p_{32}p_{42}p_{54} + p_{31}r_{12}p_{42}p_{54}$ (D) (S) (S)
$\hat{r}_{45}$	$p_{54} + p_{42}p_{32}p_{53} + p_{42}r_{12}p_{31}p_{53}$ (D) (S) (S)

Once all of the reproduced correlations have been obtained for a path model (see Table 8.2), they are displayed adjacent to the observed correlations. Those reproduced correlations that have a difference greater than .05 from the empirical correlations are indicated with an asterisk (see Table 8.3). Any differences that are substantially larger than the .05 criterion indicate that the model is not consistent with the empirical data and revisions to the model are warranted prior to describing any of the causal effects. This method of testing for model fit is only possible when there are one or more missing paths in the model—if all possible paths are included, the reproduced correlations will *always* be exactly equivalent to the observed correlations (Tate, 1992)—by definition, the fit of the model will be perfect. *Recall that if one goal of any analysis is a parsimonious solution, we should always have some missing paths in a model.*

If it is determined that a model does not "fit" the data, consideration should be given to retaining included paths and incorporating excluded paths. This is accomplished by first testing all missing paths for each endogenous variable in the model (Tate, 1992). In our working example, we originally regressed z_4 on z_2 but chose to exclude the regression of z_4 on z_1 and z_3. In order to test the missing paths for z_4, we must regress z_4 on *all* of its direct causal determinants (z_1, z_2, and z_3). Similarly, we would then regress z_5 on z_1, z_2, z_3, and z_4. Support for adding any originally excluded paths is indicated by a significant path coefficient (β) in the computer output. Support for the original model is indicated by any nonsignificant path coefficients. Second, we would want to examine empirical support for all paths that we initially chose to include. This is also accomplished by examining the significance of each path coefficient—significance denotes that the model (at least, that particular coefficient) is supported by the data. If a path coefficient is not statistically significant, one should consider dropping it from the model *unless there is strong theoretical support for its inclusion* (Tate, 1992).

Table 8.2 Calculations of Reproduced Correlations for the Initial Model (Male Life Expectancy) Shown in Figure 8.3.

$\hat{r}_{13}$ $=$ p_{31} $+$ $r_{12}p_{32}$

$= (-.395) + (-.621)(-.123) = \mathbf{-.319}$
    (D)   (U)

$\hat{r}_{14}$ $=$ $r_{12}p_{42}$

$= (-.621)(-.596) = \mathbf{.370}$
        (U)

$\hat{r}_{15}$ $=$ $p_{31}p_{53}$ $+$ $r_{12}p_{32}p_{53}$ $+$ $r_{12}p_{42}p_{54}$

$= (-.395)(-.454) + (-.621)(-.123)(-.454) + (-.621)(-.596)(.573) = \mathbf{.356}$
        (I)   (U)   (U)

- -

$\hat{r}_{23}$ $=$ p_{32} $+$ $r_{12}p_{31}$

$= (-.123) + (-.621)(-.395) = \mathbf{.122}$
    (D)   (U)

$\hat{r}_{24}$ $=$ p_{42}

$= (-.596) = \mathbf{-.596}$
    (D)

$\hat{r}_{25}$ $=$ $p_{32}p_{53}$ $+$ $p_{42}p_{54}$ $+$ $r_{12}p_{31}p_{53}$

$= (-.123)(-.454) + (-.596)(.573) + (-.621)(-.395)(-.454) = \mathbf{-.397}$
        (I)   (I)   (U)

- -

$\hat{r}_{34}$ $=$ $p_{32}p_{42}$ $+$ $p_{31}r_{12}p_{42}$

$= (-.123)(-.596) + (-.395)(-.621)(-.596) = \mathbf{-.073}$
        (S)   (S)

$\hat{r}_{35}$ $=$ p_{53} $+$ $p_{32}p_{42}p_{54}$ $+$ $p_{31}r_{12}p_{42}p_{54}$

$= (-.454) + (-.123)(-.596)(.573) + (-.395)(-.621)(-.596)(.573) = \mathbf{-.495}$
    (D)   (S)   (S)

- -

$\hat{r}_{45}$ $=$ p_{54} $+$ $p_{42}p_{32}p_{53}$ $+$ $p_{42}r_{12}p_{31}p_{53}$

$= (.573) + (-.596)(-.123)(-.454) + (-.596)(-.621)(-.395)(-.454) = \mathbf{.606}$
    (D)   (S)   (S)

In assessing the fit of our model in Figure 8.3, it can be seen from Table 8.3 that six of the ten reproduced correlations have differences greater (and substantially so!) than .05. Upon examination of the significance tests for missing paths resulting from the supplemental regression runs as described in the previous paragraph, it was determined that several paths should be added—specifically, the paths from z_1 to z_4 (p_{41}), from z_2 to z_5 (p_{52}), and from z_3 to z_4 (p_{43}). Additionally, because its beta coefficient was not significant, it was decided that the path from z_2 to z_3 (p_{32}) be removed from the model. The resulting revised path diagram, including path coefficients, is presented in Figure 8.4.

Once a model has been revised, the fit should again be reassessed. The path decompositions for our revised model are shown in Table 8.4. Calculation of the subsequent reproduced correlations is presented in Table 8.5. Reproduced correlations for the revised model are once again compared to the empirical correlations (see Table 8.3). This model obviously results in a much better fit than the initial model—only one of the reproduced correlations exceeds the .05 criterion. Had the fit *not* substantially improved, this process would continue until an adequate fit of the model to the empirical data has been achieved.

Table 8.3 Observed and Reproduced Correlations for the Initial (Figure 8.3) and the Revised (Figure 8.4) Models (Male Life Expectancy).

	z_1	z_2	z_3	z_4	z_5
	Observed Correlations				
z_1	1.000				
z_2	-.621	1.000			
z_3	-.318	.125	1.000		
z_4	.648	-.596	-.655	1.000	
z_5	.592	-.564	-.835	.870	1.000
	Reproduced Correlations (Initial Model)				
z_1	1.000				
z_2	-.621	1.000			
z_3	-.319	.122	1.000		
z_4	.370*	-.596	-.073*	1.000	
z_5	.356*	-.397*	-.495*	.606*	1.000
	Reproduced Correlations (Revised Model)				
z_1	1.000				
z_2	-.621	1.000			
z_3	-.318	.197*	1.000		
z_4	.648	-.643	-.688	1.000	
z_5	.578	-.630	-.866	.906	1.000

* Difference between reproduced and observed correlation is greater than 0.05.

At this point, we are satisfied with the fit of our model to the associated empirical data and can describe the causal effects of the variables and their correlations. A table that summarizes the causal effects of a model is typically presented in published research. Such a summary for our revised model is presented in Table 8.6. The reader should note that the indirect effects listed in the table are simply the sum of all indirect effects as identified in the path decompositions. The total effects consist of the sum of the direct and indirect effects.

Figure 8.4 Path Diagram for the Revised Model (Male Life Expectancy), Including Path Coefficients.

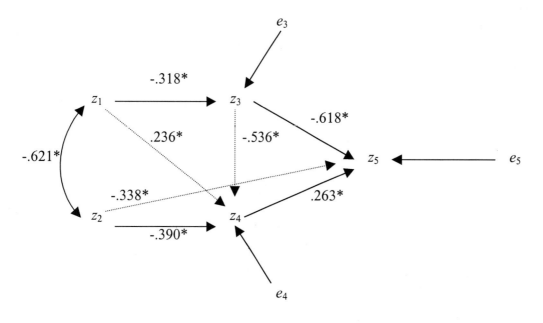

* Significant at the .05 level.

<u>Note</u>. Revised paths are shown with dashed arrows.

z_1 = *region*
z_2 = *develop*
z_3 = *deathrat*
z_4 = *lndocs*
z_5 = *lifexpm*

Interpretation of Results

As somewhat indicated, interpretation of the SPSS output for a path analysis is quite extensive since it requires several hand calculations. Once you have conducted all the regression analyses for the initial path model, the path (β) coefficients with the respective level of significance should be noted within the path model. These coefficients indicate the estimated change in the respective endogenous variables and are used to calculate the reproduced correlations through path decomposition. Reproduced correlations are then compared to the empirical correlations to test the model fit. If any reproduced correlations exhibit more than a .05 difference from the empirical correlations, the model is not consistent with the empirical data and should be revised. Once a consistent model has been generated, the specific causal effects for each endogenous variable are determined with respect to direct, indirect and total effects. Utilizing the path decompositions is imperative in this process, since both the direct and indirect effects are identified for each path. The reader should note that a path may have several indirect effects; consequently the sum of these indirect values represents the overall indirect effect for the path. The total effect is also calculated by adding the direct and indirect effects for each path. Finally, R^2 is interpreted to indicate the amount of variance in each endogenous variable that is explained by its structural model.

Table 8.4 Path Decompositions for the Revised Model (Male Life Expectancy) Shown in Figure 8.4.

Reproduced Correlation	Path Decomposition
$\hat{r}_{13}$	$\underset{(D)}{p_{31}}$
$\hat{r}_{14}$	$\underset{(D)}{p_{41}} + \underset{(U)}{r_{12}p_{42}} + \underset{(I)}{p_{31}p_{43}}$
$\hat{r}_{15}$	$\underset{(I)}{p_{31}p_{53}} + \underset{(U)}{r_{12}p_{42}p_{54}} + \underset{(I)}{p_{41}p_{54}} + \underset{(S)}{p_{31}p_{43}p_{54}} + \underset{(U)}{r_{12}p_{52}}$
$\hat{r}_{23}$	$\underset{(U)}{r_{12}p_{31}}$
$\hat{r}_{24}$	$\underset{(D)}{p_{42}} + \underset{(U)}{r_{12}p_{41}} + \underset{(U)}{r_{12}p_{31}p_{43}}$
$\hat{r}_{25}$	$\underset{(D)}{p_{52}} + \underset{(I)}{p_{42}p_{54}} + \underset{(U)}{r_{12}p_{31}p_{53}} + \underset{(U)}{r_{12}p_{31}p_{43}p_{54}} + \underset{(U)}{r_{12}p_{41}p_{54}}$
$\hat{r}_{34}$	$\underset{(D)}{p_{43}} + \underset{(I)}{p_{31}p_{41}} + \underset{(S)}{p_{31}r_{12}p_{42}}$
$\hat{r}_{35}$	$\underset{(D)}{p_{53}} + \underset{(S)}{p_{43}p_{54}} + \underset{(S)}{p_{31}p_{41}p_{54}} + \underset{(S)}{p_{31}r_{12}p_{52}} + \underset{(S)}{p_{31}r_{12}p_{42}p_{54}}$
$\hat{r}_{45}$	$\underset{(D)}{p_{54}} + \underset{(S)}{p_{43}p_{53}} + \underset{(S)}{p_{41}p_{31}p_{53}} + \underset{(S)}{p_{41}r_{12}p_{52}} + \underset{(S)}{p_{42}p_{52}} + \underset{(S)}{p_{42}r_{12}p_{31}p_{53}} + \underset{(S)}{p_{43}p_{31}r_{12}p_{52}}$

Continuing with our example (see Figure 8.1) that seeks to investigate the causal effects among the variables "region of the world," "status as a developing nation," "number of deaths," "number of doctors," and "male life expectancy," we screened data for missing cases and outliers. Outliers were identified by calculating Mahalanobis distance and conducting **Explore**. Figure 8.5 presents these results and indicates that cases #29 and #56 should be eliminated from further analysis since they exceed the chi square criterion of 20.516 (df=5). Variables were then evaluated for normality by creating a scatterplot matrix (see Figure 8.6). Since the variable of doctors (*docs*) is curvilinearly related to male life expectancy, it was transformed to *lndocs* by taking its natural log. Next, multivariate normality and homoscedascticity were assessed by creating a residuals plot (see Figure 8.7). The fairly consistent spread of residuals indicates that the test assumptions are fulfilled. Prior to conducting the regression analyses, a correlation matrix (see Figure 8.8) was created since these empirical correlations will be needed later to test model fit. Applying the initial model, the first series of regression analyses was conducted. Since this model has three endogenous variables, the following analyses were run: z_3 on z_1 and z_2 (see Figure 8.9); z_4 on z_2 (see Figure 8.10); and z_5 on z_3 and z_4 (see Figure 8.11). Prior to interpreting the path coefficients, one should review the tolerance statistic for each exogenous variable included in each regression analysis in order to determine if multicollinearity can be assumed. If tolerance is greater than .1, one may proceed with interpreting the path coefficients. For our example, tolerance statistics were all adequate. Path (beta) coefficients from the output were transferred to the path diagram of the

Table 8.5 Calculations of Reproduced Correlations for the Revised Model (Male Life Expectancy) Shown in Figure 8.4.

$\hat{r}_{13}$ $=$ p_{31}

   $= (-.318) = \mathbf{-.318}$
   (D)

$\hat{r}_{14}$ $=$ p_{41} + $r_{12}p_{42}$ + $p_{31}p_{43}$

   $= (.236) + (-.621)(-.390) + (-.318)(-.536) = \mathbf{.648}$
   (D)   (U)   (I)

$\hat{r}_{15}$ $=$ $p_{31}p_{53}$ + $r_{12}p_{42}p_{54}$ + $p_{41}p_{54}$ + $p_{31}p_{43}p_{54}$ + $r_{12}p_{52}$

   $= (-.318)(-.618) + (-.621)(-.390)(.263) + (.236)(.263) + (-.318)(-.536)(.263) + (-.621)(-.338) = \mathbf{.578}$
   (I)   (U)   (I)   (I)   (U)

$\hat{r}_{23}$ $=$ $r_{12}p_{31}$

   $= (-.621)(-.318) = \mathbf{.197}$
   (U)

$\hat{r}_{24}$ $=$ p_{42} + $r_{12}p_{41}$ + $r_{12}p_{31}p_{43}$

   $= (-.390) + (-.621)(.236) + (-.621)(-.318)(-.536) = \mathbf{-.643}$
   (D)   (U)   (U)

$\hat{r}_{25}$ $=$ p_{52} + $p_{42}p_{54}$ + $r_{12}p_{31}p_{53}$ + $r_{12}p_{31}p_{43}p_{54}$ + $r_{12}p_{41}p_{54}$

   $= (-.338) + (-.390)(.263) + (-.621)(-.318)(-.618) + (-.621)(-.318)(-.536)(.263) + (-.621)(.236)(.263) = \mathbf{-.630}$
   (D)   (I)   (U)   (U)   (U)

$\hat{r}_{34}$ $=$ p_{43} + $p_{31}p_{41}$ + $p_{31}r_{12}p_{42}$

   $= (-.536) + (-.318)(.236) + (-.318)(-.621)(-.390) = \mathbf{-.688}$
   (D)   (I)   (S)

$\hat{r}_{35}$ $=$ p_{53} + $p_{43}p_{54}$ + $p_{31}p_{41}p_{54}$ + $p_{31}r_{12}p_{52}$ + $p_{31}r_{12}p_{42}p_{54}$

   $= (-.618) + (-.536)(.263) + (-.318)(.236)(.263) + (-.318)(-.621)(-.338) + (-.318)(-.621)(-.390)(.263) = \mathbf{-.866}$
   (D)   (I)   (S)   (S)   (S)

$\hat{r}_{45}$ $= p_{54} + p_{43}p_{53}$ + $p_{41}p_{31}p_{53}$ + $p_{41}r_{12}p_{52}$ + $p_{42}p_{52}$ + $p_{42}r_{12}p_{31}p_{53}$ + $p_{43}p_{31}r_{12}p_{52}$

   $= (.263) + (-.536)(-.618) + (.236)(-.318)(-.618) + (.236)(-.621)(-.338) + (-.390)(-.338) + (-.390)(-.621)(-.318)(-.618) + (-.536)(-.318)(-.621)(-.338) = \mathbf{.906}$
   (D)   (I)   (S)   (S)   (I)   (S)   (S)

initial model as seen in Figure 8.3. All coefficients were significant with the exception of p_{32}. Reproduced correlations were then calculated through the path decompositions (see Table 8.2) and resulted in seven of the reproduced correlations differing from the empirical correlations by more than .05 (see Table 8.3). Since this initial model did not fit the empirical data, regression analyses were conducted on the following missing paths: z_4 on z_1, z_2, and z_3 (see Figure 8.12), and z_5 on z_1, z_2, z_3, and z_4 (see Figure 8.13). Evaluation of the path coefficients and respective levels of significance indicate that only the following paths were significant: z_3 on z_1; z_4 on z_1, z_2, and z_3; and z_5 on z_2, z_3, and z_4. Consequently, regression analyses were conducted again to include only those significant paths for each endogenous variable in order to obtain "final" path coefficients (non-significant paths are excluded, therefore changing the values of the significant paths). These analyses are presented in Figures 8.14 and 8.15. The revised model with respective path coefficients is displayed in Figure 8.4. Calculation of reproduced correlations through path decompositions (see Table 8.5) and subsequent comparison to the empirical correlations (see Table 8.3) indicate the revised model fits the empirical data. Utilizing calculations for the direct and indirect effects from Table 8.5, we summarize the causal effects of the revised model in Table 8.6. In addition, R^2 is noted for each endogenous variable within this summary table. R^2 can be found in the final (accepted) regression analyses for each endogenous variable (see Figures 8.12, 8.14, and 8.15). For example, causal effects of *region, develop, deathrat,* and *lndocs,* explain 93.2% (R^2=.932) of variance in *lifeexpm.*

Table 8.6 Summary of Causal Effects for Revised Model (Male Life Expectancy).

Outcome	Determinant	Causal Effects		
		Direct	Indirect	Total
Death Rate	Region	-.318*	–	-.318
(R^2 = .101)	Developing Status	–	–	– [+]
Doctors	Region	.236*	.170	.406 [+]
(R^2 = .738)	Developing Status	-.390*	–	-.390 [+]
	Death Rate	-.536*	-.075	-.611
Male Life Expectancy	Region	–	.304	.304 [+]
(R^2 = .932)	Developing Status	-.338*	-.103	-.441 [+]
	Death Rate	-.618*	-.141	-.759 [+]
	Doctors	.263*	.463	.726

* Direct effect is significant at the .05 level.

[+] Total effect may be incomplete due to unanalyzed components.

Figure 8.5 Outliers Determined by Mahalanobis Distance.

Extreme Values

			Case Number	Value
MAH_1	Highest	1	56	24.06622
		2	29	20.82071
		3	72	20.00161
		4	30	19.16860
		5	43	16.08589
	Lowest	1	35	.82904
		2	63	.86684
		3	40	.90204
		4	33	1.04412
		5	42	1.12388

Cases #56 and #29 exceed the $\chi^2(5)$=20.516 criteria and should be eliminated from further analysis.

Figure 8.6 Scatterplot for Model (Male Life Expectancy) Variables.

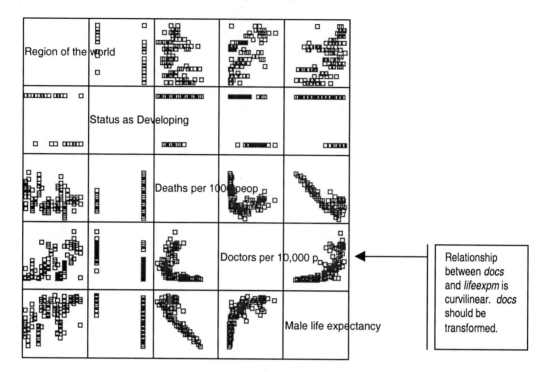

Figure 8.7 Residuals Plot for Model (Male Life Expectancy) Variables.

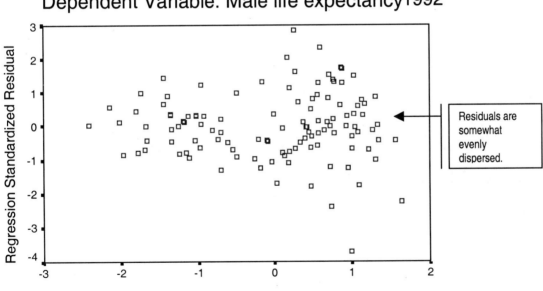

Figure 8.8 Correlation Matrix for Model (Male Life Expectancy) Variables.

Correlations

		Region of the world	Status as Developing Country	Deaths per 1000 people, 1992	Natural log of doctors per 10000	Male life expectancy 1992
Region of the world	Pearson Correlation	1.000	-.621**	-.318**	.648**	.592**
	Sig. (2-tailed)	.	.000	.000	.000	.000
	N	120	120	119	119	120
Status as Developing Country	Pearson Correlation	-.621**	1.000	.125	-.596**	-.564**
	Sig. (2-tailed)	.000	.	.176	.000	.000
	N	120	120	119	119	120
Deaths per 1000 people, 1992	Pearson Correlation	-.318**	.125	1.000	-.655**	-.835**
	Sig. (2-tailed)	.000	.176	.	.000	.000
	N	119	119	119	118	119
Natural log of doctors per 10000	Pearson Correlation	.648**	-.596**	-.655**	1.000	.870**
	Sig. (2-tailed)	.000	.000	.000	.	.000
	N	119	119	118	119	119
Male life expectancy 1992	Pearson Correlation	.592**	-.564**	-.835**	.870**	1.000
	Sig. (2-tailed)	.000	.000	.000	.000	.
	N	120	120	119	119	120

**. Correlation is significant at the 0.01 level (2-tailed).

Figure 8.9 Regression Output for *deathrat* (z_3) on *region* (z_1) and *develop* (z_2).

Model Summary

Model	R	R Square	Adjusted R Square	Std. Error of the Estimate
1	.332[a]	.110	.095	4.40

a. Predictors: (Constant), DEVELOP, REGION

Coefficients[a]

Model		Unstandardized Coefficients B	Std. Error	Standardized Coefficients Beta	t	Sig.	Collinearity Statistics Tolerance	VIF
1	(Constant)	14.499	1.697		8.542	.000		
	REGION	-.356	.101	-.395	-3.513	.001	.607	1.648
	DEVELOP	-1.352	1.237	-.123	-1.093	.277	.607	1.648

a. Dependent Variable: DEATHRAT

Path coefficient of *deathrat* on *develop* is NOT significant.

217

Figure 8.10 Regression Output for *lndocs* (z_4) on *develop* (z_2).

Model Summary

Model	R	R Square	Adjusted R Square	Std. Error of the Estimate
1	.596[a]	.356	.350	1.2596

a. Predictors: (Constant), DEVELOP

Coefficients[a]

Model		Unstandardized Coefficients		Standardized Coefficients	t	Sig.	Collinearity Statistics	
		B	Std. Error	Beta			Tolerance	VIF
1	(Constant)	3.202	.242		13.207	.000		
	DEVELOP	-2.216	.276	-.596	-8.036	.000	1.000	1.000

a. Dependent Variable: LNDOCS

Path coefficient is significant.

Figure 8.11 Regression Output for *lifeexpm* (z_5) on *deathrat* (z_3) and *lndocs* (z_4).

Model Summary

Model	R	R Square	Adjusted R Square	Std. Error of the Estimate
1	.935[a]	.874	.872	3.55

a. Predictors: (Constant), DEATHRAT, LNDOCS

Coefficients[a]

Model		Unstandardized Coefficients		Standardized Coefficients	t	Sig.	Collinearity Statistics	
		B	Std. Error	Beta			Tolerance	VIF
1	(Constant)	67.029	1.342		49.959	.000		
	LNDOCS	3.631	.277	.573	13.093	.000	.571	1.752
	DEATHRAT	-1.002	.097	-.454	-10.370	.000	.571	1.752

a. Dependent Variable: LIFEEXPM

Path coefficients are significant.

218

Figure 8.12 Regression Output of Missing Paths: *lndocs* (z_4) on *region* (z_1), *develop* (z_2), *deathrat* (z_3).

Model Summary

Model	R	R Square	Adjusted R Square	Std. Error of the Estimate
1	.859[a]	.738	.731	.8131

a. Predictors: (Constant), REGION, DEATHRAT, DEVELOP

Coefficients[a]

Model		Unstandardized Coefficients B	Unstandardized Coefficients Std. Error	Standardized Coefficients Beta	t	Sig.	Collinearity Statistics Tolerance	Collinearity Statistics VIF
1	(Constant)	3.904	.403		9.681	.000		
	REGION	7.172E-02	.020	.236	3.643	.000	.550	1.818
	DEVELOP	-1.450	.230	-.390	-6.310	.000	.601	1.663
	DEATHRAT	-.187	.018	-.536	-10.563	.000	.892	1.121

a. Dependent Variable: LNDOCS

All path coefficients are significant.

Figure 8.13 Regression Output of Missing Paths: *lifeexpm* (z_5) on *region* (z_1), *develop* (z_2), *deathrat* (z_3) and *lndocs* (z_4).

Model Summary

Model	R	R Square	Adjusted R Square	Std. Error of the Estimate
1	.966[a]	.933	.930	2.62

a. Predictors: (Constant), LNDOCS, DEVELOP, REGION, DEATHRAT

Coefficients[a]

Model		Unstandardized Coefficients		Standardized Coefficients	t	Sig.	Collinearity Statistics	
		B	Std. Error	Beta			Tolerance	VIF
1	(Constant)	79.189	1.757		45.083	.000		
	REGION	6.446E-02	.067	.033	.960	.339	.493	2.030
	DEVELOP	-7.677	.861	-.326	-8.913	.000	.446	2.244
	DEATHRAT	-1.366	.080	-.618	-17.011	.000	.451	2.218
	LNDOCS	1.571	.302	.248	5.199	.000	.262	3.815

a. Dependent Variable: LIFEEXPM

Path coefficient of *lifeexpm* on *region* is not significant and should not be included.

Figure 8.14 Regression Output of Significant Paths: *deathrat* (z_3) on *region* (z_1).

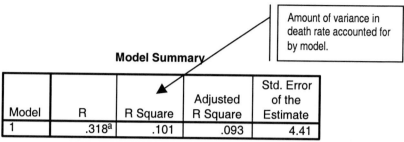

Model Summary

Model	R	R Square	Adjusted R Square	Std. Error of the Estimate
1	.318[a]	.101	.093	4.41

a. Predictors: (Constant), REGION

Coefficients[a]

Model		Unstandardized Coefficients		Standardized Coefficients	t	Sig.	Collinearity Statistics	
		B	Std. Error	Beta			Tolerance	VIF
1	(Constant)	12.857	.790		16.273	.000		
	REGION	-.286	.079	-.318	-3.626	.000	1.000	1.000

a. Dependent Variable: DEATHRAT

Figure 8.15 Regression Output of Significant Paths: *lifeexpm* (z_5) on *develop* (z_2), *deathrat* (z_3) and *lndocs* (z_4).

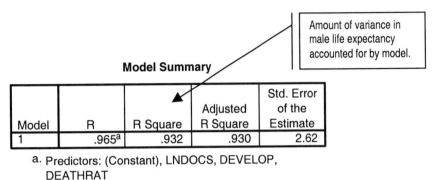

Amount of variance in male life expectancy accounted for by model.

Model Summary

Model	R	R Square	Adjusted R Square	Std. Error of the Estimate
1	.965^a	.932	.930	2.62

a. Predictors: (Constant), LNDOCS, DEVELOP, DEATHRAT

Coefficients^a

Model		Unstandardized Coefficients		Standardized Coefficients	t	Sig.	Collinearity Statistics	
		B	Std. Error	Beta			Tolerance	VIF
1	(Constant)	79.811	1.632		48.899	.000		
	DEVELOP	-7.963	.808	-.338	-9.852	.000	.506	1.977
	DEATHRAT	-1.364	.080	-.618	-17.002	.000	.451	2.217
	LNDOCS	1.665	.286	.263	5.823	.000	.293	3.417

a. Dependent Variable: LIFEEXPM

Final path coefficients.

Writing Up Results

Since the process of path analysis typically revises an initial model that was generated, the summary of results should first discuss the initial model. The path diagram with the path coefficients should be presented. Significant coefficients should be indicated with an asterisk. Within the results summary, you should state that reproduced correlations were calculated to check the model fit as well as indicate how many of the reproduced correlations were not consistent with the empirical correlations. This narration should also include a description of how the revised model was derived (i.e., theory, logic, analysis of missing paths). You should then present the revised model in narrative and pictorial form (path diagram). Your summary should also discuss the significance of the revised model path coefficients. A table that compares the empirical and reproduced correlations for the initial and revised models should be presented. Since the revised model should be a good fit, indicate that reproduced correlations are consistent with empirical correlations in the narrative. A final component of a path analysis results section is a summary table of the causal effects for the revised model. This table should include the direct, indirect, and total effects for each endogenous variable. A flag is used to indicate total effects that may be incomplete due to unanalyzed components. This table should also present R^2 for each endogenous variable. The amount of total effect (direct and indirect) for each endogenous variable should be discussed in the results summary, beginning with the endogenous variable of most interest.

Figure 8.16 summarizes the components of the results narrative for path analysis as well as how it is supported by numerous tables and figures.

Figure 8.16 Steps for Presenting Path Analysis Results.

Results Narrative	**Tables/Figures**

---

1. Present initial model: variables and flow. ⟶ Summarize initial model in path diagram.

2. Describe any data elimination and/or transformation.

3. Discuss significance of path coefficients. ⟶ Present path coefficients in path diagram.

4. Describe how reproduced correlations were not consistent with empirical correlations. ⟶ Create table that compares empirical correlations to reproduce correlations for the initial model.

5. Describe process of revising model.

6. Present revised model: variables, flow, and significant path coefficients. ⟶ Summarize revised model in path diagram (including path coefficients).

7. Describe how reproduced correlations were consistent with empirical correlations. ⟶ Create table that compares empirical correlations to reproduce correlations for the revised model.

8. Discuss causal effects for each endogenous variable: total causal effects and R^2. ⟶ Create table of causal effects (direct, indirect, and total) for each endogenous variable.

The following results narrative applies the output from Figures 8.5 – 8.15. Due to space constraints, we will utilize applicable figures and tables that were previously presented in the text.

A path analysis was conducted to determine the causal effects among the variables of region of the world *(region, z_1)*, status as a developing nation *(develop, z_2)*, number of deaths per 1,000 people *(deathrat, z_3)*, number of doctors per 10,000 people *(docs, z_4)*, and male life expectancy *(lifexpm, z_5)*. Prior to the analysis, two outliers were removed. In addition, the variable of *docs* was transformed by taking its natural log. The initial model, presented in Figure 8.3, was not consistent with the empirical data. More specifically, six of the reproduced correlations exceeded a difference of .05. Tests of the missing paths in the initial model indicated that three additional paths would significantly contribute to the model: *region* on *docs, develop* on *life-expm,* and *deathrat* on *docs.* In addition, the non-significant path of *develop* on *deathrat* was removed from the model. Thus, a revised model was generated and is presented in Figure 8.4. Recomputation of reproduced correlations for the revised model indicated consistency with the empirical correlations as only one reproduced correlation exceeding a difference of .05 (see Table 8.3). All path coefficients were significant at the .05 level. The direct, indirect, and total causal effects of the revised model are presented in Table 8.6. The outcome of primary interest was male life expectancy; the determinant with the largest total causal effect was death rate (-.759). The remaining determinants of male life expectancy as indicated by total causal effect were number of doctors (.726), status as a developing nation (-.441), and region (.304). This model explained approximately 93% of variance in male life expectancy. The primary determinant of the number of doctors was death rate (-.611) with region (.406) and status as a developing nation (-.390) following. Approximately 74% of variance in the number of doctors was explained by the model. The primary determinant of death rate was region (-.318), which explained approximately 10% of variance in death rate.

SECTION 8.4 SAMPLE STUDY AND ANALYSIS

This section provides a complete example that applies the entire process of conducting path analysis: development of model and research questions, data screening methods, test methods, interpretation of output, and presentation of results. The SPSS data set of *country.sav* is utilized.

Problem

In this example, we are interested in developing a causal model for explaining infant mortality. More specifically, we will investigate the causal effects among the following variables: number of doctors per 10,000 people (*docs, z_1*), gross domestic product (*gdp, z_2*), death rate per 1,000 people (*deathrat, z_3*), birth rate per 1,000 (*birthrat, z_4*), and infant mortality per 1,000 live births (*infmr, z_4*). Utilizing logic and theory, we develop the path model shown in Figure 8.17. Specific research questions generated are:

(1) Is our model—which describes the causal effects among the variables "number of doctors," "gross domestic product," "death rate per 1,000," "birth rate per 1,000," and "infant mortality per 1,000 live births"—consistent with our observed correlations among these variables?

(2) If our model is consistent, what are the estimated direct, indirect, and total causal effects among the variables?

Method, Output, and Interpretation

Since path analysis requires a great deal of interpretation throughout the process of conducting the analysis, we have combined discussion of methods, output, and interpretation in this section.

Figure 8.17 Path Diagram for the Initial Model (Infant Mortality).

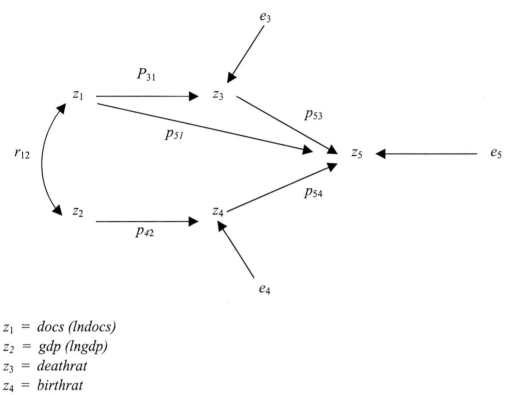

$z_1 = docs \ (lndocs)$
$z_2 = gdp \ (lngdp)$
$z_3 = deathrat$
$z_4 = birthrat$
$z_5 = infmr$

Figure 8.18 Outliers Determined by Mahalanobis Distance.

Extreme Values

			Case Number	Value
MAH_1	Highest	1	72	14.74252
		2	101	14.71959
		3	37	12.30083
		4	91	12.15138
		5	81	12.02676
	Lowest	1	30	.89494
		2	50	.98185
		3	53	1.18165
		4	45	1.26494
		5	93	1.30022

No cases exceed $\chi^2(5)=20.516$.

Figure 8.19 Scatterplot for Model (Infant Mortality) Variables.

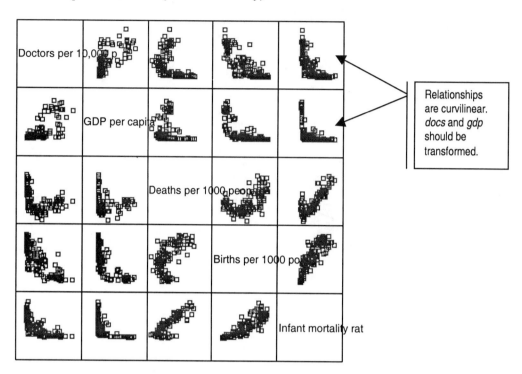

Relationships are curvilinear. *docs* and *gdp* should be transformed.

Figure 8.20 Residuals Plot for Model (Infant Mortality)Variables.

Scatterplot

Dependent Variable: Infant mortality rate 1992

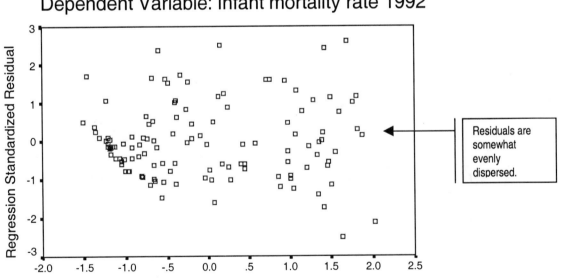

Residuals are somewhat evenly dispersed.

Figure 8.21 Correlation Matrix for Model (Infant Mortality Variables)

Correlations

		LNDOCS	LNGDP	DEATHRAT	BIRTHRAT	INFMR
LNDOCS	Pearson Correlation	1.000	.824**	-.643**	-.821**	-.831**
	Sig. (2-tailed)		.000	.000	.000	.000
	N	121	121	120	120	121
LNGDP	Pearson Correlation	.824**	1.000	-.512**	-.803**	-.809**
	Sig. (2-tailed)	.000	.	.000	.000	.000
	N	121	122	121	121	122
DEATHRAT	Pearson Correlation	-.643**	-.512**	1.000	.568**	.780**
	Sig. (2-tailed)	.000	.000	.	.000	.000
	N	120	121	121	121	121
BIRTHRAT	Pearson Correlation	-.821**	-.803**	.568**	1.000	.862**
	Sig. (2-tailed)	.000	.000	.000	.	.000
	N	120	121	121	121	121
INFMR	Pearson Correlation	-.831**	-.809**	.780**	.862**	1.000
	Sig. (2-tailed)	.000	.000	.000	.000	.
	N	121	122	121	121	122

**. Correlation is significant at the 0.01 level (2-tailed).

Once the path model was generated, all model variables were screened for missing data outliers and tested for assumptions. Identification of outliers was done by conducting a preliminary **Regression** to calculate Mahalanobis distance. The **Explore** procedure was completed to determine if any cases exceeded the chi square criterion of 20.516 (df=5). No outliers were found (see Figure 8.18). Test assumptions were assessed by creating a scatterplot matrix and a residuals plot. The scatterplot matrix indicated that the variables *docs* and *gdp* were curvilinear (see Figure 8.19). These variables were transformed by taking the natural log of each. The residuals plot was then created with the transformed variables and demonstrated fair dispersion (see Figure 8.20). With normality and homoscedasticity assumed, a correlation matrix was then created for all the model variables (see Figure 8.21). Finally, the following series of **Regression** analyses were conducted for the three endogenous variables: z_3 on z_1 (see Figure 8.22); z_4 on z_2 (see Figure 8.23), and z_5 on z_1, z_3, and z_4 (see Figure 8.24). All tolerance statistics were greater than .1. Path coefficients can be seen in the path diagram (see Figure 8.25); coefficients were then used to calculate the reproduced correlations through the path decompositions, which are displayed respectively in Tables 8.7 and 8.8. A comparison of the reproduced correlations to the empirical correlations shows that six of the reproduced correlations differ by more than .05 from the empirical correlations (see Table 8.9). Consequently, we concluded that our initial model was not consistent with the empirical data. Analysis of missing paths, which essentially include all possible paths for each endogenous variable, were conducted for the following: z_3 on z_1, z_2, z_4 (see Figure 8.26); z_4 on z_1, z_2, z_3 (see Figure 8.27), and z_5 on z_1, z_2, z_3, and z_4 (see Figure 8.28). Analysis of missing paths for *deathrat* (z_3) reveals no additional paths that would contribute to the model. Evaluation of missing paths for *birthrat* (z_4) indicates that the path from *lndocs* (z_1) would significantly contribute to the model. Finally, analysis of missing paths for *infmr* (z_5) indicates two revisions: removal of the path from *lndocs* and the addition of the path from *lngdp*. In order to obtain the accurate path coefficients for our revised model, regression analysis must be conducted again utilizing only the appropriate paths: z_3 on z_1; z_4 on z_1 and z_2 (see Figure 8.29), and z_5 on z_2, z_3, and z_4 (see Figure 8.30). Since paths did not change for z_3, the results from the original analysis may be used (see Figure 8.22). The revised model is presented in Figure 8.31. Reproduced correlations were calculated, as defined by the path decompositions (Tables 8.10 & 8.11), and were compared to the empirical correlations (see Table 8.9). Only one reproduced correlation exceeded the criterion of a .05 difference. Thus, we concluded that the revised model is consistent with empirical data. The final step was to calculate the direct, indirect, and total effects for each endogenous variable. These causal effects are presented in Table 8.12.

Figure 8.22 Regression Output for *deathrat* (z_3) on *lndocs* (z_1).

Model Summary

Model	R	R Square	Adjusted R Square	Std. Error of the Estimate
1	.643[a]	.414	.409	3.48

a. Predictors: (Constant), LNDOCS

Coefficients[a]

Model		Unstandardized Coefficients		Standardized Coefficients	t	Sig.	Collinearity Statistics	
		B	Std. Error	Beta			Tolerance	VIF
1	(Constant)	13.127	.439		29.875	.000		
	LNDOCS	-1.874	.205	-.643	-9.131	.000	1.000	1.000

a. Dependent Variable: DEATHRAT

Path coefficient is significant.

Figure 8.23 Regression Output for *birthrat* (z_4) on *lngdp* (z_2).

Model Summary

Model	R	R Square	Adjusted R Square	Std. Error of the Estimate
1	.803[a]	.645	.642	7.92

a. Predictors: (Constant), Natural log of GDP

Coefficients[a]

Model		Unstandardized Coefficients B	Unstandardized Coefficients Std. Error	Standardized Coefficients Beta	t	Sig.	Collinearity Statistics Tolerance	Collinearity Statistics VIF
1	(Constant)	81.791	3.511		23.296	.000		
	Natural log of GDP	-6.977	.475	-.803	-14.697	.000	1.000	1.000

a. Dependent Variable: Births per 1000 population, 1992

Path coefficient is significant.

Figure 8.24 Regression Output for *infmr* (z_5) on *lndocs* (z_1), *deathrat* (z_3), and *birthrat* (z_4).

Model Summary

Model	R	R Square	Adjusted R Square	Std. Error of the Estimate
1	.937[a]	.879	.876	15.196

a. Predictors: (Constant), BIRTHRAT, DEATHRAT, LNDOCS

Coefficients[a]

Model		Unstandardized Coefficients B	Unstandardized Coefficients Std. Error	Standardized Coefficients Beta	t	Sig.	Collinearity Statistics Tolerance	Collinearity Statistics VIF
1	(Constant)	-28.326	9.337		-3.034	.003		
	LNDOCS	-4.104	1.713	-.148	-2.395	.018	.273	3.662
	DEATHRAT	3.770	.402	.396	9.379	.000	.585	1.710
	BIRTHRAT	1.720	.187	.521	9.202	.000	.326	3.067

a. Dependent Variable: INFMR

All path coefficients are significant.

Table 8.7 Path Decompositions for the Initial Model (Infant Mortality) Shown in Figure 8.17.

Reproduced Correlation	Path Decomposition
$\hat{r}_{13}$	p_{31} (D)
$\hat{r}_{14}$	$r_{12}p_{42}$ (U)
$\hat{r}_{15}$	$p_{51} + p_{31}p_{53} + r_{12}p_{42}p_{54}$ (D) (I) (U)
$\hat{r}_{23}$	$r_{12}\,p_{31}$ (U)
$\hat{r}_{24}$	p_{42} (D)
$\hat{r}_{25}$	$p_{42}p_{54} + r_{12}p_{31}p_{53} + r_{12}p_{51}$ (I) (U) (U)
$\hat{r}_{34}$	$p_{31}r_{12}p_{42}$ (S)
$\hat{r}_{35}$	$p_{53} + p_{31}p_{51} + p_{31}r_{12}p_{42}p_{54}$ (D) (I) (S)
$\hat{r}_{45}$	$p_{54} + p_{42}r_{12}p_{51} + p_{42}r_{12}p_{31}p_{53}$ (D) (S) (S)

Table 8.8 Path Decompositions and Calculation of Reproduced Correlations for the Initial Model (Infant Mortality) Shown in Figure 8.17.

$\hat{r}_{13} = p_{31}$
$\quad = (-.643) = \textbf{-.643}$
$\qquad$ (U)

$\hat{r}_{14} = r_{12}p_{42}$
$\quad = (.824)(-.803) = \textbf{-.662}$
$\qquad$ (U)

$\hat{r}_{15} = p_{51} + p_{31}p_{53} + r_{12}p_{42}p_{54}$
$\quad = (-.148) + (-.643)(.396) + (.824)(-.803)(.521) = \textbf{-.748}$
$\qquad$ (D) $\qquad\qquad$ (I) $\qquad\qquad\qquad$ (U)

- -

$\hat{r}_{23} = r_{12} \, p_{31}$
$\quad = (.824)(-.643) = \textbf{-.530}$
$\qquad$ (U)

$\hat{r}_{24} = p_{42}$
$\quad = (-.803) = \textbf{-.803}$
$\qquad$ (D)

$\hat{r}_{25} = p_{42}p_{54} + r_{12}p_{31}p_{53} + r_{12}p_{51}$
$\quad = (-.803)(.521) + (.824)(-.643)(.396) + (.824)(-.148) = \textbf{-.750}$
$\qquad$ (I) $\qquad\qquad\qquad$ (U) $\qquad\qquad\qquad$ (U)

- -

$\hat{r}_{34} = p_{31}r_{12}p_{42}$
$\quad = (-.643)(.824)(-.803) = \textbf{.425}$
$\qquad\qquad$ (S)

$\hat{r}_{35} = p_{53} + p_{31}p_{51} + p_{31}r_{12}p_{42}p_{54}$
$\quad = (.396) + (-.643)(-.148) + (-.643)(.824)(-.803)(.521) = \textbf{.713}$
$\qquad$ (D) $\qquad\qquad$ (I) $\qquad\qquad\qquad$ (S)

- -

$\hat{r}_{45} = p_{54} + p_{42}r_{12}p_{51} + p_{42}r_{12}p_{31}p_{53}$
$\quad = (.521) + (-.803)(.824)(-.148) + (-.803)(.824)(-.643)(.396) = \textbf{.787}$
$\qquad$ (D) $\qquad\qquad\qquad$ (S) $\qquad\qquad\qquad$ (S)

Table 8.9 Empirical and Reproduced Correlations for the Initial Model (Figure 8.17) and the Revised Model (Figure 8.31)

	z_1	z_2	z_3	z_4	z_5
			Observed Correlations		
z_1	1.000				
z_2	.824	1.000			
z_3	-.643	-.512	1.000		
z_4	-.821	-.803	.568	1.000	
z_5	-.831	-.809	.780	.862	1.000
			Reproduced Correlations (Initial Model)		
z_1	1.000				
z_2	.824	1.000			
z_3	-.643	-.530	1.000		
z_4	-.662*	-.803	.425*	1.000	
z_5	-.748*	-.750*	.713*	.787*	1.000
			Reproduced Correlations (Revised Model)		
z_1	1.000				
z_2	.824	1.000			
z_3	-.643	-.530	1.000		
z_4	-.820	-.803	.528	1.000	
z_5	-.824	-.817	.768	.738*	1.000

* Difference between reproduced and observed correlation is greater than 0.05.

Figure 8.25 Path Diagram for the Initial Model (Infant Mortality), Including Path Coefficients

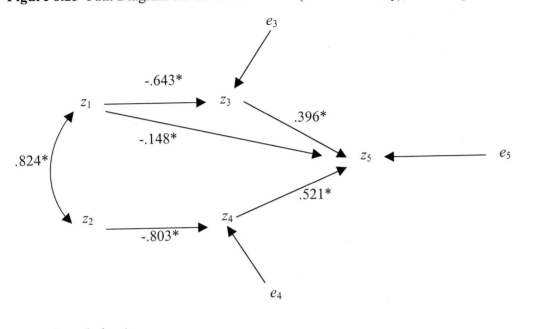

z_1 = *docs (lndocs)*
z_2 = *gdp (lngdp)*
z_3 = *deathrat*
z_4 = *birthrat*
z_5 = *infmr*

Figure 8.26 Regression Output of Missing Paths: *deathrat* (z_5) on *lndocs* (z_1), *lngdp* (z_2), and *birthrat* (z_4).

Model Summary

Model	R	R Square	Adjusted R Square	Std. Error of the Estimate
1	.647[a]	.419	.404	3.50

a. Predictors: (Constant), Births per 1000 population, 1992, Natural log of GDP, Natural log of doctors per 10000

Coefficients[a]

Model		Unstandardized Coefficients B	Unstandardized Coefficients Std. Error	Standardized Coefficients Beta	t	Sig.	Collinearity Statistics Tolerance	Collinearity Statistics VIF
1	(Constant)	9.503	3.689		2.576	.011		
	LNDOCS	-1.899	.412	-.652	-4.605	.000	.250	4.001
	LNGDP	.346	.404	.116	.856	.394	.272	3.682
	BIRTHRAT	3.700E-02	.047	.107	.792	.430	.277	3.616

a. Dependent Variable: DEATHRAT

Path coefficients for *lngdp* on *deathrat* and *birthrat* on *deathrat* are NOT signficant and should not be included.

Figure 8.27 Regression Output of Missing Paths: *birthrat* (z_4) on *lndocs* (z_1), *lngdp* (z_2), and *deathrat* (z_3).

Model Summary

Model	R	R Square	Adjusted R Square	Std. Error of the Estimate
1	.851[a]	.725	.718	6.93

a. Predictors: (Constant), Deaths per 1000 people, 1992, Natural log of GDP, Natural log of doctors per 10000

Coefficients[a]

Model		Unstandardized Coefficients		Standardized Coefficients	t	Sig.	Collinearity Statistics	
		B	Std. Error	Beta			Tolerance	VIF
1	(Constant)	60.082	5.040		11.921	.000		
	LNDOCS	-3.851	.814	-.459	-4.732	.000	.252	3.966
	LNGDP	-3.425	.738	-.399	-4.639	.000	.320	3.126
	DEATHRAT	.145	.184	.050	.792	.430	.584	1.712

a. Dependent Variable: BIRTHRAT

Path coefficient for *deathrat* on *birthrat* is NOT significant and should not be included.

235

Figure 8.28 Regression Output of Missing Paths: *infmr* (z_5) on *lndocs* (z_1), *lngdp* (z_2), *deathrat* (z_5), and *birthrat* (z_4).

Model Summary

Model	R	R Square	Adjusted R Square	Std. Error of the Estimate
1	.946[a]	.895	.891	14.202

a. Predictors: (Constant), Births per 1000 population, 1992, Deaths per 1000 people, 1992, Natural log of GDP, Natural log of doctors per 10000

Coefficients[a]

Model		Unstandardized Coefficients		Standardized Coefficients	t	Sig.	Collinearity Statistics	
		B	Std. Error	Beta			Tolerance	VIF
1	(Constant)	25.204	15.399		1.637	.104		
	LNDOCS	-.452	1.820	-.016	-.248	.805	.211	4.732
	LNGDP	-6.946	1.646	-.245	-4.219	.000	.270	3.706
	DEATHRAT	3.896	.377	.410	10.338	.000	.581	1.721
	BIRTHRAT	1.402	.190	.425	7.374	.000	.275	3.636

a. Dependent Variable: INFMR

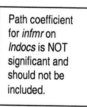

Path coefficient for *infmr* on *lndocs* is NOT significant and should not be included.

236

Figure 8.29 Regression Output for Significant Paths: *birthrat* (z_4) on *lndocs* (z_1) and *lngdp* (z_2).

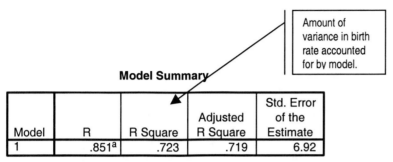

Amount of variance in birth rate accounted for by model.

Model Summary

Model	R	R Square	Adjusted R Square	Std. Error of the Estimate
1	.851[a]	.723	.719	6.92

a. Predictors: (Constant), Natural log of doctors per 10000, Natural log of GDP

Coefficients[a]

Model		Unstandardized Coefficients		Standardized Coefficients	t	Sig.	Collinearity Statistics	
		B	Std. Error	Beta			Tolerance	VIF
1	(Constant)	61.796	4.545		13.597	.000		
	LNDOCS	-4.149	.720	-.494	-5.761	.000	.321	3.116
	LNGDP	-3.393	.736	-.396	-4.610	.000	.321	3.116

a. Dependent Variable: BIRTHRAT

Final path coefficients for revised model.

Figure 8.30 Regression Output for Significant Paths: *infmr* (z_5) on *lngdp* (z_2), *deathrat* (z_3), and *birthrat* (z_4).

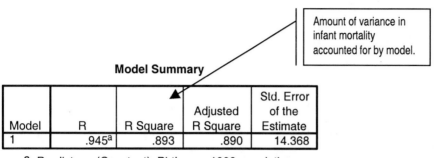

Model Summary

Model	R	R Square	Adjusted R Square	Std. Error of the Estimate
1	.945[a]	.893	.890	14.368

a. Predictors: (Constant), Births per 1000 population, 1992, Deaths per 1000 people, 1992, Natural log of GDP

Coefficients[a]

Model		Unstandardized Coefficients B	Std. Error	Standardized Coefficients Beta	t	Sig.	Collinearity Statistics Tolerance	VIF
1	(Constant)	30.347	15.347		1.977	.050		
	LNGDP	-7.512	1.455	-.264	-5.163	.000	.351	2.852
	DEATHRAT	3.792	.344	.407	11.012	.000	.669	1.495
	BIRTHRAT	1.374	.175	.419	7.864	.000	.322	3.106

a. Dependent Variable: INFMR

Final path coefficients for revised model.

Figure 8.31 Path Diagram for the Revised Model (Infant Mortality), Including Path Coefficients.

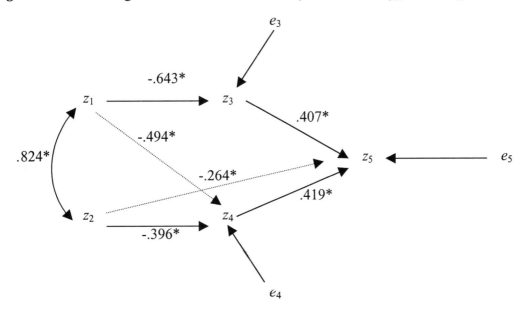

* Significant at the .05 level.

Note. Revised paths are shown with dashed arrows.

z_1 = *docs (lndocs)*
z_2 = *gdp (lngdp)*
z_3 = *deathrat*
z_4 = *birthrat*
z_5 = *infmr*

Presentation of Results

The following summary of results applies the output from Figures 8.18 – 8.31. The reader should note that due to space constraints, we have referenced appropriate figures and tables that were previously presented in the text.

A path analysis was conducted to determine the causal effects among the variables of number of doctors per 10,000 people (*docs*, z_1), gross domestic product (*gdp*, z_2), number of deaths per 1,000 people (*deathrat*, z_3), birth rate per 1,000 people (*birthrat*, z_4) and infant mortality per 1,000 live births (*infmr*, z_5).   Prior to the analysis, the variables of *docs* and *gdp* were transformed by taking the natural log. The initial model, presented in Figure 8.17, was not consistent with the empirical data. More specifically, six of the reproduced correlations exceeded a difference of .05. Tests of the missing paths in the initial model indicated that two additional paths would significantly contribute to the model: *birthrat* on *docs* and *infmr* on *gdp*. In addition, the nonsignificant path of *infmr* on *docs* was removed from the model. Thus, a revised model was generated and is presented in Figure 8.31. Computation of reproduced correlations for the revised model indicated consistency with the empirical correlations as only one reproduced correlation exceeded a difference of .05 (see Table 8.9). All path coefficients were significant at the

.05 level. The direct, indirect, and total causal effects of the revised model are presented in Table 8.12. The outcome of primary interest was infant mortality; the determinant with the largest total causal effect was number of doctors (–.469). The remaining determinants of infant mortality, as indicated by total causal effect, were gross domestic product (.430), birth rate (.419), and death rate (.407). This model explained approximately 89.3% of variance in infant mortality. The primary determinant of the birth rate was number of doctors (.494) followed by gross domestic product (.396). Approximately 72.3% of variance in the birth rate was explained by the model. The primary determinant of death rate was the number of doctors (–.643), which explained approximately 41.4% of variance in death rate.

Table 8.10 Path Decompositions for the Revised Model (Infant Mortality) Shown in Figure 8.31.

Reproduced Correlation	Path Decomposition
$\hat{r}_{13}$	p_{31} (D)
$\hat{r}_{14}$	$p_{41} + r_{12}p_{42}$ (D) (U)
$\hat{r}_{15}$	$p_{31}p_{53} + p_{41}p_{54} + r_{12}p_{52} + r_{12}p_{42}p_{54}$ (I) (I) (U) (U)
$\hat{r}_{23}$	$r_{12}\,p_{31}$ (U)
$\hat{r}_{24}$	$p_{42} + r_{12}p_{41}$ (D) (U)
$\hat{r}_{25}$	$p_{52} + p_{42}p_{54} + r_{12}p_{31}p_{53} + r_{12}p_{41}p_{54}$ (D) (I) (U) (U)
$\hat{r}_{34}$	$p_{31}p_{41} + p_{31}r_{12}p_{42}$ (S) (S)
$\hat{r}_{35}$	$p_{53} + p_{31}p_{41}p_{54} + p_{31}r_{12}p_{52} + p_{31}r_{12}p_{42}p_{54}$ (D) (I) (S) (S)
$\hat{r}_{45}$	$p_{54} + p_{42}p_{52} + p_{42}r_{12}p_{31}p_{53} + p_{41}p_{31}p_{53}$ (D) (S) (S) (S)

Table 8.11   Path Decompositions and Calculation of Reproduced Correlations for the Revised Model Shown in Figure 8.31.

$\hat{r}_{13} =$ p_{31}
   $= (-.643) = \mathbf{-.643}$
        (D)

$\hat{r}_{14} =$ p_{41} $+$ $r_{12}p_{42}$
   $= (-.494) + (.824)(-.396) = \mathbf{-.820}$
        (D)   (U)

$\hat{r}_{15} =$ $p_{31}p_{53}$ $+$ $p_{41}p_{54}$ $+$ $r_{12}p_{52}$ $+$ $r_{12}p_{42}p_{54}$
   $= (-.643)(.407) + (-.494)(.419) + (.824)(-.264) + (.824)(-.396)(.419) = \mathbf{-.824}$
            (I)   (I)   (U)   (U)

- -

$\hat{r}_{23} =$ $r_{12}\,p_{31}$
   $= (.824)(-.643) = \mathbf{-.530}$
        (U)

$\hat{r}_{24} =$ p_{42} $+$ $r_{12}p_{41}$
   $= (-.396) + (.824)(-.494) = \mathbf{-.803}$
        (D)   (U)

$\hat{r}_{25} =$ p_{52} $+$ $p_{42}p_{54}$ $+$ $r_{12}p_{31}p_{53}$ $+$ $r_{12}p_{41}p_{54}$
   $= (-.264) + (-.396)(.419) + (.824)(-.643)(.407) + (.824)(-.494)(.419) = \mathbf{-.817}$
        (D)   (I)   (U)   (U)

- -

$\hat{r}_{34} =$ $p_{31}p_{41}$ $+$ $p_{31}r_{12}p_{42}$
   $= (-.643)(-.494) + (-.643)(.824)(-.396) = \mathbf{.528}$
            (S)   (S)

$\hat{r}_{35} =$ p_{53} $+$ $p_{31}p_{41}p_{54}$ $+$ $p_{31}r_{12}p_{52}$ $+$ $p_{31}r_{12}p_{42}p_{54}$
   $= (.407) + (-.643)(-.494)(.419) + (-.643)(.824)(-.264) + (-.643)(.824)(-.396)(.419) = \mathbf{.768}$
        (D)   (S)   (S)   (S)

- -

$\hat{r}_{45} =$ p_{54} $+$ $p_{42}p_{52}$ $+$ $p_{42}r_{12}p_{31}p_{53}$ $+$ $p_{41}p_{31}p_{53}$
   $= (.419) + (-.396)(-.264) + (-.396)(.824)(-.643)(.407) + (-.494)(-.643)(.407) = \mathbf{.738}$
        (D)   (S)   (S)   (S)

Table 8.12 Summary of Causal Effects for Revised Model (Infant Mortality) Shown in Figure 8.31.

		Causal Effects		
Outcome	Determinant	Direct	Indirect	Total
Death Rate	Doctors	-.643*	—	-.643
$(R^2 = .414)$	GDP	—	—	—[+]
Birth Rate	Doctors	-.494*	—	-.494 [+]
$(R^2 = .723)$	GDP	-.396*	—	-.396 [+]
	Death Rate	—	—	—[+]
Infant Mortality	Doctors	—	-.469	-.469 [+]
$(R^2 = .893)$	GDP	-.264*	-.166	-.430 [+]
	Death Rate	.407*	—	.407 [+]
	Birth Rate	.419*	—	.419 [+]

* Direct effect is significant at the .05 level.

[+] Total effect may be incomplete due to unanalyzed components.

SECTION 8.5 SPSS "How To"

This section presents the steps for examining and conducting path analysis using linear **Regression**. For an extensive discussion on multiple regression, the reader is referred to Chapter 7. To open the Regression Dialogue Box, select the following:

```
Analyze
      Regression
            Linear
```

Linear Regression Dialogue Box (see Figure 8.32)

Once in this box, click the first endogenous variable to be analyzed and move it to the Dependent Box. For our example, the first endogenous variable is *deathrat*. Click each exogenous variable that has been identified as having a causal path to the specific endogenous variable and move to the Independent(s) Box. For our initial model, we only predicted that *lndocs* would have causal effect on *deathrat*. For method, select Enter; this is the default. Next, click **Statistics**.

Linear Regression Statistics Dialogue Box (see Figure 8.33)

Within this box, select **Model Fit** and **Collinearity Diagnostics**. Then click **Continue**. Click **OK**.

Figure 8.32 Linear Regression Dialogue Box.

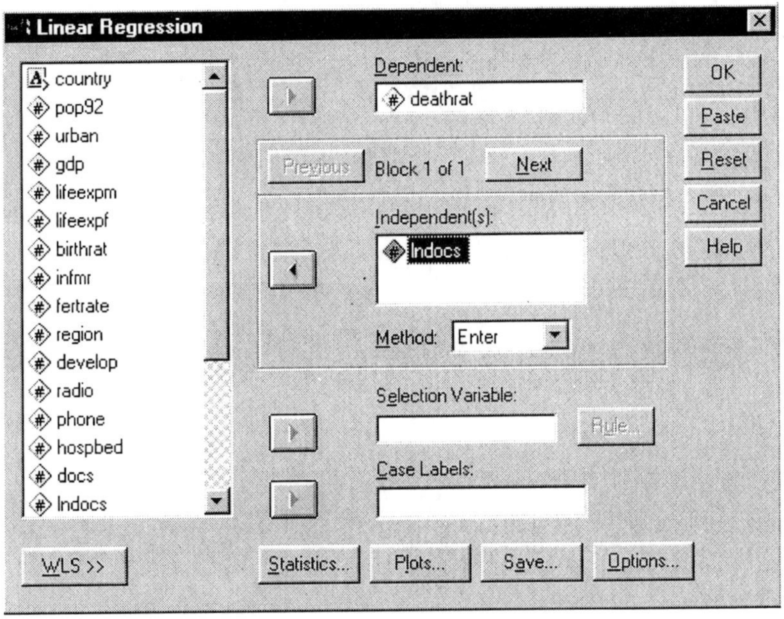

Figure 8.33 Linear Regression Statistics Dialogue Box.

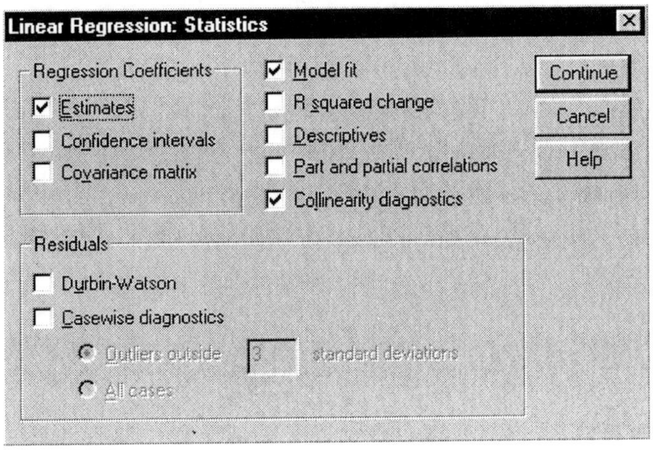

This process of regression analysis would be conducted for each endogenous variable within the initial model. For our initial model for infant mortality, the following three analyses were conducted:

Analysis	Endogenous Variable	Exogenous Variables
1	*deathrat*	*lndocs**
2	*birthrat*	*lngdp**
3	*infmr*	*lndocs*, deathrat*, birthrat**

*Indicates significant path coefficients at the .05 level.

Since the initial model was not consistent with the empirical data, the following analyses were conducted to explore the significance of paths missing from the initial model:

Analysis	Endogenous Variable	Exogenous Variables
4	*deathrat*	*lndocs*, lngdp, birthrat*
5	*birthrat*	*lndocs*, lngdp*, deathrat*
6	*infmr*	*lndocs, lngdp*, deathrat*, birthrat**

*indicates significant path coefficients at the .05 level

Since these analyses revealed that only some of the paths were significant, the following analyses were conducted to determine path coefficients for those paths that were significant:

Analysis	Endogenous Variable	Exogenous Variables
7	*birthrat*	*lndocs*, lngdp**
8	*infmr*	*lngdp*, deathrat*, birthrat**

*indicates significant path coefficients at the .05 level

Since the very first analysis produced the path coefficient of *deathrat* on *lndocs*, this analysis did not need to be repeated. The reader should note that this example required eight regression analyses to create an appropriate path model—this is quite common in path analysis.

Summary

Path analysis allows the researcher to determine causal effects among numerous variables. This technique is not exploratory in nature; rather, the researcher is testing the legitimacy of a causal model that has been based upon logic, theory, and/or experience. This causal model is depicted in a path diagram, in which effects between variables are represented by arrows. A straight line with a single arrowhead represents a direct effect, while a curved line with two arrowheads represents the bivariate correlation between two variables. An indirect effect occurs when a variable intervenes between the effect of two variables. Although a path model seeks to explain the causal determinants (i.e., IVs or, in path analysis, referred to as exogenous variables) of one variable (i.e., the DV or the endogenous variable), a model may examine several endogenous variables due to indirect effects.

Once a causal model has been developed, numerous regression analyses are conducted to determine path (beta) coefficients. To test the model fit, one must calculate the reproduced correlations for each path. Reproduced correlations are calculated through the development and application of path decompositions. If several reproduced correlations differ from empirical correlations by more than .05, then the model is not consistent with the empirical data. To revise the model, one examines missing paths within the model. Utilizing only paths that are significant, the model is revised and once again tested by comparing the empirical and reproduced correlations. Once a model has very few reproduced correlations that significantly differ from the empirical, the model is said to be consistent with empirical data. Figure 8.34 provides a checklist for conducting path analysis.

Figure 8.34 Checklist for Conducting Path Analysis.

I.   Develop Path Model
    a.   Create path diagram.
    b.   Develop path decompositions.

II. Screen Data
    a.   Missing Data?
    b.   Multivariate Outliers?
        ❑ Run preliminary Regression to calculate Mahalanobis' Distance.
            1. 🖱 **Analyze...Regression...Linear.**
            2. Identify a variable that serves as a case number and move to Dependent Variable box.
            3. Identify all appropriate quantitative variables and move to Independent(s) box.
            4. 🖱 **Save.**
            5. Check **Mahalanobis'.**
            6. 🖱 **Continue.**
            7. 🖱 **OK.**
            8. Determine chi square χ^2 critical value at $p<.001$.
        ❑ Conduct **Explore** to test outliers for Mahalanobis chi square χ^2.
            1. 🖱 **Analyze...Descriptive Statistics...Explore**
            2. Move *mah_1* to Dependent Variable box.
            3. Leave Factor box empty.
            4. 🖱 **Statistics.**
            5. Check **Outliers.**
            6. 🖱 **Continue.**
            7. 🖱 **OK.**
        ❑ Delete outliers for subjects when χ^2 exceeds critical χ^2 at $p<.001$.
    c.   Linearity, Normality, Homoscedasticity?
        ❑ Create Scatterplot Matrix of all model variables.
            ❑ Scatterplot shapes are not close to elliptical shapes→reevaluate univariate normality and consider transformations.
        ❑ Run Normality Plots with Tests within **Explore.**
        ❑ Run preliminary Regression to create residual plot.
            1. 🖱 **Analyze...Regression...Linear**
            2. Move primary endogenous variable (DV) to Dependent Variable box.
            3. Move exogenous variables (IVs) to Independent(s) Variable box.
            4. 🖱 **Plot.**
            5. Select ZRESID for y-axis.
            6. Select ZPRED for x-axis.
            7. 🖱 **Continue.**
            8. 🖱 **OK.**
        ❑ If residuals are clustered at the top, bottom, left, or right area in plot→ reevaluate univariate normality and consider transformations.

(Figure 8.34 is continued on the next page.)

Figure 8.34 Checklist for Conducting Path Analysis. (*Continued*)

III. Conduct Multiple Regression Analyses for Path Analysis

    a.   Run Regression using **Linear Regression** for each endogenous variable.

         1. ⍅ **Analyze...**⍅ **Regression...**⍅ **Linear**

         2. Move endogenous variable (DV) to Dependent Variable box

         3. Move exogenous variables (IVs) to Independent(s) box

         4. Select **Enter**.

         5. ⍅ **Statistics.**

         6. Check **Model Fit** and **Collinearity Diagnostics**.

         7. ⍅ **Continue**

         8. ⍅ **OK.**

    b.   Interpret tolerance.

    c.   If tolerance for each exogenous variable is greater than .1, interpret path (beta) coefficient for each path.

    d.   Transfer path coefficients to path diagram.

    e.   Calculate reproduced correlation coefficients through path decompositions.

    f.   Compare reproduced correlations to empirical correlations.

    g.   If only a few reproduced correlations differ from the empirical correlations by more than .05, your model is fairly consistent with the empirical data. Proceed with step IV.

    h.   If several reproduced correlations differ from the empirical correlations by more than .05, evaluate missing paths to determine if other paths may significantly contribute to the model. Analyze missing paths by following the steps beginning with III.a.

    i.   Once significant paths have been determined, conduct regression analysis using only the significant paths. Analyze significant paths by following the steps beginning with III.a.

IV. Summarize Results

    a.   Describe path model.

    b.   Present path diagram.

    c.   Describe any data elimination or transformation.

    d.   Present path coefficients in path diagram.

    e.   Describe how reproduced correlations were not consistent with empirical correlations.

    f.   Describe process for revising model.

    g.   Present revised path diagram with path coefficients.

    h.   Describe how reproduced correlations were consistent with empirical correlations.

    i.   Present table that compares empirical correlations to reproduced correlations for both the initial and revised models.

    j.   Discuss causal effects for each endogenous variable: total causal effects and R^2.

    k.   Present table of causal effects (direct, indirect, and total) for each endogenous variable.

Exercises for Chapter 8

The following exercises utilize the path model depicted below as well as data set, *county.sav*. Specifically, the variables of *lndocs* (z_1), *lngdp* (z_2), *deathrat* (z_3), *birthrat* (z_4), and *lifeexpf* (z_5) will be utilized.

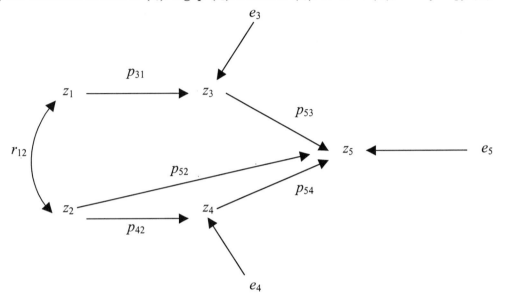

1. Determine the path decompositions for the model. Be sure to label which are direct (D), indirect (I), unanalyzed (U), and spurious (S).

2. Identify the regression analyses necessary for testing this initial model.

3. Create a correlation matrix that includes all model variables. Conduct the regression analyses identified in #2. What are the following path coefficients?

    a. $r_{12} =$

    b. $p_{31} =$

    c. $p_{42} =$

    d. $p_{52} =$

    e. $p_{53} =$

f $\quad p_{54} =$

4. Applying the path decompositions from #1, calculate the reproduced correlations.

5. Which reproduced correlations differ from the empirical correlations by more than .05?

6. Is this model consistent with empirical data? If not, what would you recommend to revise the model?

CHAPTER 9

FACTOR ANALYSIS

In this chapter, we shift the focus of our attention—this time to the third group of advanced/multivariate statistical techniques previously discussed in this text. Specifically, we present a discussion of a procedure known as *factor analysis*, which is used to describe the underlying structure that "explains" a set of variables. It is a technique—similar to correlation and regression—that capitalizes on shared variability. Factor analysis has many practical applications within social science settings, as the reader will see as a result of our discussion.

SECTION 9.1 PRACTICAL VIEW

Purpose

Generally speaking, factor analysis is a procedure used to determine the extent to which *measurement overlap* (Williams, 1992)—that is, shared variance—exists among a set of variables. Its underlying purpose is to determine if measures for different variables are, in fact, measuring something in common. The mathematical procedure essentially takes the variance, as defined by the intercorrelations among a set of measures, and attempts to allocate it in terms of a smaller number of underlying hypothetical variables (Williams, 1992). These underlying, hypothetical—and unobservable—variables are called *factors*. Factor analysis, then, is essentially a process by which the number of variables is reduced by determining which variables "cluster" together, and factors are the groupings of variables that are measuring some common entity or construct.

The main set of results obtained from a factor analysis consists of *factor loadings*. A factor loading is interpreted as the Pearson correlation coefficient of an original variable with a factor. Like correlations, loadings range in value from -1.00 (representing a perfect negative association with the factor) through 0 to +1.00 (indicating perfect positive association). Variables typically will have loadings on all factors but will usually have high loadings on only one factor (Aron & Aron, 1999).

Another index provided in the results of a factor analysis is the list of *communalities* for each variable. Communalities represent the proportion of variability for a given variable that is explained by the factors (Agresti & Finlay, 1997) and allows the researcher to examine how individual variables reflect the sources of variability (Williams, 1992). Communalities may also be interpreted as the squared multiple correlation of the variable as predicted from the combination of factors, or as the sum of squared loadings across all factors for that variable.

The process by which the factors are determined from a larger set of variables is called *extraction*. There are actually several types of factor extraction techniques, although the most commonly used empirical approaches are principal components analysis and factor analysis (Stevens, 1992). (It should be noted that the term "factor analysis" is *commonly* used to represent the *general* process of variable reduction, regardless of the actual method of extraction utilized. For a detailed description of the various additional extraction techniques available, including maximum likelihood, unweighted least squares, generalized least squares, image factoring, and alpha factoring, the reader should refer to Tabachnick and Fidell, 1996.) In both principal components analysis and factor analysis, linear combinations (the factors) of original variables are produced, and a small number of these combinations typically account for the majority of the variability within the set of intercorrelations among the original variables.

In ***principal components analysis***, *all* sources of variability—unique, shared, and error variability—are analyzed for each observed variable. However, in ***factor analysis***, only *shared* variability is analyzed—both unique and error variability are ignored. This is based on the belief that unique and error variance serve only to "confuse" the picture of the underlying structure of a set of variables (Tabachnick & Fidell, 1996). In other words, principal components analysis analyzes variance; factor analysis analyzes covariance. Principal components analysis is usually the preferred method of factor extraction, especially when the focus of an analysis searching for underlying structure is truly exploratory, which is typically the case. Its goal is to extract the maximum variance from a data set, resulting in a few orthogonal (uncorrelated) components. When principal components analysis is used for extraction, the resulting linear combinations are often referred to as "components," as opposed to "factors." For the remainder of this chapter, we will limit our discussion to principal components analysis.

Since principal components analysis is an exploratory procedure, the first—and probably most important—decision required by the researcher is deciding how many components or factors to retain and, thus, interpret. The most widely accepted criterion was developed in 1960 by Kaiser, and has become known appropriately as "Kaiser's rule." The rule states that only those components whose eigenvalues are greater than 1 should be retained. An ***eigenvalue*** is defined as the amount of total variance explained by each factor, with the total amount of variability in the analysis equal to the number of original variables in the analysis (i.e., each variable contributes one unit of variability to the total amount due to the fact that the variance has been standardized).

A second, graphical method for determining the number of components is called the *scree test* and involves the examination of a scree plot. A ***scree plot*** is a graph of the magnitude of each eigenvalue (vertical axis) plotted against their ordinal numbers (horizontal axis). In order to determine the appropriate number of components to retain and interpret, one should look for the "knee," or bend, in the line. A typical scree plot will show the first one or two eigenvalues to be relatively large in magnitude, with the magnitude of successive eigenvalues dropping off rather drastically. At some point, the line will appear to level off. This is indicative of the fact that these successive eigenvalues are relatively small and, for the most part, of equal size. The recommendation is to retain all components with eigenvalues in the sharp descent of the line *before* the first one where the leveling effect occurs (Stevens, 1992). (If you are curious about the origin of the name for this type of plot, "scree" is formally defined as the rock debris located at the bottom of a cliff—an image one could envision in an actual scree plot.)

A third criterion used to determine the number of factors to keep in a factor or principal components analysis is to retain and interpret as many factors as will account for a certain amount of total variance. A general rule of thumb is to retain the factors that account for at least 70% of the total variability (Stevens, 1992), although there may be situations where the researcher will desire an even greater amount of variability to be accounted for by the components. However, this may not always be feasible. For example, assume we wanted to reduce the number of variables in an analysis containing 20 original variables. If it takes 15 components (or factors) to achieve the 70% criterion, we have not gained much with respect to variable reduction and some underlying structure. Realize that in this situation, some factors will undoubtedly be variable-specific (i.e., only one variable will load on a given factor). Therefore, we have not uncovered any underlying structure for the *combination* of original variables.

A final criterion for retaining components is the assessment of model fit. Recall in Chapter 8 that we discussed the assessment of model fit for a path analysis model. The assessment of model fit involved the computation of the reproduced correlations (i.e., those that would occur assuming the model represents reality) and comparing them to the original, observed correlations. If the number of correlations that are reasonably close (again, within .05 of each other) is small, it can be assumed that the model is consistent with the empirical data. One advantage of all factor analytic procedures over path analysis is that computer analysis programs—including SPSS—will calculate the reproduced correlations for you, and even provide a percentage of the total that exceed the cutoff value of .05.

With four different criteria to evaluate, how can one be sure of the number of components to retain and interpret, especially if examination of the four criteria results in different decisions regarding the number of components? Stevens (1992) offers several suggestions when faced with this often occurring dilemma. He states that Kaiser's rule has been shown to be quite accurate when the number of original variables is < 30 and the communalities are > .70, or when $N > 250$ and the mean communality is ≥ .60 (p. 379). In other situations, use of the scree test with an $N > 250$ will provide fairly accurate results, provided that most of the communalities are somewhat large (i.e., > .30). Our recommendation is to examine all four criteria for alternative factor solutions and weigh them against the overall goal of any multivariate analysis—parsimony. *It is our belief that the principle of parsimony is more important in factor or principal components analysis than in any other analysis technique.*

Let us examine these various criteria for deciding how many components to keep through the development of an example that we will submit to a principal components analysis. Assume we wanted to determine what, if any, underlying structure exists for measures on ten variables, consisting of:

- male life expectancy (*lifexpm*),
- female life expectancy (*lifexpf*),
- births per 1,000 people (*birthrat*),
- infant mortality rate (*infmr*),
- fertility rate per woman (*fertrate*),
- natural log of doctors per 10,000 people (*lndocs*),
- natural log of radios per 100 people (*lnradio*),
- natural log of telephones per 100 people (*lnphone*),
- natural log of gross domestic product (*lngdp*), and
- natural log of hospital beds per 10,000 people (*lnbeds*).

We would first examine the number of eigenvalues greater than 1.00. The table of total variance accounted for in the initial factor solution for these ten variables is shown in Figure 9.1. With an eigenvalue equal to 8.161, only the first component has an eigenvalue that exceeds the criterion value of 1.00. The second component (.590) does not even approach the criterion value. Additionally, the first component accounts for nearly 82% of the total variability in the original variables, while the second component only accounts for 6%.

Figure 9.1   Initial Eigenvalues and Percentage of Variance Explained by Each Component for Initial Example.

Component	Initial Eigenvalues			Extraction Sums of Squared Loadings		
	Total	% of Variance	Cumulative %	Total	% of Variance	Cumulative %
1	8.161	81.608	81.608	8.161	81.608	81.608
2	.590	5.902	87.510			
3	.372	3.717	91.227			
4	.338	3.384	94.611			
5	.246	2.460	97.071			
6	.168	1.677	98.748			
7	5.491E-02	.549	99.297			
8	3.319E-02	.332	99.629			
9	2.978E-02	.298	99.927			
10	7.316E-03	7.316E-02	100.000			

Examination of the scree plot for this solution (see Figure 9.2) provides us with similar results. The first component is much larger than subsequent components in terms of eigenvalue magnitude; eigenvalues of successive components drop off quite drastically. Clearly, the line begins to level off at the second component. Based on this plot, it appears that we should retain and interpret only the first component.

Although we do not provide the output here, the reproduced correlation matrix indicates that 18 (40%) of the residuals (i.e., differences between observed and reproduced correlations) have absolute values greater than .05. Since this indicates that the one-component model does not fit the empirical correlations very well, we might want to investigate the possibility of a two-component model.

The reader should realize that when attempting a revised model with a different number of factors, the values for the initial eigenvalues, percent of variance explained, and the appearance of the scree plot will not change—these are based solely on the original correlations. The only substantive difference can be noticed in the numbers of residuals that exceed the criterion value. In the two-component model, only 11 (24%) of the residuals exceed our .05 criterion—a substantial improvement over our previous percentage, equal to 40%. We now compare the two possible models side-by-side in order to determine which we will interpret.

Figure 9.2 Scree Plot (Example number 1).

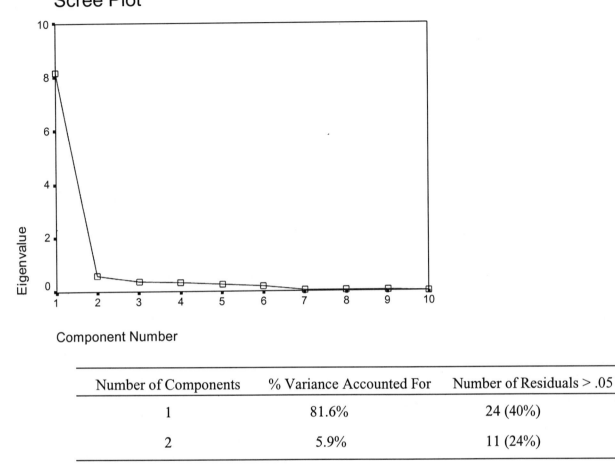

Scree Plot

Number of Components	% Variance Accounted For	Number of Residuals > .05
1	81.6%	24 (40%)
2	5.9%	11 (24%)

Based on the variance explained and the scree plot, it would appear that one component should be interpreted; however, this may be an oversimplification of the reduction of our original data. Furthermore, the addition of a second component certainly improved the fit of the model to our empirical data. For this latter reason, we will proceed with the interpretation of the two-component solution.

Before we attempt to interpret the components based on the values of the loadings, it is imperative that we discuss the topic of factor (component) rotation. **Rotation** is a process by which a factor solution is made more interpretable without altering the underlying mathematical structure. Rotation is a complex mathematical procedure and it is sometimes helpful to consider it from a geometric perspective. For the sake of simplicity, let us assume that we have four variables we have submitted to a factor analysis. The analysis returned to us two components and the associated hypothetical loadings are presented below:

Variable	Loading on Component 1	Loading on Component 2
A	.850	.120
B	.700	.210
C	−.250	.910
D	.210	−.750

If we were to plot these combinations of loadings in a scatterplot of Component 1 by Component 2, the result would appear as in Figure 9.3. Notice that the possible loadings on each component range from -1.00 to +1.00 and that the locations of the four variables in that geometric space have been plotted according to the combination of loadings on the two components in Part (a) of Figure 9.3—for example, Variable A is located at X=.850, Y=.120. Although the points are generally near the lines, if we were able to rotate the axes, we would notice a better "fit" of the loadings to the actual components. In Part (b), we now see how three of the four loadings line up nearly perfectly with the two components. This process alters the original values of the component loadings without changing their mathematical properties. This gives us the ability to name the components with greater ease since three of the variables correlate nearly perfectly with the two components, and the rotated factor loadings would change accordingly.

The researcher must decide whether to use an orthogonal or oblique rotation. ***Orthogonal rotation*** is a rotation of factors that results in factors being uncorrelated with each other; the resultant computer output is a *loading* matrix (i.e., a matrix of correlations between all observed variables and factors) where the size of the loading reflects the extent of the relationship between each observed variable and each factor. Since the goal of factor analysis is to obtain underlying factors which are uncorrelated (thereby representing some *unique* aspect of the underlying structure), it is recommended that orthogonal rotation be used instead of oblique. There are three types of orthogonal rotation procedures—varimax, quartimax, and equamax—of which varimax is the most commonly used. Orthogonal rotation methods are described further in Section 9.5.

Oblique rotation results in factors being correlated with each other. Several matrices are produced from an oblique rotation: a *factor correlation* matrix (i.e., a matrix of correlations between all factors); a loading matrix that is separated into a *structure* matrix (i.e., correlations between factors and variables); and a *pattern* matrix (i.e., unique relationships with no overlapping among factors and each observed variable). The interpretation of factors is based on loadings found in the pattern matrix. One would use an oblique rotation only if there were some prior belief that the underlying factors are correlated. Several types of oblique rotations exist, including direct oblimin, direct quartimin, orthoblique, and promax. Direct oblimin is arguably the most frequently used form of oblique rotation. Oblique rotation methods are described further in Section 9.5.

Once we have rotated the initial solution, we are ready to *attempt* interpretation. We emphasize the *attempt* at interpretation because, by its very nature, interpretation of components or factors involves much subjective decision making on the part of the researcher (Williams, 1992). The rotated component loadings for our working example are presented in Figure 9.4. Initially, the reader should notice that each variable has a loading on each component, although in most cases each has a high loading on only one component. Some of the variables in our example have loaded relatively high on both components, but we will "assign" a given variable to the component with the higher loading (as shown by the shaded boxes) and attempt to interpret them in that fashion.

Figure 9.3   Illustration of Geometric Interpretation of Rotation of Components.

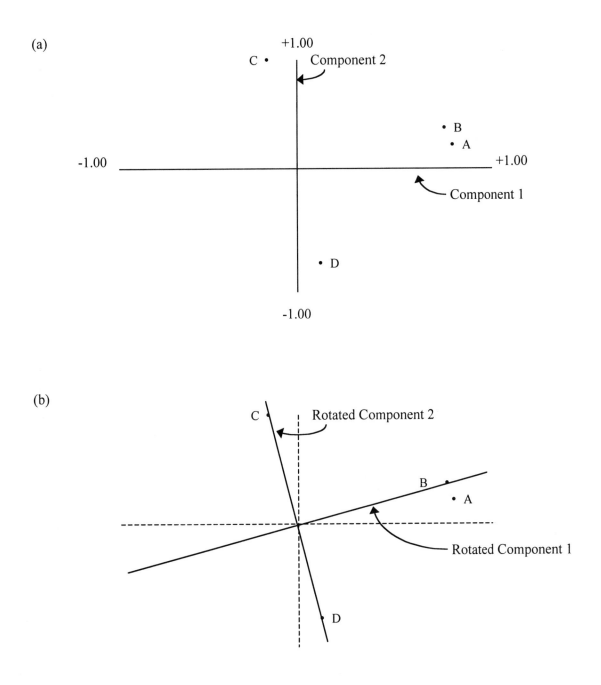

Figure 9.4   Component Loadings for the Rotated Solution.

	Component	
	1	2
Fertility rate per woman 1990	-.879	-.305
Births per 1,000 population 1992	-.858	-.393
Male life expectancy 1992	.846	.450
Infant mortality rate 1992 (per 1,000 live births)	-.839	-.461
Female life expectancy 1992	.829	.509
Natural log of doctors per 10,000	.763	.502
Natural log of radios per 100 people	.279	.879
Natural log hospital beds per 10,000	.480	.719
Natural log of GDP	.627	.688
Natural log of phones per 100 people	.675	.678

When interpreting or naming components, one should pay particular attention to two aspects—the size and direction—of each loading. The name attached to each component should reflect the relative sizes and directions of the loadings. Notice that Component 1 contains both high positive and high negative loadings; this is referred to as a *bipolar* factor. In other words, the name we assign to Component 1 should *primarily* reflect the strong positive loadings for male life expectancy (*lifexpm*) and female life expectancy (*lifexpf*), and subsequently the loading for *lndocs*. Due to negative loadings, the name should also reflect the *opposite* of fertility rate (*fertrat*), birth rate (*birthrat*), and infant mortality rate (*infmr*). Bipolar components or factors are usually more difficult to interpret. Since this component seems to address the *end* of an individual's life span as opposed to the beginning of that lifetime, and factoring in the number of docs (*lndocs*), we might consider attaching the label of *Healthy Lifespan* to this component. The second component—on which all variables have positive loadings—addresses the numbers of radios, hospital beds, and phones, as well as the nation's gross domestic product. We might consider attaching the label of *Economic Stature* to this component.

It is important to note that principal components analysis may be used as a variable reducing scheme for further analyses (Stevens, 1992). We have already examined the main application of the analysis—to determine empirically how many underlying constructs account for the majority of the variability among a set of variables. Principal components analysis may also be used as a precursor to multiple regression as a means of reducing the number of predictors, especially if the number of predictor variables is quite large relative to the number of subjects. Additionally, components analysis may be used to reduce the number of criterion variables in a multivariate analysis of variance. It is often recommended that a large number of DVs not be used in a MANOVA procedure; if you have a large number of DVs that you are interested in analyzing, reducing the overall number of variables through a principal components analysis would allow you to accomplish this task rather efficiently.

If a principal components analysis is to be followed by a subsequent analytic procedure, factor scores are often used. *Factor scores* are estimates of the scores subjects would have received on each of the factors had they been measured directly (Tabachnick & Fidell, 1996). Many different procedures may be used to estimate factor scores, the most basic of which is to simply sum the values across all variables that load on a given factor or component. Alternatively, a mean could be calculated across all variables that would then represent the score on a factor. Factor scores could also be estimated or predicted through a regression analysis. Those values are then entered into the subsequent analysis as if they were "raw" variables.

Finally, there are two basic types of factor analytic procedures, based on their overall intended function: exploratory and confirmatory factor analyses. In *exploratory factor analysis*, the goal is to describe and summarize data by grouping together variables that are correlated. The variables included in the analysis may or may not have been chosen with these underlying structures in mind (Tabachnick & Fidell, 1996). Exploratory factor analysis usually occurs during the early stages of research, when it often proves useful to consolidate numerous variables.

Confirmatory factor analysis is much more advanced and sophisticated than exploratory factor analysis. It is often used to test a theory about latent (i.e., underlying, unobservable) processes that might occur among variables. A major difference between confirmatory and exploratory factor analyses is that in a confirmatory analysis, variables are painstakingly and specifically chosen in order to adequately represent the underlying processes (Tabachnick & Fidell, 1996). The main purpose of confirmatory factor analysis is to confirm—or disconfirm—some *a priori* theory. LISREL, as previously discussed in Chapter 8, is often used as the analytical computer program in such studies.

Sample Research Questions

Building on the example we began discussing in the previous section, we now specify the main research questions to be addressed by our principal components analysis. Using the ten original variables, the appropriate research questions would be:

(1)   How many reliable and interpretable components are there among the following ten variables: male life expectancy, female life expectancy, births per 1,000 people, infant mortality rate, fertility rate per woman, number of doctors per 10,000 people, number of radios per 100 people, number of telephones per 100 people, gross domestic product, and number of hospital beds per 10,000 people?

(2)   If reliable components are identified, how might we interpret those components?

(3)   How much variance in the original set of variables is accounted for by the components?

SECTION 9.2 ASSUMPTIONS AND LIMITATIONS

If principal components analysis and factor analysis are being used in a descriptive fashion as a method of summarizing the relationships among a large set of variables, assumptions regarding the distributions of variables in the population are really not in force and, therefore, do not need to be assessed (Tabachnick & Fidell, 1996). This is usually the case since, as previously mentioned, principal components and factor analyses are almost always exploratory and descriptive in nature. It should be noted, however, that if the variables are normally distributed, the resultant factor solution will be enhanced; to the extent to which normality fails, the solution is degraded (although it still may be worthwhile) (Tabachnick & Fidell, 1996).

Previous versions of SPSS provided a statistical test of model fit—a value for a test statistic was provided and subsequently evaluated using a chi-square criterion. In situations where this test statistic was evaluated—in an inferential manner—and used to determine the number of factors or components, assessment of model assumptions takes on much greater importance. Since recent revisions of SPSS have omitted the chi-square test of model fit, this criterion can obviously no longer be used to determine the number of factors to interpret. Therefore, it is not necessary to test the assumptions of multivariate normality and linearity. However, we recommend that both of these assumptions be evaluated and any necessary transformations be made—*ensuring the quality of data will only improve the quality of the resulting factor or component solution.*

As a reminder to the reader, these two aforementioned assumptions are formally stated as follows:

(1)   All variables, as well as all linear combinations of variables, must be normally distributed (assumption of multivariate normality).
(2)   The relationships among all pairs of variables must be linear.

Factor analyses, in general, are subject to a potentially severe limitation. Recall that the basis for any underlying structure that is obtained from a factor analysis are the relationships among all original variables in the analysis. Correlation coefficients have a tendency to be less reliable when estimated from small samples. If unreliable—or at least, *less* reliable—correlations exist among variables, and those variables are subjected to a factor analysis, the resultant factors will also not be very reliable. Tabachnick and Fidell (1996) offer the following guidelines for sample sizes and factor analyses:

Approximate Sample Size	Estimated Reliability
50	very poor
100	poor
200	fair
300	good
500	very good
1000	excellent

As a general rule of thumb, they suggest that a data set include *at least* 300 cases for a factor analysis to return reliable factors. If a solution contains several high-loading variables (> .80), a smaller sample (e.g., $n = 150$) would be sufficient.

Stevens (1992) has offered a different, although somewhat similar, set of recommendations, based on the number of variables (with minimum/maximum loadings) per component (p. 384). Specifically, these recommendations are as follows:

(1)   Components with four or more loadings above .60 in absolute value (i.e., |.60|) are reliable, regardless of sample size.
(2)   Components with about 10 or more low loadings (i.e., < |.40|) are reliable as long as the sample size is greater than 150.

(3)   Components with only a few low loadings should not be interpreted unless the sample size is at least 300.

It should be noted that the above constitute *general* guidelines, *not* specific criteria which *must* be met by the applied researcher. If researchers are planning a factor analysis with small sample sizes, it is recommended that they apply *Bartlett's sphericity test*. This procedure tests the null hypothesis that the variables in the population correlation matrix are uncorrelated. If one fails to reject this hypothesis, there is no reason to do a principal components analysis since the variables are already uncorrelated (i.e., they have nothing in common) (Stevens, 1992).

Methods of Testing Assumptions

The reader will recall that the assessment of multivariate normality is not easily accomplished through the use of standard statistical software packages. The most efficient method of assessing multivariate normality is to assess univariate normality—recall that univariate normality is a necessary condition for multivariate normality. Normality among individual variables may be evaluated by examining the values for skewness and kurtosis, normal probability plots, and/or bivariate scatterplots.

The assumption of linearity is best tested through the inspection of bivariate scatterplots obtained for each pair of original variables. The reader will recall that if a relationship is in fact linear, a general elliptical shape should be apparent in the scatterplot.

SECTION 9.3 PROCESS AND LOGIC

The Logic Behind Factor Analysis

The underlying, mathematical objective in principal components analysis is to obtain uncorrelated linear combinations of the original variables that account for as much of the total variance in the original variables as possible (Johnson & Wichern, 1998). These uncorrelated linear combinations are referred to as the ***principal components***. The logic behind principal components analysis involves the partitioning of this total variance by initially finding the first principal component (Stevens, 1992). The ***first principal component*** is the linear combination that accounts for the *maximum* amount of variance and is defined by the equation:

$$PC_1 = a_{11}x_1 + a_{12}x_2 + a_{13}x_3 + ... + a_{1p}x_p \qquad \text{(Equation 9.1)}$$

where PC_1 is the first principal component, x_i refers to the measure on the original variable, and a_{1i} refers to the weight assigned to a given variable for the first principal component (the first subscript following the *a* identifies the specific principal component, and the second subscript identifies the original variable)—e.g., the term $a_{11}x_1$ refers to the product of the weight for variable 1 on PC_1 and the original value for an individual on variable 1. The subscript $_p$ is equal to the total number of original variables. This linear combination, then, accounts for the maximum amount of variance within the original set of variables—the variance of the first principal component is equal to the largest eigenvalue (i.e., the eigenvalue for the first component).

The analytic operation then proceeds to find the second linear combination— *uncorrelated* with the first linear combination—that accounts for the next largest amount of variance (after that which has been attributed to the first component has been removed). The resulting equation would be:

$$PC_2 = a_{21}x_1 + a_{22}x_2 + a_{23}x_3 + \ldots + a_{2p}x_p \qquad \text{(Equation 9.2)}$$

It is important to note that the extracted principal components are not related. In other words,

$$r_{PC_1 \bullet PC_2} = 0$$

The third principal component is constructed so that it is uncorrelated with the first two and accounts for the next largest amount of variance in the system of original variables, after the two largest amounts have been removed. This process continues until all variance has been accounted for by the extracted principal components.

Interpretation of Results

The process of interpreting factor analysis results focuses on the determination of the number of factors to retain. As mentioned earlier, there are several methods/criteria to utilize in this process:

1) Eigenvalue—Components with eigenvalues greater than 1 should be retained. This criteria is fairly reliable when the number of variables is < 30 and communalities are $> .70$, or the number of individuals is > 250 and the mean communality is $\geq .60$.
2) Variance—Retain components that account for at least 70% total variability.
3) Scree Plot—Retain all components within the sharp descent, before eigenvalues level off. This criteria is fairly reliable when the number of individuals is > 250 and communalities are $> .30$.
4) Residuals—Retain the components generated by the model if only a few residuals (the difference between the empirical and reproduced correlations) exceed .05. If several reproduced correlations differ, you may want to include more components.

Since the sample size and number of variables can impact the number of factors generated in the analysis as well as the assessment of these four criteria, we recommend utilizing all four. Another reason to examine all four criteria is that within an exploratory factor analysis, the eigenvalue is the default criteria for determining the number of factors, which can lead to an inaccurate number of factors retained. For example, if an analysis determines that only two components have eigenvalues greater than 1, the model generated will include only those two components. However, the researcher may examine the other three criteria and determine that one more component should be included. In such an instance, the analysis would have to be conducted again to override the eigenvalue criteria so that three components instead of two would be generated.

Once criteria have been evaluated and you have determined the appropriate number of components to retain (the reader should note that this decision may lead to further analysis in order to include the appropriate number of components), the nature of each component must be assessed in order to interpret/name it. This is done by noting positive and negative loadings, ordering variables with respect to loading strength, and examining the content of variables that composes each component.

Although this interpretation process has been somewhat applied to our initial example, we will describe this process more in-depth in conjunction with the output. Our example seeks to determine what, if any, underlying structure exists for measures on the following ten variables: male life expectancy (*lifexpm*), female life expectancy (*lifexpf*), births per 1,000 people (*birthrat*), infant mortality rate (*infmr*), fertility rate per woman (*fertrate*), natural log of doctors per 10,000 people (*lndocs*), natural log of radios per 100 people (*lnradio*), natural log of telephones per 100 people (*lnphone*), natural log of gross domestic product (*lngdp*), and natural log of hospital beds per 10,000 people (*lnbeds*). Data were first screened for missing data and outliers. No outliers were found when utilizing Mahalanobis distance. Univariate linearity and normality were analyzed by creating a scatterplot matrix (see Figure 9.5). The elliptical shapes indicate normality and linearity. The reader should note that the following variables were previously transformed variables by taking the natural log: number of doctors per 10,000 people, number of radios per 100 people, number of telephones per 100 people, gross domestic product, and number of hospital beds per 10,000 people. A factor analysis was then conducted using **Data Reduction**, which utilized the eigenvalue criteria and varimax rotation. Applying the four methods of interpretation, we first examined the eigenvalues in the table of total variance (see Figure 9.6). Only one component had an eigenvalue greater than 1; however, the eigenvalue criteria is only reliable when the number of variables is less than 30 and communalities are greater than .7. Figure 9.7 presents the communalities and indicates that two variables have values less than .7. Consequently, the application of the eigenvalue criteria is questionable. The next criteria to assess is variance, also displayed in Figure 9.6. The first component accounts for nearly 82% of the total variance in the original variables, whereas the second component accounts for only 5.9%. The reader should note that since only one component was retained, the factor solution was not rotated. The scree plot was then assessed and indicates that the eigenvalues after the first component drop off drastically (see Figure 9.2). These last two methods imply that only the first component should be retained. However, evaluation of residuals (differences between the reproduced correlations and the original correlations) indicate that two components should be investigated. Figure 9.8 presents the reproduced correlations as well as the residuals. Of the 45 residuals, 18 (40%) exceed the .05 criteria. Since two of the four criteria are in question, we will investigate retaining two components to improve model fit.

Figure 9.5 Scatterplot Matrix of Variables (Example number 1).

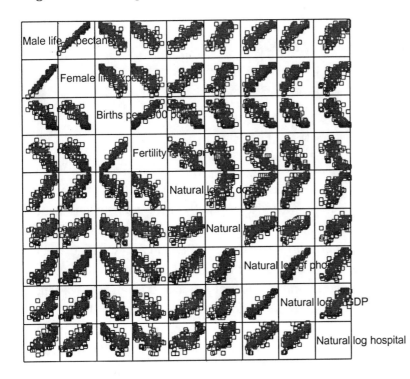

Figure 9.6 Table of Total Variance for One Component Solution (Example number 1).

Total Variance Explained

Component	Initial Eigenvalues			Extraction Sums of Squared Loadings		
	Total	% of Variance	Cumulative %	Total	% of Variance	Cumulative %
1	8.161	81.608	81.608	8.161	81.608	81.608
2	.590	5.902	87.510			
3	.372	3.717	91.227			
4	.338	3.384	94.611			
5	.246	2.460	97.071			
6	.168	1.677	98.748			
7	5.491E-02	.549	99.297			
8	3.319E-02	.332	99.629			
9	2.978E-02	.298	99.927			
10	7.316E-03	7.316E-02	100.000			

Extraction Method: Principal Component Analysis.

Figure 9.7 Communalities for One-Component Solution (Example number 1).

Communalities

	Initial	Extraction
LIFEEXPM	1.000	.894
LIFEEXPF	1.000	.936
BIRTHRAT	1.000	.846
INFMR	1.000	.894
FERTRATE	1.000	.779
LNDOCS	1.000	.830
LNRADIO	1.000	.573
LNPHONE	1.000	.899
LNGDP	1.000	.839
LNBEDS	1.000	.670

Eigenvalue criteria is questionable since two communalities are less than .7.

Extraction Method: Principal Component Analysis.

A factor analysis was conducted again, this time eliminating the eigenvalue criteria and indicating that two factors should be retained. Varimax rotation was also applied. Since we have eliminated the eigenvalue criteria and found the scree plot and variance criteria both to indicate retaining only one component, we will immediately move to the fourth criteria—assessment of residuals (see Figure 9.9). This time only 11 residuals were greater than .05; consequently, the model has been improved. Because two components were retained, the model was rotated to improve model fit. The table of total variance displays the amount of variance for each component before and after rotation (see Figure 9.10). One should note that factor rotation does not effect the total variance accounted for by the model but does change how the variance is distributed among the retained components. Prior to rotation, the first component accounted for 81.61% and the second for only 5.9%. However, once rotated, the first component accounted for 53.54% and the second for 33.97%. Figure 9.11 displays how variables were loaded into the components after rotation. Assessment of component loadings is necessary to name each component. Component number 1 was composed of both negative and positive loadings, which somewhat complicates matters. Positive loadings included the variables of male life expectancy, female life expectancy, and the number of doctors. Negative loadings included fertility rate, birth rate, and infant mortality. One should also note the variables with the highest loadings were fertility rate (-.879) followed by birth rate (-.859).   Since these variables all seemed to relate to the health of one's life, this component will be named *Healthy Lifespan*. Component number 2 included all positive loadings and addressed the number of radios, number of hospital beds, gross domestic product, and the number of phones, respectively. This component will be labeled *Economic Stature*.

Figure 9.8 Reproduced Correlations and Residuals for One-Component Solution (Example number 1).

Reproduced Correlations

		LIFEEXPM	LIFEEXPF	BIRTHRAT	INFMR	FERTRATE	LNDOCS	LNRADIO	LNPHONE	LNGDP	LNBEDS
Reproduced Correlation	LIFEEXPM	.894[b]	.915	-.870	-.894	-.834	.861	.716	.896	.866	.774
	LIFEEXPF	.915	.936[b]	-.890	-.915	-.854	.881	.732	.917	.886	.792
	BIRTHRAT	-.870	-.890	.846[b]	.870	.812	-.838	-.696	-.872	-.843	-.753
	INFMR	-.894	-.915	.870	.894[b]	.835	-.862	-.716	-.897	-.866	-.774
	FERTRATE	-.834	-.854	.812	.835	.779[b]	-.804	-.668	-.837	-.809	-.723
	LNDOCS	.861	.881	-.838	-.862	-.804	.830[b]	.690	.864	.835	.746
	LNRADIO	.716	.732	-.696	-.716	-.668	.690	.573[b]	.718	.694	.620
	LNPHONE	.896	.917	-.872	-.897	-.837	.864	.718	.899[b]	.869	.776
	LNGDP	.866	.886	-.843	-.866	-.809	.835	.694	.869	.839[b]	.750
	LNBEDS	.774	.792	-.753	-.774	-.723	.746	.620	.776	.750	.670[b]
Residual[a]	LIFEEXPM		7.300E-02	3.658E-02	-6.64E-02	2.648E-02	1.548E-02	-7.40E-02	-1.949E-02	-4.82E-02	-8.83E-02
	LIFEEXPF	7.300E-02		3.122E-02	-5.68E-02	2.157E-02	-3.42E-03	-4.29E-02	-1.958E-02	-4.45E-02	-5.12E-02
	BIRTHRAT	3.658E-02	3.122E-02		-1.67E-02	.146	2.105E-02	7.335E-02	5.575E-02	3.509E-02	3.658E-02
	INFMR	-6.638E-02	-5.68E-02	-1.668E-02		-3.83E-03	2.955E-02	6.251E-02	3.408E-02	5.469E-02	5.166E-02
	FERTRATE	2.648E-02	2.157E-02	.146	-3.83E-03		3.657E-02	9.574E-02	6.335E-02	6.850E-02	6.203E-02
	LNDOCS	1.548E-02	-3.42E-03	2.105E-02	2.955E-02	3.657E-02		-6.38E-02	1.062E-03	-9.95E-03	-3.69E-02
	LNRADIO	-7.400E-02	-4.29E-02	7.335E-02	6.251E-02	9.574E-02	-6.38E-02		2.086E-02	1.551E-02	2.879E-02
	LNPHONE	-1.949E-02	-1.96E-02	5.575E-02	3.408E-02	6.335E-02	1.062E-03	2.086E-02		7.333E-02	-3.53E-03
	LNGDP	-4.818E-02	-4.45E-02	3.509E-02	5.469E-02	6.850E-02	-9.95E-03	1.551E-02	7.333E-02		1.684E-02
	LNBEDS	-8.831E-02	-5.12E-02	3.658E-02	5.166E-02	6.203E-02	-3.69E-02	2.879E-02	-3.529E-03	1.684E-02	

Extraction Method: Principal Component Analysis.

a. Residuals are computed between observed and reproduced correlations. There are 18 (40.0%) nonredundant residuals with absolute values > 0.05.

b. Reproduced communalities

Figure 9.9 Reproduced Correlations and Residuals for Two-Component Solution (Example number 1).

Reproduced Correlations

		LIFEEXPM	LIFEEXPF	BIRTHRAT	INFMR	FERTRATE	LNDOCS	LNRADIO	LNPHONE	LNGDP	LNBEDS
Reproduced Correlation	LIFEEXPM	.919[b]	.931	-.903	-.917	-.881	.872	.632	.876	.840	.730
	LIFEEXPF	.931	.946[b]	-.911	-.930	-.884	.888	.679	.904	.870	.764
	BIRTHRAT	-.903	-.911	.890[b]	.900	.873	-.852	-.585	-.845	-.808	-.694
	INFMR	-.917	-.930	.900	.915[b]	.877	-.871	-.639	-.878	-.843	-.734
	FERTRATE	-.881	-.884	.873	.877	.865[b]	-.824	-.513	-.800	-.761	-.641
	LNDOCS	.872	.888	-.852	-.871	-.824	.834[b]	.654	.855	.824	.727
	LNRADIO	.632	.679	-.585	-.639	-.513	.654	.851[b]	.785	.780	.767
	LNPHONE	.876	.904	-.845	-.878	-.800	.855	.785	.915[b]	.890	.812
	LNGDP	.840	.870	-.808	-.843	-.761	.824	.780	.890	.866[b]	.796
	LNBEDS	.730	.764	-.694	-.734	-.641	.727	.767	.812	.796	.748[b]
Residual[a]	LIFEEXPM		5.692E-02	6.987E-02	-4.34E-02	7.294E-02	4.886E-03	9.570E-03	6.809E-04	-2.22E-02	-4.42E-02
	LIFEEXPF	5.692E-02		5.256E-02	-4.20E-02	5.134E-02	-1.02E-02	1.060E-02	-6.652E-03	-2.79E-02	-2.29E-02
	BIRTHRAT	6.987E-02	5.256E-02		-4.72E-02	8.397E-02	3.510E-02	-3.75E-02	2.898E-02	6.287E-04	-2.19E-02
	INFMR	-4.337E-02	-4.20E-02	-4.720E-02		-4.64E-02	3.926E-02	-1.41E-02	1.558E-02	3.088E-02	1.122E-02
	FERTRATE	7.294E-02	5.134E-02	8.397E-02	-4.64E-02		5.618E-02	-5.90E-02	2.600E-02	2.041E-02	-1.97E-02
	LNDOCS	4.886E-03	-1.02E-02	3.510E-02	3.926E-02	5.618E-02		-2.85E-02	9.578E-03	1.010E-03	-1.82E-02
	LNRADIO	9.570E-03	1.060E-02	-3.754E-02	-1.41E-02	-5.90E-02	-2.85E-02		-4.632E-02	-7.10E-02	-.118
	LNPHONE	6.809E-04	-6.65E-03	2.898E-02	1.558E-02	2.600E-02	9.578E-03	-4.63E-02		5.246E-02	-3.90E-02
	LNGDP	-2.221E-02	-2.79E-02	6.287E-04	3.088E-02	2.041E-02	1.010E-03	-7.10E-02	5.246E-02		-2.88E-02
	LNBEDS	-4.420E-02	-2.29E-02	-2.195E-02	1.122E-02	-1.97E-02	-1.82E-02	-.118	-3.899E-02	-2.88E-02	

Extraction Method: Principal Component Analysis.

a. Residuals are computed between observed and reproduced correlations. There are 11 (24.0%) nonredundant residuals with absolute values > 0.05.

b. Reproduced communalities ·

Figure 9.10 Table of Total Variance for Two-Component Solution (Example number 1).

Total Variance Explained

Component	Initial Eigenvalues			Extraction Sums of Squared Loadings			Rotation Sums of Squared Loadings		
	Total	% of Variance	Cumulative %	Total	% of Variance	Cumulative %	Total	% of Variance	Cumulative %
1	8.161	81.608	81.608	8.161	81.608	81.608	5.354	53.542	53.542
2	.590	5.902	87.510	.590	5.902	87.510	3.397	33.968	87.510
3	.372	3.717	91.227						
4	.338	3.384	94.611						
5	.246	2.460	97.071						
6	.168	1.677	98.748						
7	5.491E-02	.549	99.297						
8	3.319E-02	.332	99.629						
9	2.978E-02	.298	99.927						
10	7.316E-03	7.316E-02	100.000						

Extraction Method: Principal Component Analysis.

Figure 9.11 Factor Loadings for Rotated Components (Example number 1).

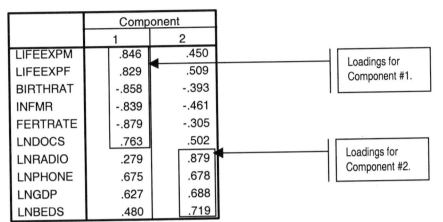

Rotated Component Matrix[a]

	Component 1	Component 2
LIFEEXPM	.846	.450
LIFEEXPF	.829	.509
BIRTHRAT	-.858	-.393
INFMR	-.839	-.461
FERTRATE	-.879	-.305
LNDOCS	.763	.502
LNRADIO	.279	.879
LNPHONE	.675	.678
LNGDP	.627	.688
LNBEDS	.480	.719

Loadings for Component #1.

Loadings for Component #2.

Extraction Method: Principal Component Analysis.
Rotation Method: Varimax with Kaiser Normalization.

a. Rotation converged in 3 iterations.

Writing Up Results

Once again, the results narrative should always describe the elimination of subjects and transformation of variables. The summary should then describe the type of factor analysis conducted and indicate if any rotation method was utilized. The interpretation process is then presented in conjunction with the results. In general, it is only necessary to indicate the number of factors retained and the criteria that led to that decision. The results of the final solution are then presented. One should summarize each component by presenting the following: the percent of variance, the number and names of variables loaded into the component, and the component loadings. This is often displayed in table format, depending upon the number of components and variables. Finally, the researcher indicates the names of components. The following summary applies the output presented in Figures 9.5 – 9.11.

Factor analysis was conducted to determine what, if any, underlying structure exists for measures on the following ten variables: male life expectancy (*lifexpm*), female life expectancy (*lifexpf*), births per 1,000 people (*birthrat*), infant mortality rate (*infmr*), fertility rate per woman (*fertrate*), doctors per 10,000 people (*docs*), radios per 100 people (*radio*), telephones per 100 people (*phone*), gross domestic product (*gdp*), and hospital beds per 10,000 people (*beds*). Prior to the analysis, evaluation of linearity and normality led to the natural log transformation of *docs, radios, phones, gdp,* and *beds*. Principal components analysis was conducted utilizing a varimax rotation. The initial analysis retained only one component. Four criteria were used to determine the appropriate number of components to retain: eigenvalue, variance, scree plot, and residuals. Criteria indicated that retaining two components should be investigated. Thus, principal components analysis was conducted to retain two components and apply the varimax rotation. Inclusion of two components increased the model fit as it decreased the number of residuals exceeding the .05 criteria.

After rotation, the first component accounted for 53.54% and the second for 33.97%. Component number 1 included items with both negative and positive loadings. Positive loadings include the variables of male life expectancy, female life expectancy, and the number of doctors. Negative loadings include fertility rate, birth rate, and infant mortality. Items with the highest loadings were fertility rate and birth rate. Component number 1 was named *Healthy Lifespan*. Component number 2 included the number of radios, number of hospital beds, gross domestic product, and the number of phones, respectively. This component was labeled *Economic Stature*. (See Table 1.)

Table 1 Component Loadings

	Loading
Component 1: Healthy Lifespan	
Fertility rate	−.879
Birth rate per 1,000 people	−.858
Male life expectancy	.846
Infant mortality	−.839
Female life expectancy	.829
Number of doctors per 10,000	.763
Component 2: Economic Stature	
Number of radios per 100 people	.879
Number of hospital beds per 10,000 people	.719
Gross domestic product	.688
Number of phones per 100 people	.678

SECTION 9.4 SAMPLE STUDY AND ANALYSIS

This section provides a complete example of the process of factor analysis. This process includes the development of research questions, data screening, test methods, interpretation of output, and presentation of results. The example utilizes the data set *schools.sav* from the SPSS Web site.

Problem

We are interested in determining what, if any, underlying structures exist for measures on the following twelve variables: % graduating in 1993 (*grad93*); % graduating in 1994 (*grad94*); average ACT score in 1993 (*act93*); average ACT score in 1994 (*act94*); 10th grade average score in 1993 for math (*math93*), reading (*read93*), science (*scienc93*); % meeting or exceeding state standards in 1994 for math (*math94me*), reading (*read94me*), science (*sci94me*); and % limited English proficiency in 1993 (*lep93*) and 1994 (*lep94*).

Method, Output, and Interpretation

Since factor analysis requires a great deal of interpretation throughout the process of analysis, we have combined the discussion of methods, output, and interpretation in this section.

Figure 9.12 Outliers for Mahalanobis Distance (Example number 2).

Extreme Values

			Case Number	Value
MAH_1	Highest	1	37	40.30603
		2	64	38.07910
		3	46	23.28122
		4	6	23.09387
		5	39	22.68936
	Lowest	1	16	2.35371
		2	27	3.40542
		3	19	3.47370
		4	61	3.95236
		5	22	4.13483

Outliers exceeding the $\chi2(12)=32.909$ at p=.001.

Figure 9.13 Scatterplot Matrix (Example number 2).

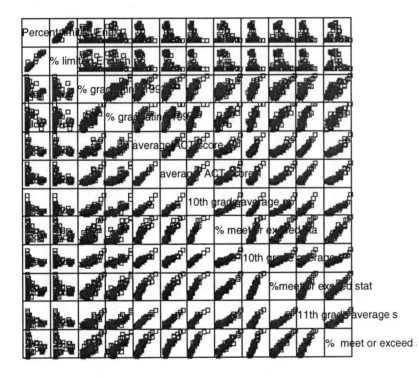

Data were evaluated to screen for outliers and assess normality and linearity. Using Mahalanobis distance, two outliers (cases number 37 and number 64) were found and eliminated (see Figure 9.12). A scatterplot matrix reveals fairly normal distributions and linear relationship among variables (see Figure 9.13). A factor analysis was then conducted using **Data Reduction**, which utilized the eigenvalue criteria and varimax rotation. Applying the four methods of interpretation, we first examine the eigenvalues in the table of total variance (see Figure 9.14). Two components were retained since they have eigenvalues greater than 1. In this example, the application of the eigenvalue criteria seems appropriate since the number of variables is less than 30 and all communalities are greater than .7 (see Figure 9.15). Evaluation of variance is done by referring back to Figure 9.14. After rotation, the first component accounts for 74.73% of the total variance in the original variables, while the second component accounts for 17.01%. The scree plot was then assessed and indicates that the eigenvalues after three components levels off (see Figure 9.16). Evaluation of residuals indicate that only three residuals are greater than .05 (see Figure 9.17). Although the scree plot suggests that the inclusion of the third component may improve the model, the residuals reveal that any model improvement would be minimal. Consequently, two components were retained.

Figure 9.14 Table of Total Variance for Two-Component Solution (Example number 2).

Total Variance Explained

Component	Initial Eigenvalues Total	% of Variance	Cumulative %	Extraction Sums of Squared Loadings Total	% of Variance	Cumulative %	Rotation Sums of Squared Loadings Total	% of Variance	Cumulative %
1	8.969	74.745	74.745	8.969	74.745	74.745	8.968	74.733	74.733
2	2.040	17.000	91.746	2.040	17.000	91.746	2.042	17.013	91.746
3	.407	3.390	95.136						
4	.205	1.705	96.841						
5	.136	1.132	97.973						
6	7.253E-02	.604	98.577						
7	5.635E-02	.470	99.047						
8	4.654E-02	.388	99.435						
9	2.780E-02	.232	99.667						
10	2.202E-02	.183	99.850						
11	1.197E-02	9.977E-02	99.950						
12	6.027E-03	5.022E-02	100.000						

Extraction Method: Principal Component Analysis.

Figure 9.15 Communalities (Example number 2).

Communalities

	Initial	Extraction
GRAD93	1.000	.701
GRAD94	1.000	.800
ACT94	1.000	.905
ACT93	1.000	.936
MATH93	1.000	.964
MATH94ME	1.000	.944
READ93	1.000	.932
READ94ME	1.000	.950
SCIENC93	1.000	.948
SCI94ME	1.000	.945
LEP93	1.000	.993
LEP94	1.000	.991

Eigenvalue criteria is acceptable since all communalities are greater than .7 and less than 30 factors are analyzed.

Extraction Method: Principal Component Analysis.

Figure 9.16   Scree Plot (Example number 2).

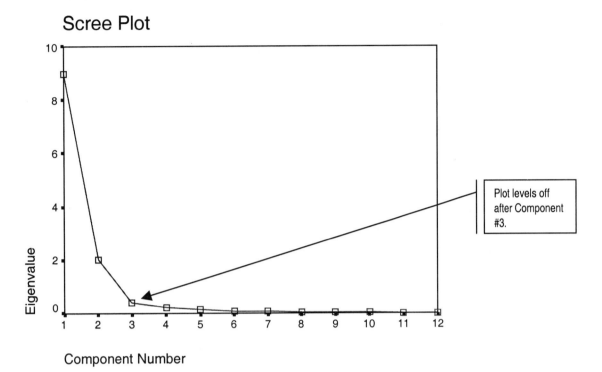

Figure 9.17   Reproduced Correlations and Residuals (Example number 2).

Reproduced Correlations

		GRAD93	GRAD94	ACT94	ACT93	MATH93	MATH94ME	READ93	READ94ME	SCIENC93	SCI94ME	LEP93	LEP94
Reproduced Correlation	GRAD93	.701[b]	.742	.783	.804	.806	.780	.796	.792	.794	.778	-.182	-.172
	GRAD94	.742	.800[b]	.850	.865	.876	.859	.863	.867	.867	.858	-7.54E-02	-6.45E-02
	ACT94	.783	.850	.905[b]	.919	.934	.919	.918	.926	.925	.919	-3.18E-02	-2.02E-02
	ACT93	.804	.865	.919	.936[b]	.947	.927	.933	.936	.937	.926	-9.69E-02	-8.51E-02
	MATH93	.806	.876	.934	.947	.964[b]	.950	.947	.956	.955	.950	-2.15E-02	-9.60E-03
	MATH94ME	.780	.859	.919	.927	.950	.944[b]	.932	.946	.944	.944	6.759E-02	7.930E-02
	READ93	.796	.863	.918	.933	.947	.932	.932[b]	.939	.939	.931	-4.20E-02	-3.03E-02
	READ94ME	.792	.867	.926	.936	.956	.946	.939	.950[b]	.949	.946	2.449E-02	3.630E-02
	SCIENC93	.794	.867	.925	.937	.955	.944	.939	.949	.948[b]	.944	6.128E-03	1.794E-02
	SCI94ME	.778	.858	.919	.926	.950	.944	.931	.946	.944	.945[b]	7.579E-02	8.750E-02
	LEP93	-.182	-7.54E-02	-3.18E-02	-9.69E-02	-2.15E-02	6.759E-02	-4.20E-02	2.449E-02	6.128E-03	7.579E-02	.993[b]	.992
	LEP94	-.172	-6.45E-02	-2.02E-02	-8.51E-02	-9.60E-03	7.930E-02	-3.03E-02	3.630E-02	1.794E-02	8.750E-02	.992	.991[b]
Residual[a]	GRAD93		6.407E-02	-5.45E-02	-2.35E-02	-3.25E-02	-4.69E-02	-1.52E-02	-2.87E-02	-6.046E-02	-5.00E-02	2.880E-02	3.362E-02
	GRAD94	6.407E-02		-6.92E-02	-3.97E-02	-2.27E-02	-2.88E-03	-2.41E-02	-9.59E-03	-4.452E-02	-2.70E-02	3.885E-03	1.825E-02
	ACT94	-5.45E-02	-6.92E-02		4.827E-02	-1.02E-02	-1.00E-02	-2.34E-02	-1.58E-02	3.731E-02	-9.30E-03	2.403E-03	-9.56E-03
	ACT93	-2.35E-02	-3.97E-02	4.827E-02		-3.47E-03	-1.09E-02	-2.18E-02	-2.20E-02	1.505E-02	-1.11E-02	3.867E-03	-2.40E-03
	MATH93	-3.25E-02	-2.27E-02	-1.02E-02	-3.47E-03		1.139E-02	2.071E-02	-3.84E-03	-7.250E-03	4.158E-03	-5.01E-03	-2.63E-03
	MATH94ME	-4.69E-02	-2.88E-03	-1.00E-02	-1.09E-02	1.139E-02		-1.83E-02	1.268E-02	-6.073E-03	7.046E-03	-6.21E-03	-9.15E-03
	READ93	-1.52E-02	-2.41E-02	-2.34E-02	-2.18E-02	2.071E-02	-1.83E-02		6.196E-03	-5.925E-03	3.737E-03	-5.08E-03	1.387E-03
	READ94ME	-2.87E-02	-9.59E-03	-1.58E-02	-2.20E-02	-3.84E-03	1.268E-02	6.196E-03		3.071E-04	5.325E-03	-2.94E-03	-8.03E-03
	SCIENC93	-6.05E-02	-4.45E-02	3.731E-02	1.505E-02	-7.25E-03	-6.07E-03	-5.92E-03	3.071E-04		6.884E-03	-4.41E-03	-8.25E-03
	SCI94ME	-5.00E-02	-2.70E-02	-9.30E-03	-1.11E-02	4.158E-03	7.046E-03	3.737E-03	5.325E-03	6.884E-03		-1.03E-02	-6.67E-03
	LEP93	2.880E-02	3.885E-03	2.403E-03	3.867E-03	-5.01E-03	-6.21E-03	-5.08E-03	-2.94E-03	-4.412E-03	-1.03E-02		-1.03E-04
	LEP94	3.362E-02	1.825E-02	-9.56E-03	-2.40E-03	-2.63E-03	-9.15E-03	1.387E-03	-8.03E-03	-8.250E-03	-6.67E-03	-1.03E-04	

Extraction Method: Principal Component Analysis.

a. Residuals are computed between observed and reproduced correlations. There are 5 (7.0%) nonredundant residuals with absolute values > 0.05.

b. Reproduced communalities

The next step was to interpret each component. Figure 9.18 presents the factor loadings for the rotated components. Component number 1 consisted of ten of the twelve variables: *scienc93, read94me, math 93, sci94me, read93, math94me, act94, grad94,* and *grad93*. These variables had positive loadings and addressed *Academic Achievement*. The second component included the remaining two variables of percent of limited English proficiency in 1994 (*lep94*) and 1993 (*lep93*). Both variables had positive loadings. Component number 2 was named *Limited English Proficiency*.

Figure 9.18   Factor Loadings for Rotated Components (Example number 2).

Rotated Component Matrix[a]

	Component 1	Component 2
MATH93	.982	-4.77E-03
READ94ME	.974	4.129E-02
SCIENC93	.973	2.284E-02
MATH94ME	.968	8.444E-02
SCI94ME	.968	9.268E-02
READ93	.965	-2.56E-02
ACT93	.964	-8.07E-02
ACT94	.951	-1.56E-02
GRAD94	.892	-6.04E-02
GRAD93	.820	-.169
LEP93	-1.71E-02	.996
LEP94	-4.94E-03	.996

Loadings for Component #1.

Loadings for Component #2.

Extraction Method: Principal Component Analysis.
Rotation Method: Varimax with Kaiser Normalization.
a. Rotation converged in 3 iterations.

Presentation of Results

The following summary applies the output from Figures 9.12 – 9.18.

Factor analysis was conducted to determine what, if any, underlying structures exist for measures on the following twelve variables: % graduating in 1993 (*grad93*); % graduating in 1994 (*grad94*); average ACT score in 1993 (*act93*); average ACT score in 1994 (*act94*); 10th grade average score in 1993 for math (*math93*), reading (*read93*), science (*scienc93*); % meeting or exceeding state standards in 1994 for math (*math94me*), reading (*read94me*), science (*sci94me*); and % limited English proficiency in 1993 (*lep93*) and 1994 (*lep94*). Prior to analysis, two outliers were eliminated. Principal components analysis was conducted utilizing a varimax rotation. The analysis produced a two-component solution, which was evaluated with the following criteria: eigenvalue, variance, scree plot, and residuals. Criteria indicated a two-component solution was appropriate.

After rotation, the first component accounted for 74.73% of the total variance in the original variables, while the second component accounted for 17.01%. Table 1 presents the loadings for each component. Component number 1 consisted of ten of the twelve variables: *math93, read94me, scienc93, math94me, sci94me, read93, act93, act94, grad94,* and *grad93*. These variables had positive loadings and addressed *Academic Achievement*. The second component included the remaining two variables of % of limited English proficiency in 1994 (*lep94*) and 1993 (*lep93*). Both variables had positive loadings. Component number 2 was labeled *Limited English Proficiency*. (See Table 1.)

Table 1 Component Loadings

	Loading
Component 1: Academic Achievement	
10th grade average math score (1993)	.982
% meeting/exceeding reading standards (1994)	.974
10th grade average science score	.973
% meeting/exceeding math standards (1994)	.968
% meeting/exceeding science standards (1994)	.968
10th grade average reading score (1993)	.965
Average ACT score (1993)	.964
Average ACT score (1994)	.951
% graduating (1994)	.892
% graduating (1993)	.820
Component 2: Limited English Proficiency	
% Limited English Proficiency (1994)	.996
% Limited English Proficiency (1993)	.996

SECTION 9.5 SPSS "HOW TO"

This section presents the steps for conducting factor analysis using **Data Reduction**. To open the Factor Analysis Dialogue Box, select the following:

```
Analyze
      Data Reduction
            Factor
```

Factor Analysis Dialogue Box (Figure 9.19)

Select each variable to be included in the analysis and move to the Variables box. Then click **Descriptives**.

Factor Analysis: Descriptive Dialogue Box (Figure 9.20)

Several descriptive statistics are provided in this Dialogue Box. Under Statistics, two options are provided: **Univariate Descriptives** and **Initial Solution**. **Univariate Descriptives** presents the means and standard deviations for each variable analyzed. **Initial Solution** is selected by default and will present the initial communalities, eigenvalues, and percent accounted for by each factor. For our example, we selected only **Initial Solution**. Under Correlation Matrix, the following options are frequently used:

Coefficients—Presents original correlation coefficients of variables.

Significance Levels—Indicates *p* values of each correlation coefficient.
KMO and Bartlett's Test of Sphericity—Tests both multivariate normality and sampling adequacy.
Reproduced—Presents reproduced correlation coefficients and residuals (difference between original and reproduced coefficients).

Our example only utilized the **Reproduced** option. After selecting the descriptive options, click **Continue**. Click **Extraction**.

Figure 9.19 Factor Analysis Dialogue Box.

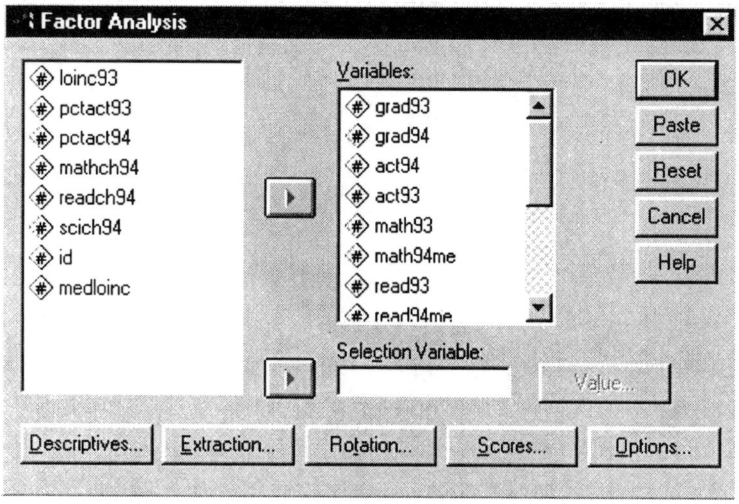

Figure 9.20 Factor Analysis: Descriptive Dialogue Box.

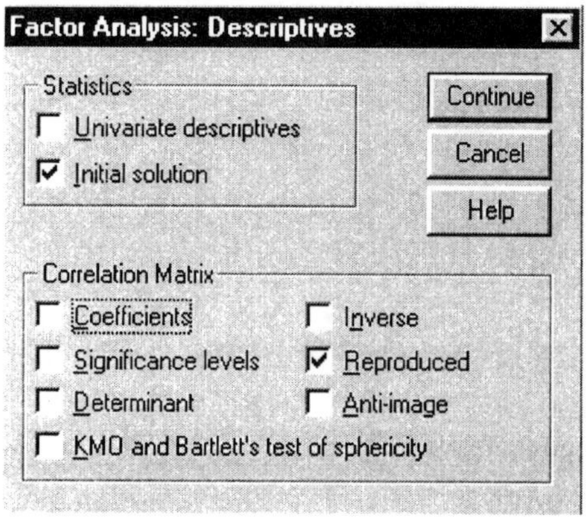

Factor Analysis: Extraction Dialogue Box (Figure 9.21)

For method, select **Principal Components**. Under Analyze, check **Correlation Matrix**. Under Display, select **Scree Plot**. Under Extract, the eigenvalue criteria of 1 is the default. Utilize the default, unless a previous analysis indicates that more components should be retained, at which time you would indicate the **Number of Factors**. Next, click **Continue**. Click **Rotation**.

Figure 9.21 Factor Analysis: Extraction Dialogue Box.

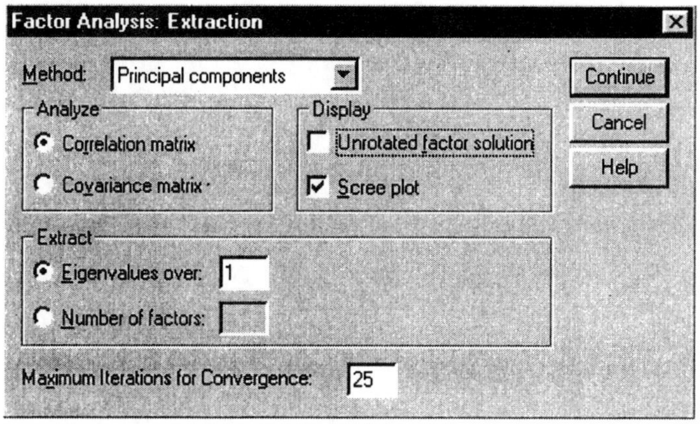

Factor Analysis: Rotation Dialogue Box (Figure 9.22)

Select the rotation method you prefer. Rotation methods available are described as follows:

Varimax—Orthogonal method that minimizes factor complexity by maximizing variance for each factor.

Direct Oblimin—Oblique method that simplifies factors by minimizing cross products of loadings.

Quartimax—Orthogonal method that minimizes factor complexity by maximizing variance loadings on each variable.

Equamax— Orthogonal method that combines both Varimax and Quartimax procedures.

Promax— Oblique method that rotates orthogonal factors to oblique positions.

Our example utilized **Varimax**. If a rotation method is indicated, check **Rotated Solution** under Display. Click **Continue**, then **Scores**.

Factor Analysis: Factor Scores Dialogue Box (Figure 9.23)

If you will be using the generated factors in future analyses, you will need to save factor scores. To do so, check **Save as Variables** and utilize the default method of **Regression**. Click **Continue**, then **Options**.

Factor Analysis: Options Dialogue Box (Figure 9.24)

Under Coefficient Display Format, check **Sorted by Size**. This will help in reading the Component Matrix. Click **Continue**. Click **OK**.

Figure 9.22 Factor Analysis: Rotation Dialogue Box.

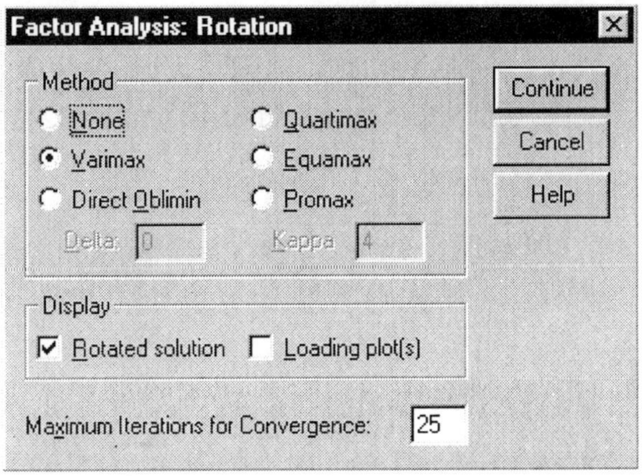

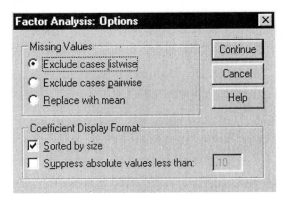

Figure 9.23 Factor Analysis: Factor Scores Dialogue Box.

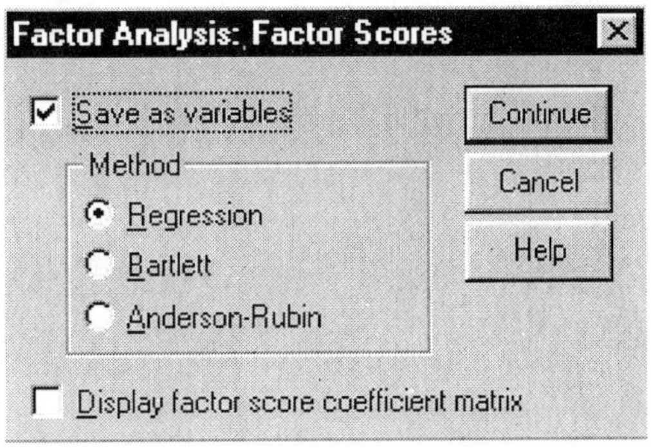

Figure 9.24 Factor Analysis: Options Dialogue Box.

Summary

Factor analysis is a technique used to identify factors that explain common variance among variables. This statistical method is often used to reduce data by grouping variables that measure a common construct. Principal components analysis is one of the most commonly used methods of extraction since this method will evaluate all sources of variability for each variable. Factors or components can also be rotated to make the components more interpretable. Orthogonal rotation methods (i.e., varimax, quartimax, equamax) result in uncorrelated factors and are the most frequently used methods. Oblique rotation methods (i.e., oblimin, promax, orthoblique) result in factors being correlated with each other.

Since principal components analysis is typically exploratory, the researcher must determine the appropriate number of components to retain. Four criteria are used in this decision-making process: 1) Eigenvalue—Components with eigenvalues greater than 1 should be retained. This criteria is fairly reliable when: the number of variables is < 30 and communalities are > .70, or the number of individuals is > 250 and the mean communality is ≥ .60; 2) Variance—Retain components that account for at least 70% total variability; 3) Scree Plot—Retain all components within the sharp descent, before eigenvalues level off. This criterion is fairly reliable when the number of individuals is > 250 and communalities are > .30; 4) Residuals—Retain the components generated by the model if only a few residuals exceed .05. If several reproduced correlations differ, you may want to include more components.

Once the appropriate number of components to retain has been determined, the researcher must then interpret/name the components by evaluating the type of variables included in each factor, the strength of factor loadings, and the direction of factor loadings. Figure 9.25 provides a checklist for conducting factor analysis.

Figure 9.25   Checklist for Conducting Factor Analysis.

I.   Screen Data
   a.   Missing Data?
   b.   Multivariate Outliers?
       ❑   Run preliminary Regression to calculate Mahalanobis' Distance.
          1. 🖰 **Analyze...Regression...Linear.**
          2. Identify a variable that serves as a case number and move to Dependent Variable box.
          3. Identify all appropriate quantitative variables and move to Independent(s) box.
          4. 🖰 **Save.**
          5. Check **Mahalanobis'.**
          6. 🖰 **Continue.**
          7. 🖰 **OK.**
          8. Determine chi square χ^2 critical value at $p<.001$.
       ❑   Conduct **Explore** to test outliers for Mahalanobis chi square χ^2.
          1. 🖰 **Analyze...Descriptive Statistics...Explore**
          2. Move *mah_1* to Dependent Variable box.
          3. Leave Factor box empty.
          4. 🖰 **Statistics.**
          5. Check **Outliers.**
          6. 🖰 **Continue.**
          7. 🖰 **OK.**
       ❑   Delete outliers for subjects when χ^2 exceeds critical χ^2 at $p<.001$.
   c.   Linearity and Normality?
       ❑   Create Scatterplot Matrix of all model variables.
          ❑   Scatterplot shapes are not close to elliptical shapes→reevaluate univariate normality and consider transformations.

II.   Conduct Factor Analysis
   a.   Run Factor Analysis using **Data Reduction**.
          1. 🖰 **Analyze...🖰 Data Reduction...🖰 Factor.**
          2. Move each studied variable to the Variables box.
          3. 🖰 **Descriptives.**
          4. Check: **Initial Solution** and **Reproduced**.
          5. 🖰 **Continue.**
          6. 🖰 **Extraction.**
          7. Check **Correlation Matrix, Unrotated Factor Solution, Scree Plot,** and **Eigenvalue.**
          8. 🖰 **Continue.**
          9. 🖰 **Rotation.**
         10. Check **Varimax** and **Rotated Solution.**
         11. 🖰 **Continue.**
         12. 🖰 **Scores.**
         13. Check **Save as Variables** and **Regression.**
         14. 🖰 **Continue.**
         15. 🖰 **Options.**
         16. Check **Sorted by Size.**
         17. 🖰 **Continue.**
         18. 🖰 **OK.**

Figure 9.25 Checklist for Conducting Factor Analysis. (*Continued*)

b. Determine appropriate number of components to retain.
    1. Eigenvalue—Components with eigenvalues greater than 1 should be retained. This criteria is fairly reliable when: the number of variables is < 30 and communalities are > .70, or the number of individuals is > 250 and the mean communality is ≥ .60;
    2.   Variance—Retain components that account for at least 70% total variability;
    3. Scree Plot—Retain all components within the sharp descent, before eigenvalues level off. This criteria is fairly reliable when the number of individuals is > 250 and communalities are > .30;
    4. Residuals—Retain the components generated by the model if only a few residuals exceed .05. If several reproduced correlations differ, you may want to include more components.
c. Conduct factor analysis again if more components should be retained.
d. Interpret components.
    1. Evaluate the type of variables loaded into each component;
    2. Note the strength and direction of loadings;
    3. Label component accordingly.

III. Summarize Results
a. Describe any data elimination or transformation.
b. Describe the initial model.
c. Describe the criteria used to determine the number of components to retain.
d. Summarize the components generated by narrating: the variables loaded into each component, the strength and direction of loadings, the component labels, and the percent of variance.
e. Create a table that summarizes each component (report component loadings).
f. Draw conclusions.

Exercises for Chapter 9

The following exercises seek to determine what underlying structure exists among the following variables: highest degree earned (*degree*), hours worked per week (*hrs1*), job satisfaction (*satjob*), years of education (*educ*), hours per day watching TV (*tvhours*), general happiness (*happy*), degree to which life is exciting *(life)*, and degree to which the lot of the average person is getting worse (*anomia5*).

1. The following output was generated for the initial analysis. Varimax rotation was utilized.

education (*educ*), hours per day watching TV (*tvhours*), general happiness (*happy*), degree to which life is exciting *(life)*, and degree to which the lot of the average person is getting worse (*anomia5*).

1. The following output was generated for the initial analysis. Varimax rotation was utilized.

Communalities

	Initial	Extraction
DEGREE	1.000	.933
HRS1	1.000	.602
SATJOB	1.000	.447
EDUC	1.000	.939
TVHOURS	1.000	.556
HAPPY	1.000	.576
LIFE	1.000	.500
ANOMIA5	1.000	.317

Extraction Method: Principal Component Analysis.

Total Variance Explained

Component	Initial Eigenvalues			Extraction Sums of Squared Loadings			Rotation Sums of Squared Loadings		
	Total	% of Variance	Cumulative %	Total	% of Variance	Cumulative %	Total	% of Variance	Cumulative %
1	2.423	30.293	30.293	2.423	30.293	30.293	1.879	23.488	23.488
2	1.426	17.822	48.115	1.426	17.822	48.115	1.734	21.677	45.165
3	1.021	12.760	60.875	1.021	12.760	60.875	1.257	15.710	60.875
4	.886	11.077	71.952						
5	.796	9.955	81.907						
6	.728	9.094	91.001						
7	.607	7.589	98.590						
8	.113	1.410	100.000						

Extraction Method: Principal Component Analysis.

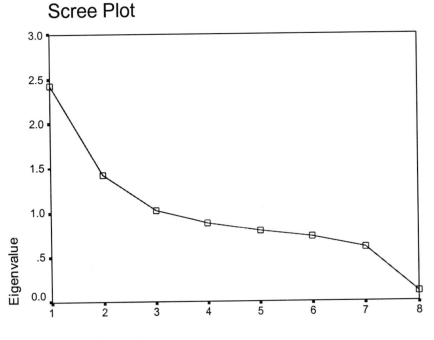

Scree Plot

Reproduced Correlations

		DEGREE	HRS1	SATJOB	EDUC	TVHOURS	HAPPY	LIFE	ANOMIA5
Reproduced Correlation	DEGREE	.933[b]	.176	-3.88E-02	.935	-.239	-.119	.230	.118
	HRS1	.176	.602[b]	-.239	.194	-.576	-7.70E-02	.141	-4.88E-02
	SATJOB	-3.88E-02	-.239	.447[b]	-6.23E-02	.214	.469	-.436	-.297
	EDUC	.935	.194	-6.23E-02	.939[b]	-.255	-.142	.252	.131
	TVHOURS	-.239	-.576	.214	-.255	.556[b]	6.629E-02	-.136	4.699E-02
	HAPPY	-.119	-7.70E-02	.469	-.142	6.629E-02	.576[b]	-.526	-.412
	LIFE	.230	.141	-.436	.252	-.136	-.526	.500[b]	.371
	ANOMIA5	.118	-4.88E-02	-.297	.131	4.699E-02	-.412	.371	.317[b]
Residual[a]	DEGREE		4.281E-03	-6.83E-02	-4.98E-02	3.196E-02	-3.87E-03	-3.42E-02	-3.74E-02
	HRS1	4.281E-03		.104	1.080E-02	.361	-3.13E-02	-4.61E-02	.112
	SATJOB	-6.83E-02	.104		-3.73E-02	-.105	-.197	.151	.158
	EDUC	-4.98E-02	1.080E-02	-3.73E-02		2.637E-02	-2.03E-03	-2.92E-02	-1.25E-02
	TVHOURS	3.196E-02	.361	-.105	2.637E-02		1.400E-02	-1.18E-02	-9.88E-02
	HAPPY	-3.87E-03	-3.13E-02	-.197	-2.03E-03	1.400E-02		.159	.177
	LIFE	-3.42E-02	-4.61E-02	.151	-2.92E-02	-1.181E-02	.159		-.217
	ANOMIA5	-3.74E-02	.112	.158	-1.25E-02	-9.880E-02	.177	-.217	

Extraction Method: Principal Component Analysis.

a. Residuals are computed between observed and reproduced correlations. There are 12 (42.0%) nonredundant residuals with absolute values > 0.05.

b. Reproduced communalities

a. Assess the eigenvalue criteria. How many components were retained? Is the eigenvalue appropriate, considering the number of factors and the communalities?

b. Assess the variance explained by the retained components. What is the total variability explained by the model? Is this amount adequate?

c. Assess the scree plot. At which component does the plot begin to level off?

d. Assess the residuals. How many residuals exceed the .05 criteria?

e. Having applied the four criteria, do you believe the number of components retained in this analysis is appropriate? If not, what is your recommendation?

2. Assume that you believe four components should be retained in the previous example. Conduct a factor analysis with varimax rotation (be sure to retain 4 components).

a. Evaluate each of the four criteria. Has the model fit improved? Explain.

b. Provide two alternatives for improving the model.

Notes:

Chapter 10

Discriminant Analysis

In Chapters 10 and 11, we give attention to the fourth, and final, group of advanced/multivariate statistical techniques. Recall from Chapter 2 that the techniques presented in these chapters have as their primary purpose the prediction of group membership—sometimes referred to as **classification**—for individuals in the sample. This is accomplished by identifying specific IVs that serve as the best predictors of membership in a particular group, as defined by the DV. In this chapter, we present discussion of discriminant analysis, which seeks to identify a combination of IVs, measured at the interval level, that best predicts membership in a particular group, as measured by a categorical DV.

SECTION 10.1 PRACTICAL VIEW

Purpose

Discriminant analysis has two basic purposes: (1) to describe major differences among groups following a MANOVA analysis, and (2) to classify subjects into groups based on a combination of measures (Stevens, 1992). Discriminant analysis was originally developed in the 1930s with the second purpose in mind—to classify subjects into one of two clearly defined groups; the technique was later expanded in order to classify subjects into any number of groups (Pedhazur, 1982). However, as of late, its main purpose has focused on the description of group differences (Tatsuoka, 1988). Both purposes are described briefly in the ensuing paragraphs.

If the discriminant analysis is being used to further describe the differences among two or more groups—often referred to as *descriptive discriminant analysis* (Stevens, 1992)—the analysis essentially involves breaking down the "between" association from a MANOVA into its additive portions. This is accomplished through the identification of uncorrelated linear combinations of the original variables. These uncorrelated linear combinations are called *discriminant functions*. Groups are subsequently differentiated along one, two, or possibly several dimensions. There exists here some carry over from factor analysis (see Chapter 9) in that the researcher is responsible for providing a meaningful name for the various discriminant functions, identified as being statistically significant.

In contrast, if the discriminant analysis is being used for purposes of prediction or classification, discriminant functions are again obtained, but their interpretation changes slightly. Instead of trying to describe dimensions on which groups differ, the goal is to determine dimensions that serve as the basis for reliably—and accurately—classifying subjects into groups. Since we have discussed several analytic techniques for investigating the existence of differences among groups, it is our intent in this chapter to focus on the use of discriminant analysis for classification purposes.

Regardless of the purpose for which it is being used, discriminant analysis is often seen as the reverse of MANOVA. You will recall that in MANOVA, the researcher takes two or more groups and compares their scores on a combination of DVs in an attempt to discover whether or not there exist significant group differences. However, in discriminant analysis, this process is reversed (Sprinthall, 2000). In MANOVA, the IVs are the grouping variables and the DVs are the predictors; whereas, in discriminant analysis, the DVs serve as the grouping variables and the IVs are the predictors (Tabachnick & Fidell, 1996). If the goal of the analysis is to describe group differences, the researcher would determine the number of dimensions (i.e., the *discriminant functions*) that maximize the differences among the groups in question. In contrast, if prediction is the goal of analysis, the researcher might use discriminant function scores in order to predict from which group subjects came. This procedure can then be used to predict membership in a particular group for "new" or future subjects from the same population.

Students oftentimes have difficulty envisioning practical uses for discriminant analysis, especially when used for classification purposes. We offer several examples from different fields of study where classifying subjects may be of interest. The following is a brief list of possible examples:

- The dean of a college of education wants to determine a process by which she can identify those students who are likely to succeed as educators and those who are not.
- Based on a comprehensive set of psychological measures, a psychologist wants to classify mental patients into one of several categories.
- A special educator wants to reliably classify exceptional students as learning disabled or mentally handicapped.
- A credit card company wants to determine a method of accurately predicting an individual's level of risk (e.g., high, moderate, low) as a potential credit customer.
- A psychiatrist wants to classify patients with respect to their level of anxiety (e.g., low and high) in social settings.

Now, let us develop our own working example to which we will refer in this chapter. Suppose that we wanted to determine a nation's status in terms of its level of development based on several measures. Specifically, those measures consist of:

- percent of the population living in urban areas (*urban*);
- gross domestic product per capita (*gdp*);
- male life expectancy (*lifexpm*);
- female life expectancy (*lifexpf*); and
- infant mortality rate per 1,000 live births (*infmr*).

Linear combinations of these five variables, which would allow us to predict a country's status as a developing nation (*develop*)—either (1) developed nation or (2) developing nation—will be determined and discussed subsequently.

The interpretation of results obtained from a discriminant analysis is fairly straightforward since they tend to parallel results that we have seen in previous analysis techniques. The main result obtained from a discriminant analysis is the summary of the discriminant functions. These functions are similar to factors or components in factor analysis—in the case when there is more than one, they represent uncorrelated linear combinations of the IVs. These combinations basically consist of regression equations—raw scores on each original variable are multiplied by assigned weights and then summed together in order to obtain a *discriminant score*, which is analogous to a factor score (discriminant functions will be discussed in greater detail in Section 10.3). The analysis returns several indices, many of which we have seen in previous chapters. An eigenvalue and percent of variance explained is provided for each discriminant function. These values are interpreted in similar fashion to their analogous counterparts in a factor or principal components analysis. A value called the *canonical correlation* is also reported. This value is equivalent to the correlation between the discriminant scores and the levels of the dependent variable. A high value for this correlation indicates a function that discriminates well between subjects; in other words, it will likely perform well in terms of classifying subjects into groups (levels of the DV). It is important for the reader to realize that when the canonical correlation is less than perfect (i.e., not equal to 1.0), some degree of error will be reflected in the assignment of individual subjects to groups (Williams, 1992). This is an important point that we will address momentarily.

Additionally, we are provided with a test of the significance of each of the discriminant functions—specifically, we are provided a value for Wilks' Lambda (Λ), similar to that which was supplied in a MANOVA output. The significance of each discriminant function is tested using a chi-square criterion—statistical significance indicates that the function discriminates well based on levels of the DV.

Another portion of the results vital to the analysis involve the actual coefficients for each discriminant function. Similar to regression coefficients in multiple regression, these coefficients serve as the weights assigned by the computer to the various original variables in the analysis. Both unstandardized and standardized coefficients are provided for interpretation by the researcher. Unstandardized coefficients are the basis for the calculation of discriminant scores, a point that will be discussed further in Section 10.3. However, since they represent various measurement scales inherent in the original variables, they cannot be used to assess the relative contributions of individual variables to the discriminant function(s) (Williams, 1992). The standardized coefficients must be used for this purpose.

These standardized coefficients, along with coefficients presented in a structure matrix, are similar to factor loadings in factor analysis. The sizes of the standardized coefficients and the function loadings indicate the degree of relationship between each variable and the discriminant function. Recall that the relative sizes and directions of these coefficients are used to attach labels to the functions for greater ease in interpretation, especially interpretation of the resultant discriminant scores. As in factor analysis, the meaning of the function is inferred from the researcher, based on the pattern of correlations, or loadings, between the function and the predictor IVs (Tabachnick & Fidell, 1996). Figure 10.1 presents the standardized and unstandardized discriminant function coefficients, as well as structure coefficients, for our working example. The reader should notice that the resulting single function is a bipolar discriminant function.

Also, one should notice that—based on the absolute values of the coefficients—the order of importance of the predictors differs when comparing the standardized and structure coefficients. *It should be noted that examination of these two indices may provide different results.* In the matrix of standardized coefficients, the most important predictor is female life expectancy (1.293), followed closely by gross domestic product (1.121), then male life expectancy (-1.117), percent urban (-.349), and infant mortality rate (-.132). However, with respect to the correlations, the most important predictor is gross domestic product (.943), followed by female life expectancy (.669), infant mortality rate (-.642), male life expectancy (.607), and finally percent urban (.462).

Figure 10.1 Unstandardized (a), Standardized (b), and Structure Coefficients (c) for "Classification as Developing Nation" Example.

(a) Unstandardized Canonical Discriminant Function Coefficients

	Function
	1
Percent urban 1992	-.016
GDP per capita	1.063
Male life expectancy 1992	-.134
Female life expectancy 1992	.142
Infant mortality rate 1992 (per 1,000 live births)	.004
(Constant)	-8.243

(b) Standardized Canonical Discriminant Function Coefficients

	Function
	1
Percent urban 1992	-.349
GDP per capita	1.121
Male life expectancy 1992	-1.117
Female life expectancy 1992	1.293
Infant mortality rate 1992 (per 1,000 live births)	.132

(c) Structure Coefficients

	Function
	1
GDP per capita	.943
Female life expectancy 1992	.669
Infant mortality rate 1992 (per 1,000 live births)	-.642
Male life expectancy 1992	.607
Percent urban 1992	.462

As previously mentioned, a nonperfect canonical correlation will result in some cases being classified incorrectly—this is, of course, almost inevitable. This situation is analogous to large residuals in multiple regression—recall that large residual values indicated that, for some individuals in the sample, the prediction equation did not accurately predict the value on the DV. Fortunately, discriminant analysis by way of statistical software programs will provide an assessment of the adequacy of classification. There are several means of accomplishing this assessment, depending on the analysis software being used by the researcher. In SPSS, the unstandardized coefficients are used to calculate the classification of cases in the original sample into DV groups. The number of correct classifications, also known as the *hit rate* (Stevens, 1992), is then compared to the actual group membership of subjects in the original sample (Williams, 1992). A table showing the *actual* group membership and *predicted* group membership is presented. This table includes the percentage of correct classifications, based on the equation resulting from the unstandardized coefficients. Figure 10.2 shows the classification results for our working example. Although there is no "rule of thumb" regarding an acceptable rate of correct classifications, one would certainly hope to achieve a high percentage. In our example, the single discriminant function resulted in nearly 93% of the cases being classified correctly.

Figure 10.2 Assessment of Adequacy of Classification Results for "Classification as Developing Nation" Example.

Classification Results[b,c]

		DEVELOP	Predicted Group Membership 0 Developed country	Predicted Group Membership 1 Developing country	Total
Original	Count	0 Developed country	24	3	27
		1 Developing country	6	87	93
	%	0 Developed country	88.9	11.1	100.0
		1 Developing country	6.5	93.5	100.0
Cross-validated [a]	Count	0 Developed country	24	3	27
		1 Developing country	7	86	93
	%	0 Developed country	88.9	11.1	100.0
		1 Developing country	7.5	92.5	100.0

[a.] Cross validation is done only for those cases in the analysis. In cross validation, each case is classified by the functions derived from all cases other than that case.

[b.] 92.5% of original grouped cases correctly classified.

[c.] 91.7% of cross-validated grouped cases correctly classified.

Stevens (1992) has suggested that assessing the accuracy of the hit rates with the *same* sample that was used to develop the discriminant function equation results in an unrealistic and mileading assessment. He suggests assessing the accuracy through an "external" classification analysis. In this manner, the data used to verify the classification are not used in the construction of the function itself. This can be accomplished in one of two ways: (1) if the sample is large enough, it may be split in half such that one half is used to construct the function and the other half is used to assess its accuracy; or (2) through the use of the *jackknife* procedure, in which each subject is classified based on a classification statistic derived from the remaining (*n* - 1) subjects. This second procedure is appropriate for small to moderate sample sizes. For more information on these two assessment procedures, the reader is referred to Stevens (1992) and Tabachnick & Fidell (1996).

Stevens (1992) has further recommended that another important consideration not be overlooked when assessing a classification procedure. Obviously, it is important that the hit rate be high—in other words, we should have mainly correct classifications. The additional consideration is the cost of misclassification—financially, morally, or ethically, etc. To the researcher, it may initially look good to have a hit rate of approximately 90%–95%, but, what about the handful of cases that are classified incorrectly? There also exist ramifications of the misclassification of an individual from Group A, for example, into Group B resulting in greater "cost" than misclassifying someone in the other direction. For example, assume we wanted to classify subjects as either low risk or high risk in terms of developing heart disease based on family history and personal habits. Obviously, labeling a person as low risk when in fact he is high risk is much more costly than identifying him as high risk when he is actually low risk. It is important to realize that these are issues that cannot be explained through statistical analyses, but must be assessed and explicated by the knowledge and intuition of the researcher.

Similar to multiple regression, there are several approaches to conducting a discriminant analysis (Tabachnick & Fidell, 1996). In *standard*, or *direct*, discriminant analysis, each predictor (IV) is entered into the equation simultaneously and is assigned only its unique association with the groups, as defined by the DV. In *sequential*, or *hierarchical*, discriminant analysis, the predictor IVs are entered into the analysis in an order specified by the researcher. The improvement in classification accuracy is assessed following the addition of each new predictor variable to the analysis. This procedure can be used to establish a priority among the predictor IVs and/or to reduce the number of possible predictors when a larger set is initially identified. Finally, in *stepwise*, or *statistical*, discriminant analysis, the order of entry of predictor variables is determined by using statistical criteria. A reduced set of predictors will be obtained by retaining only statistically significant predictors.

Sample Research Questions

Again, based on our working example, let us now proceed to the specification of a series of possible research questions for our analysis:

(1) Can status as a developing nation (i.e., develop*ed* or develop*ing*) be reliably predicted from knowledge of percent of population living in urban areas, gross domestic product, male and female life expectancy, and infant mortality rate?

(2) If developing nation status can be predicted reliably, along how many dimensions do the two groups differ? How can those dimensions be interpreted?

(3) Given the obtained classification functions, how adequate is the classification (in other words, what proportion of cases is classified correctly)?

SECTION 10.2 ASSUMPTIONS AND LIMITATIONS

Because the analytical procedures involved in discriminant analysis are so similar to those involved in MANOVA, the assumptions are basically the same. There are, of course, some "adjustments" to the assumptions, necessitated by the classification situation. The assumptions for discriminant analysis when being used for classification are as follows:

(1) The observations on the predictor variables must be randomly sampled and must be independent of one another.

(2) The sampling distribution of any linear combination of predictors is normal (multivariate normality).

(3) The population covariance matrices for the predictor variables in each group must be equal (the homogeneity of covariance matrices assumption or the assumption of homoscedasticity).

(4) The relationships among all pairs of predictors within each group must be linear.

An additional, and potentially serious, limitation of discriminant analysis is that it can be sensitive to sample size; therefore, consideration should be given to this issue. Stevens (1992) states that "unless sample size is large, relative to the number of variables, both the standardized coefficients and the correlations are very unstable" (p. 277). Further, he states that unless the ratio of total sample size to the number of variables (i.e., $\frac{N}{p}$) is quite large—approximately 20 to 1—one should use extreme caution in interpreting the results. For example, if five variables are used in a discriminant analysis, there should be a minimum of 100 subjects in order for the researcher to have confidence in interpreting the discriminant function, and in expecting to see the same function appear with another sample from the same population.

Methods of Testing Assumptions

Multivariate normality in discriminant analysis implies that the sampling distributions of the linear combinations of predictor variables are normally distributed (Tabachnick & Fidell, 1996). There exists no test of the normality of all linear combinations of sampling distributions of means of predictors; however, bivariate scatterplots are often used to examine univariate normality. It is important to note that discriminant analysis is robust to violations of multivariate normality, provided the violation is caused by skewness rather than outliers (Tabachnick & Fidell, 1996). If outliers are present, it is imperative that the researcher transform or eliminate them prior to proceeding with the discriminant analysis. Multivariate outliers can be identified using Mahalanobis distance within **Regression**. You will recall from Chapter 6 that, as a conservative suggestion, robustness is expected with samples of 20 cases in the smallest group (on the DV), even if there are five or fewer predictors. In situations where multivariate normality is subject, an alternative classification procedure exists (Stevens, 1992). This technique is called *logistic regression* and is the topic of discussion in Chapter 11.

Discriminant analysis, when used for classification, is not robust to violations of the assumption of homogeneity of variance-covariance matrices (Tabachnick & Fidell, 1996). Cases tend to be overclassified into groups with a greater amount of dispersion, resulting in greater error as evidenced by a lower hit rate. Checking for univariate normality is a good starting point for assessing possible violations of homoscedasticity. A more applicable test for homoscedasticity in a discriminant analysis involves the examination of scatterplots of scores on the first two discriminant functions produced separately for each group. Obviously, this test is only possible in solutions resulting in more than one discriminant function. Rough equality in the overall size of the scatterplots provides evidence of homogeneity of variance-covariance matrices (Tabachnick & Fidell, 1996).

Linearity is, of course, best assessed through inspection of bivariate scatterplots. If both variables in the pair of predictors for each group on the DV are normally distributed and linearly related, the shape of the scatterplot should be elliptical. If one of the variables is not normally distributed, the relationship will not be linear and the scatterplot between the two variables will not appear oval shaped. Violations of the assumption of linearity are often seen as less serious than violations of other assumptions in that it tends to lead to reduced power as opposed to an inflated Type I error rate (Tabachnick & Fidell, 1996).

SECTION 10.3 PROCESS AND LOGIC

The Logic Behind Discriminant Analysis

Discriminant analysis is a *mathematical maximization* procedure. The goal of the procedure is to find uncorrelated linear combinations of the original (predictor) variables that maximize the between-to-within association, as measured by the sum-of-squares and cross-products (SSCP) matrices (Stevens, 1992). These uncorrelated linear combinations are referred to as the *discriminant functions*. The logic behind discriminant analysis involves finding the function with the largest eigenvalue—this results in maximum discrimination among groups (Stevens, 1992). The *first discriminant function* is the linear combination that maximizes the between-to-within association and is defined by the equation:

$$DF_1 = a_{10}x_0 + a_{11}x_1 + a_{12}x_2 + a_{13}x_3 + ... + a_{1p}x_p \qquad \text{(Equation 10.1)}$$

where DF_1 is the first discriminant function, x_i refers to the measure on an original predictor variable, and a_{1i} refers to the weight (coefficient) assigned to a given variable for the first discriminant function (the first subscript following the a identifies the specific discriminant function, and the second subscript identifies the original variable)—e.g., the term $a_{11}x_1$ refers to the product of the weight for variable 1 on DF_1 and the original value for an individual on variable 1. The term $a_{10}x_0$ represents the constant within the equation. The subscript p is equal to the total number of original predictor variables. This linear combination, then, provides for the maximum separation among groups.

This equation is used to obtain a discriminant function score for each subject. Using the unstandardized coefficients as presented in Figure 10.1, we can make the appropriate substitutions and arrive at the actual equation for our data and resulting discriminant function:

$$DF_1 = (-8.243) + (-.016)x_1 + (1.063)x_2 + (-.134)x_3 + (.142)x_4 + (.004)x_5$$

where each x represents an actual score on a predictor variable (specifically, x_1 = percent urban, x_2 = gross domestic product, x_3 = male life expectancy, x_4 = female life expectancy, and x_5 = infant mortality rate).

The analytic procedure then proceeds to find the second linear combination— *uncorrelated* with the first linear combination—that serves as the next best separator of the groups. The resulting equation would be:

$$DF_2 = a_{20}x_0 + a_{21}x_1 + a_{22}x_2 + a_{23}x_3 + ... + a_{2p}x_p \qquad \text{(Equation 10.2)}$$

As in factor analysis, it is again important to note that the constructed discriminant functions are not related. In other words,

$$r_{DF_1 \bullet DF_2} = 0$$

The third discriminant function is constructed so that it is uncorrelated with the first two and so that it serves as the third best separator of the groups. This process continues until the maximum possible number of discriminant functions has been obtained. If k is the number of groups and p is the number of predictor variables, the maximum possible number of discriminant functions will be the smaller of p and $(k - 1)$ (Stevens, 1992). Thus, in a situation with 3 groups and 10 variables, the analysis would construct two discriminant functions. In any situation with only 2 groups, only one discriminant function would be constructed.

The actual classification of subjects occurs in the following manner for a situation involving two groups (additional groups simply involve more of the same calculations). Once a discriminant function has been constructed, the location *of each group* on the discriminant function ($\bar{y}_k$) is determined by multiplying the vectors of means for each subject on all predictor variables ($\bar{x}_1$) by the vector of unstandardized coefficients (a'):

$$\bar{y}_1 = a'\bar{x}_1 \qquad \text{(Equation 10.3)}$$

and

$$\bar{y}_2 = a'\bar{x}_2 \qquad \text{(Equation 10.4)}$$

The midpoint between the two groups on the discriminant function is then calculated as:

$$m = \frac{(\bar{y}_1 + \bar{y}_2)}{2} \qquad \text{(Equation 10.5)}$$

Subjects are then classified—based on their individual discriminant function score, z_i— into one of the two groups based on the following *decision rule*:

If $z_i \geq m$, then classify the subject into Group 1
If $z_i < m$, then classify the subject into Group 2

Finally, it is important to note that within statistical analysis programs, it is possible to specify the prior probabilities of being classified into a specific group. ***Prior probability*** is essentially defined as the probability of membership in group k prior to collection of the data. The default option in the programs is to assign equal *a priori* probabilities. For example, if we have two possible groups in which to classify subjects, we would assume that they have an equal (50-50) chance of being classified in either group. The researcher may decide to, based on *extensive* content knowledge, to assign different *a priori* probabilities. However, it is critical to note that this can have a substantial effect on the classification function; therefore, caution has been strongly recommended in using anything but equal prior probabilities (Stevens, 1992).

Interpretation of Results

Discriminant analysis output typically has four parts: 1) preliminary statistics that describe group differences and covariances; 2) significance tests and strength of relationship statistics for each discriminant function; 3) discriminant function coefficients; and 4) group classification. Interpretation of these four parts will be discussed subsequently.

Discriminant analysis produces a series of preliminary statistics that assist in interpreting the overall analysis results: a table of means and standard deviations for each IV by group, ANOVA analysis testing for group differences among the IVs, covariance matrices, and Box's M Test. Examination of group means and standard deviations and the ANOVA results is helpful in determining how groups differ within each IV. The ANOVA results, presented in the table of Tests of Equality of Group Means, includes Wilks' Lambda, F-test, degrees of freedom, and p values for each IV. Group differences are usually significant; if they are not, the functions generated will not be very accurate in classifying individuals. The Box's M test is also included in preliminary statistics and is an indicator of significant differences in the covariance matrices among groups. A significant F-test ($p<.001$) indicates that group covariances are not equal. Failure of the homogeneity of covariance assumption may limit the interpretation of results. However, one should keep in mind that the Box's Test is highly sensitive to non-normal distributions and therefore should be interpreted with caution.

The next section of output to interpret is the significance tests and strength of relationship statistics for each discriminant function, presented in the Eigenvalues and Wilks' Lambda Tables. The Eigenvalue Table displays the eigenvalue, percent of variance, and canonical correlation for each discriminant function. The canonical correlation represents the correlation between the discriminant function and the levels of the DV. By squaring the canonical correlation, we calculate the effect size (η^2) of the function, which indicates the percent of variability in the function explained by the different levels in the DV.

The Wilks' Lambda Table provides chi-square tests of significance for each function. Statistics displayed include Wilks' Lambda, chi-square, degrees of freedom, and level of significance. Essentially, these statistics represent the degree to which there are significant group differences in the IVs after the effects of the previous function(s) have been removed. These significance tests help in determining the number of functions to interpret. For example, if an analysis generated three functions, of which the first two functions were significant, only the first two functions would be interpreted.

Once the number of functions to interpret has been determined, each function can then be interpreted/named by examining the variables that are most related to it. Two tables are utilized for this process—Standardized Canonical Discriminant Function Coefficients and the Structure Matrix. The table of Standardized Canonical Discriminant Function Coefficients presents the standardized discriminant function coefficients, which represent the degree to which each variable contributes to each function. The Structure Matrix presents the correlation coefficients between the variables and functions.

The next step in interpreting discriminant analysis results is assessing the accuracy of the functions in classifying subjects in the appropriate groups. The table of Classification results displays the original and predicted frequency and percent of subjects within each group. The final step in this interpretation process is to determine the extent to which group differences support the functions generated. This is done by reviewing group means for each function as presented in the table of Functions at Group Centroids.

Continuing with our example that seeks to predict the status of a developing nation (*develop*) from the knowledge of percent of population living in urban areas (*urban*), gross domestic product (*gdp*), male (*lifeexpm*) and female (*lifexpf*) life expectancy, and infant mortality (*infmr*), we screened data for outliers. Utilizing Mahalanobis distance, one outlier (case number 83) was removed as it exceeded the chi-square critical value $\chi^2(5)=20.515$ at $p=.001$ (see Figure 10.3). Assessment of normality and linearity was conducted by evaluating bivariate scatterplots of the IVs. Figure 10.4 indicates that the variable of *gdp* is not normally distributed and does not linearly relate with the other variables. Consequently, the transformed (natural log) version of this variable was utilized (*lngdp*). A discriminant analysis was then conducted using **Classify**. The Enter method was applied. Figure 10.5 reveals significant ($p<.001$) group differences for each IV. Although the Box's M Test indicates that homogeneity of covariance cannot be assumed, our interpretation will continue since the Box's test is highly sensitive to non-normality (refer to Fig 10.6). Figure 10.7 presents the Eigenvalues and Wilks' Lambda tables. Since the DV consisted of two levels, only one function could be generated. The canonical correlation ($r = .748$) indicates that the function is highly related to the levels in the DV. Squaring this value produces the effect size, which reveals that 55.9% ($\eta^2=.748^2=.559$) of function variance is accounted for by the DV. The overall Wilks' Lambda was significant, $\Lambda=.441$, $\chi^2(5, N=120)=94.50$, $p<.001$, and indicates that the function of predictors significantly differentiated between countries being developed and developing. Evaluation of the standardized discriminant function coefficients reveal that female life expectancy (1.29) had the highest loading, followed by gross domestic product (1.12), male life expectancy (-1.12), urban population (-.349), and infant mortality (.132) (see Figure 10.8a). In contrast, variable correlations with the function (Figure 10.8b) indicate that gross domestic product ($r=.943$) has the strongest relationship, followed by female life expectancy ($r=.669$), infant mortality ($r=-.642$), male life expectancy ($r=.607$), and urban population ($r=.464$). These differences in function and correlation coefficients make it somewhat difficult to interpret the function, especially since only one function was generated. However, both statistics indicate that the top three variables are gross domestic product and male and female life expectancy. With these variables in mind, we have named the function, *Life and Economic Well-being*.

The next step in interpreting discriminant analysis is evaluating the accuracy of the function in classifying subjects in the appropriate groups. Figure 10.9 presents the classification results. In addition, the cross-validation procedure was utilized to "double-check" the accuracy of classification. Initial classification indicates that 92.5% of the cases were correctly classified. Cross-validation supported this level of accuracy (91.7%). Initial classification results revealed that 88.9% of the *developed* countries were correctly classified, while 93.5% of the *developing* countries were correctly classified. The means of the discriminant functions are consistent with these results (see Figure 10.10). Developed countries had a function mean of 2.07, while developing countries had a mean of -.601. These results suggest that countries with high life expectancy and gross domestic product are likely to be classified as *developed*.

Figure 10.3 Mahalanobis Distance (Example number 1).

Extreme Values

			Case Number	Value
MAH_4	Highest	1	83	29.49218
		2	122	18.24153
		3	37	15.68729
		4	81	14.39278
		5	94	12.89373
	Lowest	1	24	.28085
		2	22	.47589
		3	50	.48938
		4	30	.84003
		5	51	.91498

Only case #83 exceeds the chi-square critical value.

Figure 10.4 Scatterplot of Predictor Variables (Example number 1).

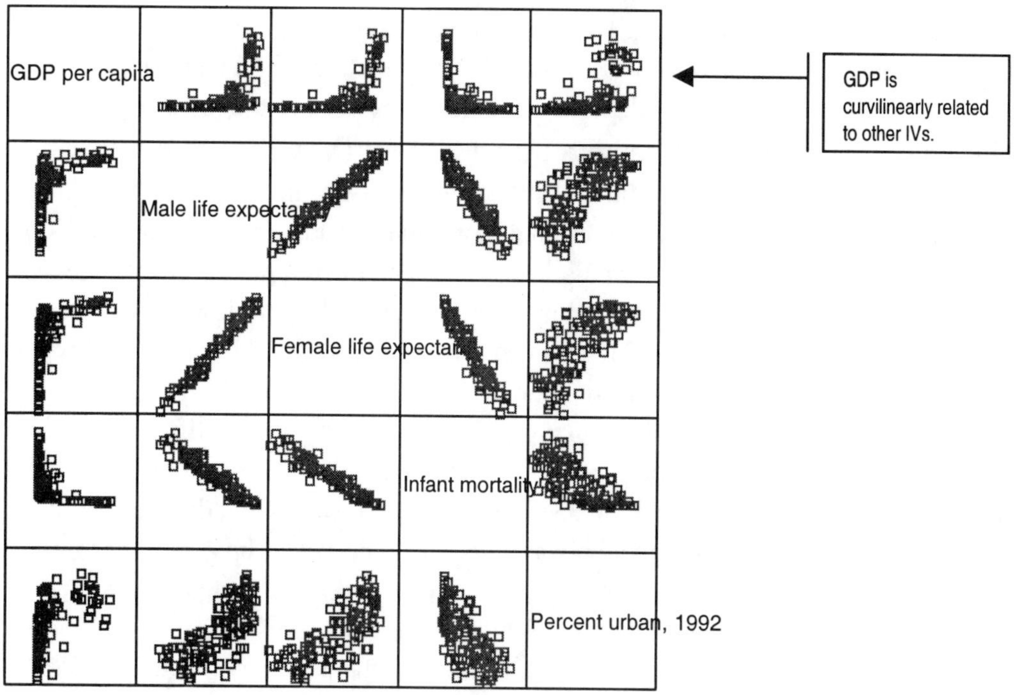

GDP is curvilinearly related to other IVs.

Figure 10.5 a) Group Statistics; b) ANOVA Summary Table (Example number 1).

(a)

Group Statistics

DEVELOP		Mean	Std. Deviation	Valid N (listwise) Unweighted	Weighted
0 Developed country	URBAN	70.0111	15.7100	27	27.000
	LIFEEXPM	72.4074	3.3542	27	27.000
	LIFEEXPF	79.0000	3.2699	27	27.000
	INFMR	11.1259	9.4002	27	27.000
	LNGDP	9.3158	.7428	27	27.000
1 Developing country	URBAN	42.7570	23.5809	93	93.000
	LIFEEXPM	58.9247	9.2492	93	93.000
	LIFEEXPF	62.6882	10.1874	93	93.000
	INFMR	71.8925	39.7841	93	93.000
	LNGDP	6.6569	1.1272	93	93.000
Total	URBAN	48.8892	24.7877	120	120.000
	LIFEEXPM	61.9583	10.0280	120	120.000
	LIFEEXPF	66.3583	11.3736	120	120.000
	INFMR	58.2200	43.5002	120	120.000
	LNGDP	7.2551	1.5317	120	120.000

(b)

Tests of Equality of Group Means

	Wilks' Lambda	F	df1	df2	Sig.
URBAN	.787	31.855	1	118	.000
LIFEEXPM	.682	54.986	1	118	.000
LIFEEXPF	.638	66.862	1	118	.000
INFMR	.657	61.641	1	118	.000
LNGDP	.470	133.008	1	118	.000

All predictor variables show significant group differences.

Figure 10.6 Box's M Test (Example number 1).

(a)   **Test Results**

Box's M		89.048
F	Approx.	5.496
	df1	15
	df2	9219.677
	Sig.	.000

Tests null hypothesis of equal population covariance matrices.

Significance indicates that homogeneity of covariance cannot be assumed.

Figure 10.7 a) Eigenvalues Table; b) Wilks' Lambda Table (Example number 1).

Eigenvalues

Function	Eigenvalue	% of Variance	Cumulative %	Canonical Correlation
1	1.266[a]	100.0	100.0	.748

a. First 1 canonical discriminant functions were used in the analysis.

Canonical correlation of function with predictor variables. Square this to calculate effect size.

(b)   **Wilks' Lambda**

Test of Function(s)	Wilks' Lambda	Chi-square	df	Sig.
1	.441	94.497	5	.000

Figure 10.8 a) Standardized Discriminant Function Coefficients Table; b) Correlation Coefficient Table (Example number 1).

(a)

Standardized Canonical Discriminant Function Coefficients

	Function
	1
URBAN	-.349
LIFEEXPM	-1.117
LIFEEXPF	1.293
INFMR	.132
LNGDP	1.121

(b)

Structure Matrix

	Function
	1
LNGDP	.943
LIFEEXPF	.669
INFMR	-.642
LIFEEXPM	.607
URBAN	.462

Pooled within-groups correlations between discriminating
variables and standardized canonical discriminant functions
Variables ordered by absolute size of correlation within function.

Figure 10.9 Classification Results (Example number 1).

Classification Results[b,c]

		DEVELOP	Predicted Group Membership 0 Developed country	Predicted Group Membership 1 Developing country	Total
Original	Count	0 Developed country	24	3	27
		1 Developing country	6	87	93
	%	0 Developed country	88.9	11.1	100.0
		1 Developing country	6.5	93.5	100.0
Cross-validated[a]	Count	0 Developed country	24	3	27
		1 Developing country	7	86	93
	%	0 Developed country	88.9	11.1	100.0
		1 Developing country	7.5	92.5	100.0

Percent of cases correctly classified for original and cross-validated results by group.

a. Cross validation is done only for those cases in the analysis. In cross validation, each case is classified by the functions derived from all cases other than that case.

b. 92.5% of original grouped cases correctly classified.

c. 91.7% of cross-validated grouped cases correctly classified.

Percent of cases correctly classified for original and cross-validated results for total sample.

Figure 10.10 Discriminant Function Means (Example number 1).

Functions at Group Centroids

DEVELOP	Function 1
0 Developed country	2.071
1 Developing country	-.601

Unstandardized canonical discriminant functions evaluated at group means

Writing Up Results

After stating any data transformations and/or outlier elimination, one should report the significance test results for each function. These results should include Wilks' Lambda, chi-square with degrees of freedom and number of subjects, and *p* value. Applying these results, you should then indicate the number of functions you chose to interpret. It is then necessary to present the standardized function coefficients and the correlation coefficients. These coefficients are typically displayed in a table, while the narration describes the primary variables associated with each function. The researcher should then indicate how the functions were labeled; the previous statistics should support this process. Classification results are presented and should include the percent of accuracy for each group and the entire sample. Finally, function means are presented in support of the functions generated. The following results statement applies the results from Figures 10.3 – 10.10.

A discriminant analysis was conducted to determine whether five variables—urban population, female life expectancy, male life expectancy, infant mortality, and gross domestic product—could predict the development (developed vs. developing) status for a country. Prior to analysis, one outlier was eliminated. Due to non-normality, the variable of gross domestic product was transformed by taking its natural log. One function was generated and was significant, $\Lambda=.441$, $\chi^2(5, N=120)=94.50$, $p<.001$, indicating that the function of predictors significantly differentiated between countries being developed and developing. Development status was found to account for 55.9% of function variance. Standardized function coefficients and correlation coefficients (see Table 1) revealed that the variables of gross domestic product, male life expectancy, and female life expectancy were most associated with the function. Based upon these results, the function was labeled *Life and Economic Well-being*. Original classification results revealed that 88.9% of the *developed* countries were correctly classified, while 93.5% of the *developing* countries were correctly classified. For the overall sample, 92.5% were correctly classified. Cross-validation derived 91.7% accuracy for the total sample. The means of the discriminant functions are consistent with these results. Developed countries had a function mean of 2.07, while developing countries had a mean of -.601. These results suggest that countries with high life expectancy and gross domestic product are likely to be classified as *developed*.

Table 1 Correlation Coefficients and Standardized Function Coefficients

	Correlation Coefficients with Discriminant Function	Standardized Function Coefficients
Female Life Expectancy	.669	1.293
Male Life Expectancy	.607	−1.117
Gross Domestic Product	.943	1.121
Urban Population	.462	−.349
Infant Mortality	−.642	.132

SECTION 10.4 SAMPLE STUDY AND ANALYSIS

This section provides a complete example of a stepwise discriminant analysis. We will develop the research question, screen data, conduct data analysis, interpret output, and summarize the results. The SPSS data set *gss.sav* is utilized.

Problem

We are interested in predicting one's life perspective (dull, routine, exciting) with the following variables: age, hours worked per week (*hrs1*), years of education (*educ*), income (*rincom91*), number of siblings (*sibs*), and number of hours spent watching TV per day (*tvhours*). The following research question is generated to address this scenario: Can life perspective be reliably predicted from knowledge of an individual's age, hours worked per week, years of education, income, number of siblings, and number of hours spent watching TV per day? The DV is *life* and has three levels of response: dull, routine, and exciting. The predictor variables are the six IVs.

Method

Prior to conducting the discriminant analysis, data were screened for outliers, normality, and linearity. Calculating Mahalanobis distance, numerous outliers were identified. Thus, cases with Mahalanobis distance exceeding the chi square critical value at $p=.001$, $\chi^2(6)=22.458$, were eliminated (see Figure 10.11). In addition, the variable of *rincom91* was transformed to eliminate cases greater than 22. This new variable was labeled *rincom2*. A bivariate scatterplot was created to evaluate normality and linearity (see Figure 10.12). In general, plots are fairly elliptical, indicating normality and linearity. A stepwise discriminant analysis was conducted using **Classify**. See the section on SPSS "How To" for an explanation of the steps necessary to generate the following output.

Output and Interpretation of Results

Figure 10.13 presents the group means and the ANOVA results for group differences. Group differences are only significant ($p<.05$) for the variables of *educ, sibs, tvhours*, and *rincom2*. The Box's M Test indicates heterogeneity of covariances (see Figure 10.14). Since a stepping procedure was utilized in this analysis, Figure 10.15 presents the variables entered at each step. Only three variables, *educ, tvhours,* and *rincom2*, were entered into the model. Figure 10.16 displays the Eigenvalues and Wilks' Lambda tables. The Eigenvalue table (Figure 10.16a) reveals that two functions were generated, with the DV accounting for only 7.2% ($\eta^2=.268^2=.072$) of the variance in the first function. The Wilks' Lambda table (Figure 10.16b) indicates that the first function is significant, $\Lambda=.920$, $\chi^2(6, N=563)=46.36$, $p<.001$. However, the second function is not significant, $\Lambda=.992$, $\chi^2(6, N=563)=4.54$, $p=.103$. Consequently, only one function will be interpreted. The standardized function coefficients and correlation coefficients presented in Figure 10.17 indicate that *educ* has the highest relationship with the function, followed by *tvhours* and *rincom2*. Although these variables are quite diverse in nature, we will label this function, *Work Success*. Classification results (see Figure 10.18) reveal that the original grouped cases were classified with only 57.2% overall accuracy. Accuracy by each group was 10.5 % for dull, 68.4% for routine, and 21.1% for exciting. The cross-

Figure 10.11 Mahalanobis Distance (Example number 2).

Extreme Values

			Case Number	Value
MAH_1	Highest	1	466	125.85957
		2	1360	113.81654
		3	406	55.96910
		4	50	53.43976
		5	121	42.28421
	Lowest	1	1032	.26375
		2	561	.34919
		3	649	.44056
		4	734	.49205
		5	1110	.52900

Since several cases exceed the $\chi 2$ critical value, cases greater than $\chi^2(6)=22.458$ were eliminated.

Figure 10.12 Scatterplot Matrix of Predictor Variables (Example number 2).

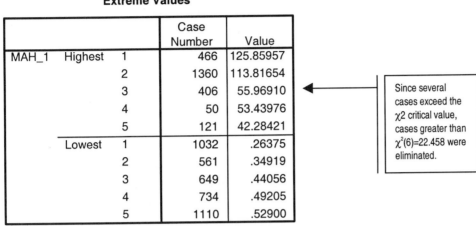

validated results supported original accuracy levels with 56.7% correctly classified overall. Group means for the function (see Figure 10.19) indicate that those who perceive life as dull had a function mean of –1.022, those who perceive life as routine had a mean of -.178, and those who perceive life as exciting had a mean of .234. These results suggest that individuals who perceive life as dull have the least amount of education, watch the most TV, and have the lowest income.

Presentation of Results

The following report summarizes the stepwise discriminant analysis results.

A stepwise discriminant analysis was conducted to determine the ability of the six variables—age, hours worked per week, years of education, income, number of siblings, and number of hours spent watching TV per day to predict one's life perspective (dull, routine, exciting). Prior to the analysis, several outliers were eliminated. In addition, the variable of income was transformed by eliminating subjects whose value exceeded 22. The analysis generated two functions; however only function one was significant, $\Lambda=.920$, $\chi^2(6, N=563)=46.36$, $p<.001$, with only 7.2% of the function variability explained by life perspective. Three variables were entered into the function: years of education, hours watching TV per day, and income, respectively. The variables of age, hours worked per week, and number of siblings were excluded. Table 1 presents the standardized function coefficients and correlation coefficients. Although these variables are quite diverse in nature, the function was labeled *Work Success*. Classification results revealed that the original grouped cases were classified with only 57.2% overall accuracy. Accuracy by each group was 10.5% for dull, 68.4% for routine, and 21.1% for exciting. The cross-validated results supported original accuracy levels with 56.7% correctly classified overall. Group means for the function indicated that those who perceive life as dull had a function mean of –1.022, those who perceive life as routine had a mean of -.178, and those who perceive life as exciting had a mean of .234. These results suggest that individuals who have the least amount of education, watch the most TV, and have the lowest income will likely perceive life as dull.

Table 1 Correlation Coefficients and Standardized Function Coefficients

	Correlation Coefficients with Discriminant Function	Standardized Function Coefficients
Education	.754	.520
Hours Watching TV per Day	–.669	–.479
Income	.660	.436

Figure 10.13 a) Group Statistics; b) ANOVA Summary Table (Example number 2).

(a)   **Group Statistics**

LIFE		Mean	Std. Deviation	Valid N (listwise) Unweighted	Weighted
1 Dull	AGE	41.5789	12.1212	19	19.000
	EDUC	12.4737	2.0647	19	19.000
	HRS1	40.1579	15.9139	19	19.000
	SIBS	4.2632	2.9785	19	19.000
	TVHOURS	3.4737	2.6535	19	19.000
	RINCOM2	8.3158	6.5833	19	19.000
2 Routine	AGE	40.5573	12.7136	262	262.000
	EDUC	13.4084	2.5785	262	262.000
	HRS1	41.4656	13.7458	262	262.000
	SIBS	3.4847	2.3370	262	262.000
	TVHOURS	2.4122	1.3889	262	262.000
	RINCOM2	12.1870	5.2068	262	262.000
3 Exciting	AGE	40.2199	11.9781	282	282.000
	EDUC	14.4397	2.6993	282	282.000
	HRS1	43.6915	15.2782	282	282.000
	SIBS	3.0957	2.5086	282	282.000
	TVHOURS	2.0851	1.3939	282	282.000
	RINCOM2	13.4291	5.6307	282	282.000
Total	AGE	40.4227	12.3119	563	563.000
	EDUC	13.8934	2.6824	563	563.000
	HRS1	42.5364	14.6276	563	563.000
	SIBS	3.3162	2.4566	563	563.000
	TVHOURS	2.2842	1.4725	563	563.000
	RINCOM2	12.6785	5.5561	563	563.000

(b)   **Tests of Equality of Group Means**

	Wilks' Lambda	F	df1	df2	Sig.
AGE	1.000	.137	2	560	.872
EDUC	.954	13.356	2	560	.000
HRS1	.993	1.838	2	560	.160
SIBS	.989	3.188	2	560	.042
TVHOURS	.965	10.083	2	560	.000
RINCOM2	.966	9.750	2	560	.000

Only *educ, sibs, tvhours,* and *rincom2* show significant group differences.

Figure 10.14 Box's M Test (Example number 2).

Test Results

Box's M		39.658
F	Approx.	3.191
	df1	12
	df2	9366.677
	Sig.	.000

Tests null hypothesis of equal population covariance matrices.

Significance indicates that homogeneity of covariance cannot be assumed.

Figure 10.15 Summary Table of Steps (Example number 2).

Variables Entered/Removed[a,b,c,d]

		Wilks' Lambda				Exact F			
Step	Entered	Statistic	df1	df2	df3	Statistic	df1	df2	Sig.
1	EDUC	.954	1	2	560.000	13.356	2	560.000	.000
2	TVHOURS	.934	2	2	560.000	9.710	4	1118.000	.000
3	RINCOM2	.920	3	2	560.000	7.876	6	1116.000	.000

At each step, the variable that minimizes the overall Wilks' Lambda is entered.

a. Maximum number of steps is 12.

b. Minimum partial F to enter is 3.84.

c. Maximum partial F to remove is 2.71.

d. F level, tolerance, or VIN insufficient for further computation.

Figure 10.16   a) Eigenvalues Table; b) Wilks' Lambda Table (Example number 2).

(a)

Eigenvalues

Function	Eigenvalue	% of Variance	Cumulativ e %	Canonical Correlation
1	.078ᵃ	90.5	90.5	.268
2	.008ᵃ	9.5	100.0	.090

a. First 2 canonical discriminant functions were used in the analysis.

Canonical correlation of function with predictor variables. Square this to calculate effect size.

(b)

Wilks' Lambda

Test of Function(s)	Wilks' Lambda	Chi-squar e	df	Sig.
1 through 2	.920	46.363	6	.000
2	.992	4.538	2	.103

Indicates that first function is significant, while second function is NOT.

Figure 10.17 a) Standardized Discriminant Function Coefficients Table; b) Correlation Coefficient Table (Example number 2).

(a)

Standardized Canonical Discriminant Function Coefficients

	Function 1	Function 2
EDUC	.520	.927
TVHOURS	-.479	.521
RINCOM2	.436	-.531

(b)

Structure Matrix

	Function 1	Function 2
EDUC	.754*	.656
TVHOURS	-.669*	.392
RINCOM2	.660*	-.352
HRS1a	.394*	-.215
SIBSa	-.193*	-.088
AGEa	.052	-.212*

Pooled within-groups correlations between discriminating variables and standardized canonical discriminant functions
Variables ordered by absolute size of correlation within function.

*. Largest absolute correlation between each variable and any discriminant function

a. This variable not used in the analysis.

Figure 10.18   Classification Results (Example number 2).

Percent of cases correctly classified for original and cross-validated results by group.

Classification Results[b,c]

		LIFE	Predicted Group Membership 1 Dull	2 Routine	3 Exciting	Total
Original	Count	1 Dull	2	13	4	19
		2 Routine	1	148	113	262
		3 Exciting	1	109	172	282
		Ungrouped cases	3	127	141	271
	%	1 Dull	10.5	68.4	21.1	100.0
		2 Routine	.4	56.5	43.1	100.0
		3 Exciting	.4	38.7	61.0	100.0
		Ungrouped cases	1.1	46.9	52.0	100.0
Cross-validated[a]	Count	1 Dull	1	13	5	19
		2 Routine	1	148	113	262
		3 Exciting	1	111	170	282
	%	1 Dull	5.3	68.4	26.3	100.0
		2 Routine	.4	56.5	43.1	100.0
		3 Exciting	.4	39.4	60.3	100.0

a. Cross validation is done only for those cases in the analysis. In cross validation, each case is classified by the functions derived from all cases other than that case.

b. 57.2% of original grouped cases correctly classified.

c. 56.7% of cross-validated grouped cases correctly classified.

Percent of cases correctly classified for original and cross-validated results for total sample.

Figure 10.19   Discriminant Function Means.

Functions at Group Centroids

LIFE	Function 1	2
1 Dull	-1.022	.350
2 Routine	-.178	-7.75E-02
3 Exciting	.234	4.842E-02

Unstandardized canonical discriminant functions evaluated at group means

SECTION 10.5 SPSS "HOW TO"

This section presents the steps for conducting a discriminant analysis using **Classify**. The preceding example from *gss.sav* will be applied. To open the Discriminant Analysis Dialogue box shown in Figure 10.20, select the following:

> **Analyze**
> > **Classify**
> > > **Discriminant**

Discriminant Analysis Dialogue Box (Figure 10.20)

Click the DV and move to the Grouping Variable box. Then click **Define Range**; indicate the lowest and the highest group values. Click each IV and move to the Independents box. Select the method by either checking **Enter Independents Together** or **Use Stepwise Method**. For our second example, we chose the stepwise. Then click **Statistics**.

Figure 10.20   Discriminant Analysis: Dialogue Box.

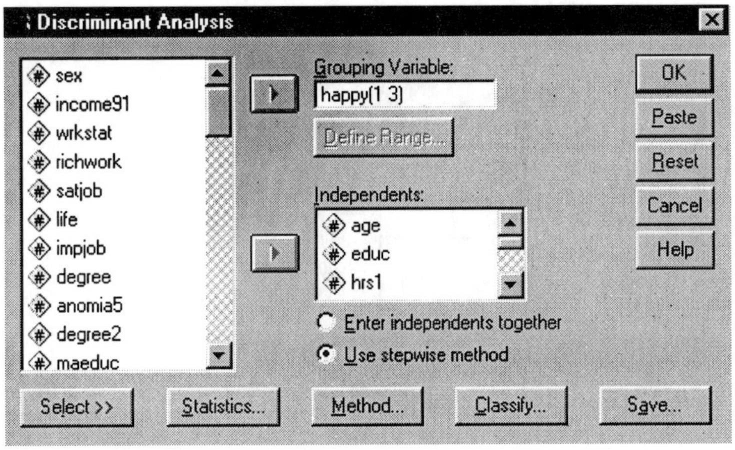

Discriminant Analysis: Statistics Dialogue Box (see Figure 10.21)

Under Descriptives, check all the options: **Means, Univariate ANOVAs,** and **Box's M**. No other options were selected within this dialogue box; however, a description of each option is as follows.

> **Fisher's**—Canonical function coefficients that maximize discrimination between levels of the DV.
>
> **Unstandardized**—Unstandardized function coefficients based on the variable raw scores.
>
> **Within-groups Correlation**—Correlation matrix of the IVs at each level of the DV.
>
> **Within-groups Covariance**— Covariance matrix of the IVs at each level of the DV.
>
> **Separate-groups Covariance**—Same as Within-groups Covariance, but a separate matrix is produced for each level of the DV.

Total Covariance—Covariance matrix for the entire sample.

After checking the Statistics options, click **Continue**, then **Method**.

Figure 10.21   Discriminant Analysis: Statistics Dialogue Box.

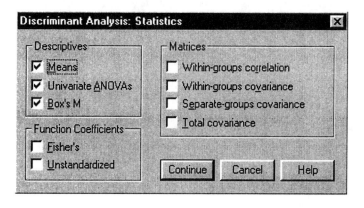

Discriminant Analysis: Stepwise Method Dialogue Box (see Figure 10.22)

This Dialogue box will appear only if you have selected the Stepwise method. Under Method, check **Wilks' Lambda**. Under Display, check **Summary of Steps**. Use the default criteria of the **F-value**. Click **Continue**, then **Classify**.

Figure 10.22 Discriminant Analysis: Stepwise Method Dialogue Box.

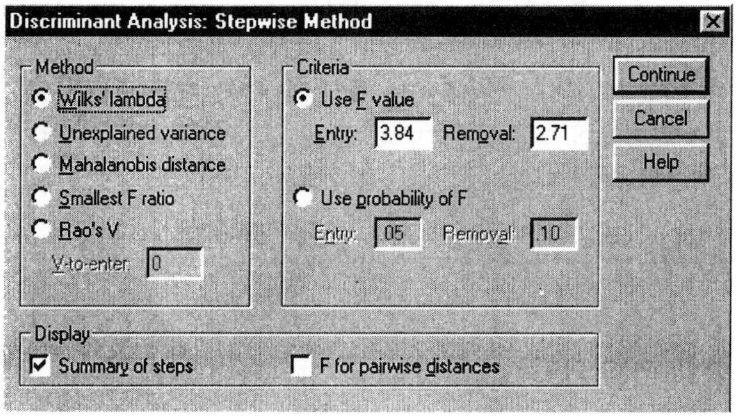

Discriminant Analysis: Classification Dialogue Box (see Figure 10.23)

If group proportions in the population are fairly equal, then select **All Groups Equal**, under Prior Probabilities. If groups are unequal, check **Compute from Group Sizes**; this is the most commonly used option and the one we chose in the example. Under Display, select **Summary Table**, which will display the number and percent of correct and incorrect classifications for each group. Also select **Leave-One-Out Classification**, which will reclassify each case based on the functions of all other cases excluding that case. Under Use Covariance Matrix, select the default **Within-groups Covariance Matrix**.

Depending upon the number of levels in the DV, you may select one or more of the options Under Plots. The **Combined-groups Plot** will create a histogram (for two groups) or a scatterplot (for three or more groups). This option is often helpful when the DV has three or more levels. The **Separate-groups Plot** creates a separate plot for each group. The **Territory Map** charts centroids and boundaries and is used only when the DV has three or more levels. For our example, we selected the **Combined-groups Plot**; however, the plot was not presented since only one function was interpreted. After you have selected the desired plots, click **Continue** then **Save**.

Figure 10.23 Discriminant Analysis: Classification Dialogue Box.

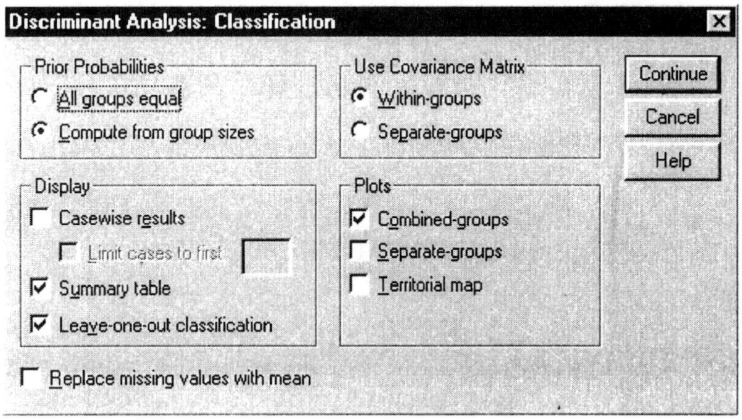

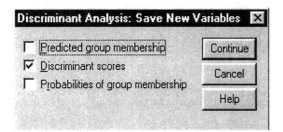

Discriminant Analysis: Save New Variables Dialogue Box (Figure 10.24)

This dialogue box provides options for saving specific results as variables for future analysis. Three options are available: **Predicted Group Membership, Discriminant Scores,** and **Probabilities of Group Membership**. It is common to save **Discriminant Scores**. Once you have selected the type(s) of variables to save, click **Continue**, then **OK**.

Figure 10.24   Discriminant Analysis: Save New Variables Dialogue Box.

Summary

Discriminant analysis is often used to predict group membership based on observed characteristics (predictor variables). The technique generates discriminant function(s) derived from linear combinations of the predictor variables that best discriminate between/among the groups. Prior to conducting discriminant analysis, data should be screened for missing data and outliers. Data should also be tested for normality, linearity, and homogeneity of covariance. The discriminant analysis procedure in SPSS generates four parts: 1) preliminary statistics that describe group differences and covariances; 2) significance tests and strength of relationship statistics for each discriminant function; 3) discriminant function coefficients; and 4) group classification. Preliminary statistics present group means and standard deviations for each IV, ANOVA results for group differences (Wilks' Lambda, F-test, degrees of freedom, and p values for each IV), and Box's M Test (test for homogeneity of covariance). Significance tests and strength of relationship statistics are presented in the Eigenvalues and Wilks' Lambda Tables. The Eigenvalue Table displays the eigenvalue, percent of variance, and canonical correlation for each discriminant function. The Wilks' Lambda Table provides chi-square tests of significance for each function (Wilks' Lambda, chi-square, degrees of freedom, and p value). Discriminant function coefficients are presented in the tables of Standardized Canonical Discriminant Function Coefficients (which represent the degree to which each variable contributes to each function) and the Structure Matrix (which displays the correlation coefficients between the variables and functions). Assessment of these coefficients assists in labeling the function(s). Group classification results are utilized to assess the accuracy of the functions in classifying subjects in the appropriate groups and are presented in a table that displays the original and predicted frequency and percent of subjects within each group. The final step in this interpretation process is to determine the extent to which group differences support the functions generated by reviewing group means for each function as presented in the table of Functions at Group Centroids. Figure 10.25 presents a checklist for conducting a discriminant analysis.

Figure 10.25 Checklist for Conducting Discriminant Analysis.

I. Screen Data
  a. Missing Data?
  b. Multivariate Outliers?
   ❑ Run preliminary Regression to calculate Mahalanobis' Distance.
      1. ⌐ᵗᴴ **Analyze…Regression…Linear.**
      2. Identify a variable that serves as a case number and move to Dependent Variable box.
      3. Identify all appropriate quantitative variables and move to Independent(s) box.
      4. ⌐ᵗᴴ **Save.**
      5. Check **Mahalanobis'.**
      6. ⌐ᵗᴴ **Continue.**
      7. ⌐ᵗᴴ **OK.**
      8. Determine chi square χ^2 critical value at $p<.001$.
   ❑ Conduct **Explore** to test outliers for Mahalanobis chi square χ^2.
      1. ⌐ᵗᴴ **Analyze…Descriptive Statistics…Explore**
      2. Move *mah_1* to Dependent Variable box.
      3. Leave Factor box empty.
      4. ⌐ᵗᴴ **Statistics.**
      5. Check **Outliers.**
      6. ⌐ᵗᴴ **Continue.**
      7. ⌐ᵗᴴ **OK.**
   ❑ Delete outliers for subjects when χ^2 exceeds critical χ^2 at $p<.001$.
  c. Linearity and Normality?
   ❑ Create Scatterplot Matrix of all IVs.
      ❑ Scatterplot shapes are not close to elliptical shapes→reevaluate univariate normality and consider transformations.

II. Conduct Discriminant Analysis
  a. Run Discriminant Analysis using **Classify**.
      1. ⌐ᵗᴴ **Analyze…**⌐ᵗᴴ **Classify…**⌐ᵗᴴ **Discriminant.**
      2. Move the DV to the Grouping Variable box.
      3. ⌐ᵗᴴ **Define Range.**
      4. Indicate the lowest and highest group value.
      5. ⌐ᵗᴴ **Continue.**
      6. ⌐ᵗᴴ **Statistics.**
      7. Check: **Means, Univariate ANOVAs** and **Box's M.**
      8. ⌐ᵗᴴ **Continue.**
      9. ⌐ᵗᴴ **Method.**
     10. Check **Wilks' Lambda, Use F-value,** and **Summary of Steps.**
     11. ⌐ᵗᴴ **Continue.**
     12. ⌐ᵗᴴ **Classification.**
     13. Check **Compute from Group Sizes, Summary Table, Leave-out-one Classification, Within-groups Matrix,** and **Combined-groups Plot.**
     14. ⌐ᵗᴴ **Continue.**
     15. ⌐ᵗᴴ **Save.**
     16. Check **Discriminant Scores.**
     17. ⌐ᵗᴴ **Continue.**
     18. ⌐ᵗᴴ **OK.**
  b. Interpret Box's M Test. If significant at $p<.001$, continue to interpret results but proceed with caution.

Figure 10.25 Checklist for Conducting Discriminant Analysis. (*Continued*)

III. Summarize Results
    a.   Describe any data elimination or transformation.
    b.   Report the significance test results for each function (Wilks' Lambda, chi-square, *df, N*, and *p*-value.
    c.   Indicate the number of functions you chose to interpret.
    d.   Present the standardized function coefficients and the correlation coefficients in narration and table.
    e.   Indicate how functions were labeled.
    f.   Present classification results, including the percent of accuracy for each group and the entire sample.
    g.   Present function means in support of the functions generated.
    h.   Draw conclusions.

Exercises for Chapter 10

This exercise utilizes the data set *schools.sav*, which can be downloaded at the SPSS Web site. Open the URL: **www.spss.com/tech/DataSets.html** in your Web browser. Scroll down until you see "Data Used in SPSS Guide to Data Analysis—8.0 and 9.0" and click on the link "dataset.exe." When the "Save As" dialogue appears, select the appropriate folder and save the file. Preferably, this should be a folder created in the SPSS folder of your hard drive for this purpose. Once the file is saved, double-click the "dataset.exe" file to extract the data sets to the folder.

Conduct a stepwise discriminant analysis with the following variables:
       IVs—*grad94, act94, pctact94, math93, math94me, read93, read94me, scienc93, sci94me*
       DV—*medlowin*

1. Develop a research question.

2. Which predictor variables show significant group differences?

3. Can homogeneity of covariance be assumed?

4. Which variables were entered into the model?

5. How many functions were generated? Why?

6. Is (are) the function(s) significant? Explain.

7. Calculate the effect size for each function.

8. Evaluating the function coefficients and correlation coefficients, what would you label the function(s)?

9. How accurate is (are) the function(s) in predicting income level for each group and the total sample?

10. What is your conclusion regarding these results?

CHAPTER 11

LOGISTIC REGRESSION

In Chapter 11, we present discussion of logistic regression—an alternative to discriminant analysis, as well as multiple regression in certain situations. Logistic regression has the same basic purpose as discriminant analysis—the classification of individuals into groups. It is, in some ways, more flexible and versatile than discriminant analysis, although mathematically, it can be quite a bit more cumbersome. In this chapter, we present discussion of logistic regression, which seeks to identify a combination of IVs—which are limited in few, if any, ways—that best predicts membership in a particular group, as measured by a categorical DV.

SECTION 11.1 PRACTICAL VIEW

Purpose

Logistic regression is basically an extension of multiple regression in situations where the DV is not a continuous or quantitative variable (George & Mallery, 2000). In other words, the DV is categorical (or discrete) and may have as few as two values. For example, in a logistic regression application, these categories might include values such as membership or nonmembership in a group, completion or noncompletion of an academic program, passing or failure to pass a course, survival or failure of a business, etc.

Due to the nature of the categorical DV in logistic regression, this procedure is also sometimes used as an alternative to discriminant analysis. Since the goal is to predict values on a DV that is categorical, [we are essentially attempting to predict membership into one of two or more "groups."] The reader should likely see the similarity between this procedure and that discussed in the previous chapter for discriminant analysis (i.e., the classification, or prediction, of subjects into groups). Although logistic regression may be used to predict values on a DV of two or more categories, our discussion will focus on binary logistic regression, in which the DV is dichotomous.

The basic concepts fundamental to multiple regression analysis—namely that several variables are regressed onto another variable using one of several selection processes—are the same for logistic regression analysis (George & Mallery, 2000), although the meaning of the resultant regression equation is considerably different. As you read in Chapter 7, a standard regression equation is composed of the sum of the products of weights and actual values on several predictor variables (IVs) in order to predict the values on the criterion variable (DV). In contrast, [the value that is being predicted in logistic regression is actually a *probability*, which ranges from 0 to 1.] More precisely, logistic regression specifies the probabilities of the particular outcomes (e.g., "pass" and "fail") for each subject or case involved. In other words, logistic regression analysis produces a regression equation that accurately predicts the probability of whether an individual will fall into one category (e.g., "pass") or the other (e.g., "fail") (Tate, 1992).

Although logistic regression is fairly similar to both multiple regression and discriminant analysis, it does provide several distinct advantages over both techniques (Tabachnick & Fidell, 1996). Unlike both discriminant analysis and multiple regression, logistic regression requires that no assumptions about the distributions of the predictor variables (IVs) need to be made by the researcher. In other words, the predictors do not have to be normally distributed, linearly related, or have equal variances within each group. It should be obvious to the reader that this fact alone makes logistic regression much more flexible than the other two techniques. Additionally, logistic regression cannot produce negative predictive probabilities, as can happen when applying multiple regression to situations involving dichotomous outcomes (Tate, 1992). In logistic regression, all probability values will be positive and will range from 0 to 1 (Tabachnick & Fidell, 1996). Another advantage is that logistic regression has the capacity to analyze predictor variables of all types—continuous, discrete, and dichotomous. Finally, logistic regression can be especially useful when the distribution of data on the criterion variable (DV) is expected or known to be *nonlinear* with one or more of the predictor variables (IVs). Logistic regression is able to produce nonlinear models, which again adds to its overall flexibility.

Let us develop an initial working example to which we will refer in this chapter. Suppose we wanted to determine a nation's status in terms of its level of development based on several measures. Specifically, those measures consist of:

- population (1992) in millions (*pop92*);
- percent of the population living in urban areas (*urban*);
- gross domestic product per capita (*gdp*);
- death rate per 1,000 people (*deathrat*);
- number of radios per 100 people (*radio*);
- number of hospital beds per 10,000 people (*hospbed*); and
- number of doctors per 10,000 (*docs*).

Combinations of these seven variables that would accurately predict the probability of a country's status as a developing nation (*develop*)—either (1) developed nation or (2) developing nation—will be determined as a result of a logistic regression analysis. Our sample analysis was conducted using a forward method of entry for the seven predictors.

The results obtained from a logistic regression analysis are somewhat different from those that we have seen accompany previous analysis techniques. There are basically three main output components to be interpreted. First, the resulting model is evaluated using goodness-of-fit tests. The table showing the results of chi-square goodness-of-fit tests for our working example is presented in Figure 11.1. One should notice that the model resulted in the inclusion of three variables from the original seven predictors—*gdp* (entered in Step 1), *hospbed* (entered in Step 2), and *urban* (entered at Step 3). At each step, this test essentially compares the actual values for cases on the DV with the predicted values on the DV. All steps resulted in significance values < .001, indicating that these three variables are significant and important predictors of the DV, *develop*. Also included in this table are the percentages of correct classification—based on the model—at each step (i.e., based on the addition of each variable). In Step 1, the model with only *gdp* correctly classified 91.96% of the cases; in Step 2, the model with *gdp* and *hospbed* correctly classified 91.96%; in Step 3, the model of *gdp, hospbed*, and *urban* correctly classified 95.54% of the cases.

Figure 11.1 Goodness-of-fit Indices for Example Number 1.

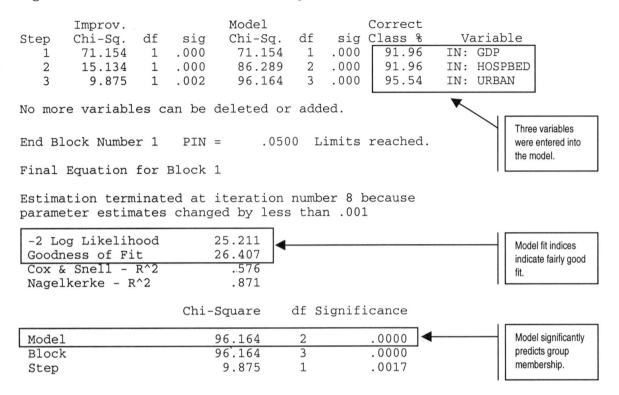

```
          Improv.                 Model             Correct
Step    Chi-Sq.  df   sig    Chi-Sq.  df   sig   Class %     Variable
  1      71.154   1  .000     71.154   1  .000   | 91.96    IN: GDP
  2      15.134   1  .000     86.289   2  .000   | 91.96    IN: HOSPBED
  3       9.875   1  .002     96.164   3  .000   | 95.54    IN: URBAN

No more variables can be deleted or added.

End Block Number 1   PIN =      .0500  Limits reached.
```

Three variables were entered into the model.

```
Final Equation for Block 1

Estimation terminated at iteration number 8 because
parameter estimates changed by less than .001

 -2 Log Likelihood        25.211
 Goodness of Fit          26.407
 Cox & Snell - R^2          .576
 Nagelkerke - R^2           .871
```

Model fit indices indicate fairly good fit.

```
               Chi-Square    df Significance

 Model            96.164      2     .0000
 Block            96.164      3     .0000
 Step              9.875      1     .0017
```

Model significantly predicts group membership.

The second table shown in Figure 11.1 includes several indices of *overall* model fit. Smaller values on the first measure, labeled ***-2 Log Likelihood***, indicate that the model fits the data better; a perfect model has a value for this measure equal to 0 (George & Mallery, 2000). The next measure, ***Goodness-of-Fit***, compares the actual values for cases on the DV with the predicted values on the DV; this measure is similar to the chi-square value in the first table. The third and fourth measures, ***Cox & Snell – R^2*** and ***Nagelkerke – R^2***, are essentially estimates of R^2 indicating the proportion of variability in the DV that may be accounted for by all predictor variables included in the equation.

The second component is a classification table for the DV. The classification table for *develop* is presented in Figure 11.2. The classification table compares the predicted values for the DV, based on the logistic regression model, with the actual observed values from the data. The predicted values are obtained by computing the probability for a particular case (this computation will be discussed later in Section 11.3) and classifies it into one of the two possible categories based on that probability. If the calculated probability is less than .50, the case is classified into the first value on the DV—in our example, the first category is *developed* nation (coded "0").

The third and final component to be interpreted is the table of coefficients for variables included in the model. The coefficients for our working example are shown in Figure 11.3. These coefficients are interpreted in similar fashion to coefficients resulting from a multiple regression. The values labeled *B* are the regression coefficients or weights for each variable used in the equation. The significance of each predictor is tested not with a *t*-test, as in multiple regression, but with a measure known as the ***Wald statistic*** and the associated significance value. The value *R* is the partial correlation coefficient between each predictor variable and the DV, holding constant all other predictors in the equation. Finally, ***Exp(B)*** provides an alternative method of interpreting the regression coefficients. The meaning of this coefficient will be explained further in Section 11.3.

315

Figure 11.2 Classification Table for Example Number 1.

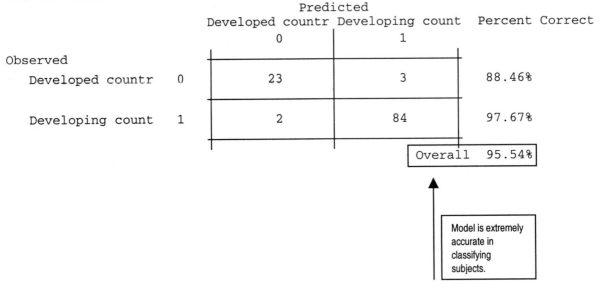

```
Classification Table for DEVELOP
The Cut Value is .50
```

Figure 11.3 Regression Coefficients for Example Number 1.

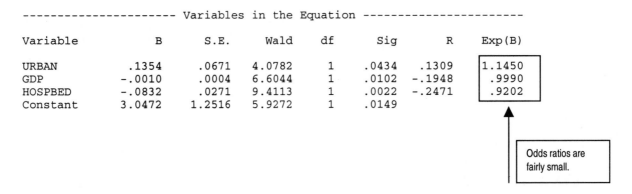

Sample Research Questions

The goal of logistic regression analysis is to correctly predict the category of outcome for individual cases. Further, attempts are made to reduce the number of predictors (in order to achieve parsimony) while maintaining a strong level of prediction. Based on our working example, let us now proceed to the specification of a series of possible research questions for our analysis:

(1)   Can status as a developing nation (i.e., develop*ed* or develop*ing*) be correctly predicted from knowledge of population; percent of population living in urban areas; gross domestic product; death rate; number of radios, hospital beds, and doctors?

(2)   If developing nation status can be predicted correctly, which variables are central in the prediction of that status? Does the inclusion of a particular variable increase or decrease the probability of the specific outcome?

(3)   How good is the model at classifying cases for which the outcome is unknown? In other words, how many developing countries are classified correctly? How many developed countries are classified correctly?

SECTION 11.2 ASSUMPTIONS AND LIMITATIONS

As mentioned earlier, logistic regression does not require the adherence to any assumptions about the distributions of predictor variables (Tabachnick & Fidell, 1996). However, if distributional assumptions are met, discriminant analysis may be a stronger analysis technique; thus, the researcher may want to opt for this procedure.

There are, however, several important issues related to the use of logistic regression. First, there is the issue of the ratio of cases to variables included in the analysis. Several problems may occur if too few cases relative to the number of predictor variables exist in the data. Logistic regression may produce extremely large parameter estimates and standard errors, especially in situations where combinations of discrete variables result in too many cells with no cases. If this situation occurs, the researcher is advised to collapse categories of the discrete variables, delete any offending categories (if patterns are evident), or simply delete any discrete variable if it is not important to the analysis (Tabachnick & Fidell, 1996). Another option open to the researcher is to increase the number of cases in the hope that this will "fill in" some of the empty cells.

Second, logistic regression relies on a goodness-of-fit test as a means of assessing the fit of the model to the data. You may recall from an earlier course in statistics that a goodness-of-fit test includes values for the expected frequencies for each cell in the data matrix formed by combinations of discrete variables. If any of the cells have expected frequencies that are too small (typically, $f_e < 5$), the analysis may have little power (Tabachnick & Fidell, 1996). All pairs of discrete variables should be evaluated to ensure that all cells have expected frequencies greater than 1 and that no more than 20% have frequencies less than 5. If either of these conditions fail, the researcher should consider accepting a lower level of power for the analysis, collapsing categories for variables with more than two levels, or deleting discrete variables so as to reduce the total number of cells (Tabachnick & Fidell, 1996).

Third, as with all varieties of multiple regression, logistic regression is sensitive to high correlations among predictor variables. This condition results in multicollinearity among predictor variables, as discussed in Chapter 7. If multicollinearity is present among variables in the analysis, one is advised to delete one or more of the redundant variables from the model in order to eliminate the multicollinear relationships (Tabachnick & Fidell, 1996).

Finally, extreme values on predictor variables should be examined carefully. As with multiple regression, resultant logistic regression models are very sensitive to outliers. A case that is actually in one outcome category may show a high probability for being in another category. Multiple cases such as this will result in a model with extremely poor fit. Standardized residuals should be examined in order to detect outliers; any identified outliers—those cases with values > |3|—should be addressed using standard methods (i.e., deletion from the sample).

SECTION 11.3 PROCESS AND LOGIC

The Logic Behind Logistic Regression

Mathematically speaking, logistic regression is based on probabilities, odds, and the logarithm of the odds (George & Mallery, 2000). Probabilities are simply the number of outcomes of a specific type expressed as a proportion of the total number of possible outcomes. For example, if we roll a single die, the probability of rolling a three would be 1 out of 6—there is only one "3" on a die and there are six possible outcomes. This ratio could also be expressed as a proportion (.167) or as a percentage (16.7%). In a logistic regression application, **odds** are defined as the ratio of the probability that an event will occur divided by the probability that the event will not occur. In other words,

$$Odds = \frac{p(X)}{1 - p(X)}$$
(Equation 11.1)

where $p(X)$ is the probability of event X occurring and $1-p(X)$ is the probability of event X not occurring. Therefore, the odds of rolling a "3" on a die are

$$Odds_{"3"} = \frac{p("3")}{1 - p("3")} = \frac{.167}{.833} = .200$$

It is important for the reader to keep in mind that probabilities will always have values that range from 0 to 1, but odds may be greater than 1. Applying the concept of odds to our working logistic regression example of classification as a developing nation would give us the following equation:

$$Odds_{developing} = \frac{p(developing)}{1 - p(developing)}$$

The effect of an IV on a dichotomous outcome is usually represented by an odds ratio. The **odds ratio**—symbolized by ψ or Exp(B)—is defined as a ratio of the odds of being classified in one category (i.e., $Y=0$ or $Y=1$) of the DV for two different values of the IV (Tate, 1992). For example, we would be interested in the odds ratio for being classified as a "developing nation" ($Y=0$) for a given increase in the value of the score on the combination of the three significant predictors of *develop*, namely *urban, gdp* and *hospbed*.

The ultimate model obtained by a logistic regression analysis is a nonlinear function (Tate, 1992). A key concept in logistic regression with which the reader must be familiar is known as a logit. A **logit** is the natural logarithm of the odds—an operation that most pocket calculators will perform. Again extending our simplified example, the logit for our odds of rolling a "3" would be

$$\ln(.200) = -1.609$$

Specifically, in logistic regression, $\hat{Y}$ is the probability of having one outcome or another based on a nonlinear model resulting from the best linear combination of predictors. We can combine the ideas of probabilities, odds, and logits into one equation:

$$\hat{Y}_i = \frac{e^u}{1 + e^u}$$
(Equation 11.2)

where $\hat{Y}_i$ is the estimated probability that the ith case is in one of the categories of the DV, and e is a constant equal to 2.718, raised to the power u, where u is the usual regression equation:

$$u = B_0 + B_1X_1 + B_2X_2 + \ldots + B_kX_k$$
(Equation 11.3)

The linear regression equation (u) is then the natural log of the probability of being in one group divided by the probability of being in the other group (Tabachnick & Fidell, 1996). The linear regression equation creates the logit or log of the odds:

$$\ln\left(\frac{\hat{Y}}{1-\hat{Y}}\right) = B_0 + B_1X_1 + B_2X_2 + \ldots + B_kX_k \qquad \text{(Equation 11.4)}$$

We tend to agree with George & Mallery (2000), who state in their text, "[These] equation[s are] probably not very intuitive to most people…it takes a lot of experience before interpreting logistic regression equations becomes intuitive." For most researchers, focus should more appropriately be directed at assessing the fit of the model, as well as its overall predictive accuracy.

Interpretation of Results

The output for logistic regression can be divided into three parts: the statistics for overall model fit, a classification table, and the summary of model variables. Although these components were introduced briefly in Section 11.1, a more detailed description will be presented here. The reader should note that the output for logistic regression looks considerably different from previous statistical methods since the output is presented in text and not in pivot tables. One should also keep in mind that the output can vary depending upon the stepping method utilized in the procedure. If a stepping method is applied, you have the option of presenting the output for each step or limiting the output to the last step. When output for each step is selected, the three components will be displayed for each step. Due to space constraints, our discussion of output and its subsequent interpretation will primarily be limited to output from the final step.

Several statistics for the overall model are presented in the first component of logistic regression output. The –2 Log Likelihood provides an index of model fit. A perfect model would have a –2 Log Likelihood of 0; consequently the lower this value, the better the model fits the data. This value actually represents the sum of the "probabilities associated with the predicted and actual outcomes for each case" (Tabachnik & Fidell, 1996, p. 582). The Goodness-of-Fit statistic is also calculated for the overall model and compares the predicted values of the subjects to their actual values. This value should also be relatively small. The next two values, Cox & Snell — R^2 and Nagelkerke — R^2, represent two different estimates of the amount of variance in the DV accounted for by the model. Chi-square statistics with levels of significance are also computed for the model, block, and step. Chi-square for the model represents the difference between the constant-only model and the model generated. When using a stepwise method, the model generated will include only selected predictors; in contrast, the enter method generates a model with all the IVs included. Consequently, this comparison varies depending on the stepping method utilized. In general, a significant model chi-square indicates that the generated model is significantly better in predicting subject membership than the constant-only model. However, the reader should note that a large sample size increases the likelihood of finding significance when a poor-fitting model may have been generated. Chi-square is also calculated for each step if a stepping method has been utilized. This value indicates the degree of model improvement when adding a selected predictor or, in other words, represents the comparison between the model generated from the previous step to the current step.

The second component of output to interpret is the classification table. This table applies the generated regression model to predicting group membership. These predictions are then compared to the actual subject values. The percent of subjects correctly classified is calculated and serves as another indicator of model fit.

The third component of output is the summary of model variables. This summary presents several statistics—*B, S.E., Wald, df, Sig., R, Exp(B)*—for each variable included in the model as well as the constant. *B*, as in multiple regression, represents the unstandardized regression coefficient and represents the effect the IV has on the DV. *S.E.* is the standard error of *B*. *Wald* is a measure of significance for *B* and represents the significance of each variable in its ability to contribute the model. Since several sources indicate that the *Wald* statistic is quite conservative (Tabachnick & Fidell, 1996), a more liberal significance level (i.e., $p<.05$ or $p<.1$) should be applied when interpreting this value. Degrees of freedom (*df*) and level of significance (*Sig.*) are also reported for the *Wald* statistic within the summary table. The partial correlation (*R*) of each IV with the DV (independent from the other model variables) is also presented. The final value presented in the summary table is *Exp(B),* which is the calculated odds ratio for each variable. The odds ratio represents the increase (or decrease if *Exp(B)* is less than 1) in odds of being classified in a category when the predictor variable increases by 1.

Figure 11.4 Tolerance Statistics for Example Number 1.

Coefficients^a

Model		B	Std. Error	Beta	t	Sig.	Tolerance	VIF
1	(Constant)	48.105	12.241		3.930	.000		
	URBAN	.116	.167	.080	.696	.488	.382	2.620
	GDP	2.099E-03	.001	.366	3.048	.003	.352	2.840
	DEATHRAT	-.755	.674	-.094	-1.121	.265	.717	1.395
	RADIO	-.197	.117	-.167	-1.683	.095	.517	1.934
	HOSPBED	2.921E-02	.129	.026	.226	.822	.368	2.719
	DOCS	1.077	.449	.339	2.399	.018	.254	3.944
	POP92	1.896E-02	.019	.073	.999	.320	.941	1.062

a. Dependent Variable: SEQUENCE

Tolerance for all variables exceeds .1; multicollinearity is not a problem.

Applying this process to our original example, we sought to investigate which IVs (population; percent of population living in urban areas; gross domestic product; death rate; number of radios, hospital beds, and doctors) are predictors of status as a developing nation (i.e., develop*ed* or develop*ing*). Since our investigation is exploratory in nature, we utilized the forward stepping method, such that only IVs that significantly predict the DV will be included in the model. Data were first screened for missing data and outliers. A preliminary multiple **Regression** was conducted to calculate Mahalanobis distance (to identify outliers) and examine multicollinearity among the seven predictors. Figure 11.4 presents the tolerance statistics for the seven predictors. Tolerance for all variables is greater than .1, indicating that multicollinearity is not a problem. The **Explore** procedure was then conducted to identify outliers (see Figure 11.5). Subjects with a Mahalanobis distance greater than $\chi^2(7)=24.322$ were eliminated. **Binary Logistic Regression** was then performed using the **Forward:LR** method. The three output components were presented in Figures 11.1 –11.3. Figure 11.1 indicates that the three variables, *gdp, hospbed,* and *urban* were entered into the overall model, which correctly classified 95.54% of the cases (see Figure 11.2). Figure 11.3 presents the summary of statistics for the model variables. Odds ratios, *Exp(B)* or e^B, indicated that as the variable urban increases by 1, subjects are 1.145 times more likely to be classified as "developing." The odds ratios for *gdp* and *hospbed* were both below 1, indicating that as *gdp* (e^B=.9990) and *hospbed* (e^B =.9202) increase by 1, the odds of being classified as "developing" decrease by the respective ratio.

Figure 11.5   Outliers for Mahalanobis Distance (Example Number 1).

Extreme Values

			Case Number	Value
MAH_11	Highest	1	68	67.13359
		2	67	42.98150
		3	84	38.17691
		4	69	24.20090
		5	83	23.28605
	Lowest	1	21	.95439
		2	40	1.09002
		3	18	1.15272
		4	9	1.30692
		5	42	1.32406

Eliminate cases that exceed $\chi2(7)=24.322$ at p=.001.

Writing Up Results

The results summary should always describe how variables have been transformed or deleted. The results for the overall model are reported within the narrative by first identifying the predictors entered into the model and addressing the following goodness of fit indices: -2 Log Likelihood, Goodness of Fit, and Model Chi-Square with degree of freedom and level of significance. The accuracy of classification should also be reported in the narrative. Finally, the regression coefficients for model variables should be presented in table and narrative format. The table should include B, *Wald, df,* level of significance, and odds ratio. The following results statement applies the example presented in Figures 11.1 – 11.3.

Forward logistic regression was conducted to determine which independent variables (population; percent of population living in urban areas; gross domestic product; death rate; number of radios, hospital beds, and doctors) were predictors of status as a developing nation (developed or developing). Data screening led to the elimination of three outliers. Regression results indicate the overall model of three predictors (gdp, hospital beds, and urban) was statistically reliable in distinguishing between developed and developing countries (-2 Log Likelihood=25.211; Goodness-of-Fit=26.407; $\chi^2(2)$=96.164, p<.0001). The model correctly classified 95.54% of the cases. Regression coefficients are presented in Table 1. Wald statistics indicated that all variables significantly predict country development. However, odds ratios for these variables indicate little change in the likelihood of country development.

Table 1. Regression Coefficients

	B	Wald	df	p	Odds Ratio
Urban	.1354	4.08	1	.0434	1.145
GDP	−.0010	6.60	1	.0102	.999
Hospital beds	−.0832	9.41	1	.0022	.920
Constant	3.0472	5.93	1	.0149	

SECTION 11.4 SAMPLE STUDY AND ANALYSIS

This section provides a complete example of the process of conducting logistic regression. This process includes the development of research questions, data screening methods, analysis methods, interpretation of output, and presentation of results. The example utilizes the data set *gss.sav* from the SPSS Web site.

Problem

In the previous chapter on discriminant analysis, we presented the second example that investigated the ability of seven IVs (age, gender, hours worked per week, years of education, income, number of siblings, and number of hours spent watching TV) to predict one's life perspective (dull, routine, exciting). For this example, we will utilize a similar scenario (*excluding the gender variable*); however, the DV will be recoded as dichotomous to fulfill the requirement of binary logistic regression. Since six IVs are being investigated, the forward stepping method will be applied. The following research question is generated to address this scenario:

> Can *life* perspective (dull/routine or exciting) be reliably predicted from the knowledge of an individuals age (*age*), hours worked per week (*hrs1*), years of education (*educ*), income (*rincom91*), number of siblings (*sibs*), number of hours spent watching TV per day (*tvhours*)?

Method

Prior to analysis, the variable of *life* was recoded as dichotomous (*life2*) and applied the following transformations: 0=missing, 1-2=0, 3=1, 8-9=missing. Data were screened for missing data and outliers. A preliminary multiple **Linear Regression** was conducted to calculate Mahalanobis' Distance and to evaluate multicollinearity among the six continuous predictors. The table of regression coefficients (see Figure 11.6) indicate that multicollinearity was not violated since tolerance statistics for all six IVs were greater than .1. **Explore** was then conducted to determine which cases exceeded the chi square criteria of $\chi^2(6)=22.458$ at $p=.001$. All subjects that exceeded this value were eliminated from the analysis (see Figure 11.7). **Binary Logistic Regression** was then conducted using Forward: LR method.

Figure 11.6 Tolerance Statistics for Example Number 2.

Coefficients^a

Model		Unstandardized Coefficients		Standardized Coefficients	t	Sig.	Collinearity Statistics	
		B	Std. Error	Beta			Tolerance	VIF
1	(Constant)	1026.574	113.922		9.011	.000		
	AGE	-1.618	1.304	-.045	-1.241	.215	.893	1.119
	HRS1	2.897	1.169	.096	2.478	.013	.763	1.310
	EDUC	-12.739	5.841	-.081	-2.181	.029	.824	1.213
	RINCOM91	-11.601	3.403	-.143	-3.409	.001	.655	1.527
	SIBS	-10.188	5.648	-.063	-1.804	.072	.947	1.056
	TVHOURS	3.277	8.660	.013	.378	.705	.918	1.089

a. Dependent Variable: ID

Tolerance for all variables exceeds .1; multicollinearity is not a problem.

Figure 11.7 Outliers for Mahalanobis Distance (Example Number 2).

Extreme Values

			Case Number	Value
MAH_1	Highest	1	466	126.80049
		2	1360	114.05523
		3	406	56.53054
		4	50	54.06832
		5	121	42.50807
	Lowest	1	649	.25234
		2	561	.26159
		3	734	.30765
		4	1032	.45789
		5	266	.48138

Eliminate cases that exceed $\chi^2(6)=22.458$ at $p=.001$.

Output and Interpretation of Results

The three components of output are presented in Figures 11.8–11.10. The statistics for overall model fit are presented in Figure 11.8 and indicate that only two variables were entered into the model: *educ* and *rincom91*. Model fit statistics were extremely large and reveal a poor-fitting model, -2Log Likelihood=748.595, Goodness of Fit=564.598. The generated model was significantly different from the constant-only model, $\chi^2(1)=33.098$, p<.0001. Figure 11.9 presents the classification table and indicates that the model correctly classified only 59.57% of subjects. The summary of model variables is displayed in Figure 11.10. Odds ratios for the *educ* (e^B=1.1609) and *rincom91* (e^B=1.0403) revealed little increase in the likelihood of perceiving life as exciting when the predictors increase by 1.

323

Figure 11.8 Goodness-of-fit Indices for Example Number 2.

```
Dependent Variable..   LIFE2

Beginning Block Number  0.  Initial Log Likelihood Function

-2 Log Likelihood   781.69271

* Constant is included in the model.


Beginning Block Number  1.  Method: Forward Stepwise (LR)
```

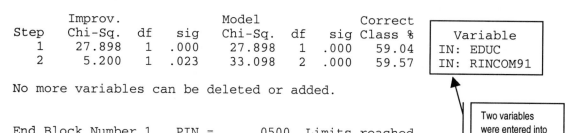

```
        Improv.              Model              Correct
Step    Chi-Sq.  df   sig    Chi-Sq.  df   sig  Class %      Variable
  1     27.898   1   .000    27.898   1   .000   59.04     IN: EDUC
  2      5.200   1   .023    33.098   2   .000   59.57     IN: RINCOM91
```

Two variables were entered into the model.

```
No more variables can be deleted or added.


End Block Number 1   PIN =      .0500  Limits reached.

Final Equation for Block 1

Estimation terminated at iteration number 3 because
Log Likelihood decreased by less than .01 percent.
```

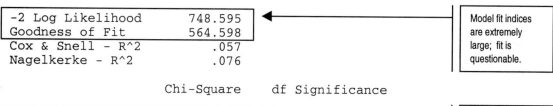

```
-2 Log Likelihood      748.595
Goodness of Fit        564.598
Cox & Snell - R^2         .057
Nagelkerke - R^2          .076
```

Model fit indices are extremely large; fit is questionable.

```
               Chi-Square    df Significance
Model            33.098      1     .0000
Block            33.098      2     .0000
Step              5.200      1     .0226
```

Model significantly predicts group membership.

Figure 11.9 Classification Table for Example Number 2.

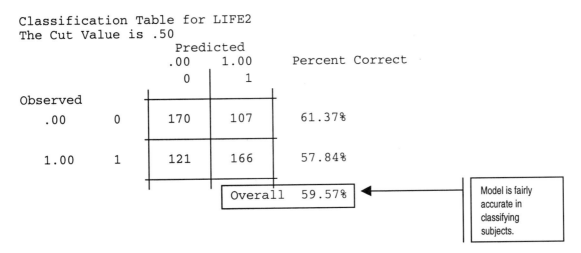

```
Classification Table for LIFE2
The Cut Value is .50
                      Predicted
                    .00    1.00      Percent Correct
                      0      1
Observed
   .00       0      170     107      61.37%

  1.00       1      121     166      57.84%

                        Overall   59.57%
```

Model is fairly accurate in classifying subjects.

Figure 11.10 Regression Coefficients for Example Number 2.

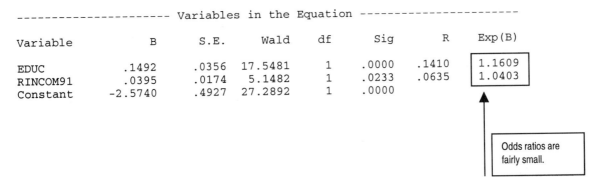

```
--------------------- Variables in the Equation ----------------------

Variable          B       S.E.     Wald    df     Sig       R      Exp(B)

EDUC            .1492    .0356   17.5481    1    .0000    .1410    1.1609
RINCOM91        .0395    .0174    5.1482    1    .0233    .0635    1.0403
Constant       -2.5740   .4927   27.2892    1    .0000
```

Odds ratios are fairly small.

Presentation of Results

Forward logistic regression was conducted to determine which independent variables (age, hours worked per week, years of education, income, number of siblings, and number of hours spent watching TV) are predictors of life perspective (dull/routine or exciting). Data screening led to the elimination of several outliers. Regression results indicated the overall model fit of two predictors (education and income) was questionable (-2 Log Likelihood=748.595, Goodness of Fit=564.598) but was statistically reliable in distinguishing between life perspective; $\chi^2(1)=33.098$, p<.0001). The model correctly classified only 59.57% of the cases. Regression coefficients are presented in Table 1. *Wald* statistics indicated that education and income significantly predict life perspective. However, odds ratios for these variables indicated little change in the likelihood of perceiving life as exciting.

Table 1. Regression Coefficients

	B	Wald	df	p	Odds Ratio
Education	.1492	17.55	1	<.0001	1.1609
Income	.0395	5.15	1	.0233	1.0403
Constant	−2.5740	27.29	1	<.0001	

SECTION 11.5 SPSS "HOW TO"

This section demonstrates the steps for conducting binary logistic regression with Example Number 2 of this chapter. Prior to conducting binary logistic regression, be sure to dichotomize (0,1) your DV. To conduct **Binary Logistic Regression**, select the following menus:

Analyze
> **Regression**
>> **Binary Logistic**

Logistic Regression Dialogue Box (see Figure 11.11)

Once in this dialogue box, identify the DV (*life2*) and move it to the Dependent box. Identify each IV and move each to the Covariates box. Next, select the desired regression method. SPSS provides seven different methods, five of which are described as follows:

Enter—Enters all the IVs at once into the model, regardless of significant contribution. This method is useful if you have previously tested the IVs and want all of them to be entered.

Forward: LR—One of the most common methods. Enters IVs, one at a time. The likelihood-ratio is used to determine variable selection.

Forward: Wald— Enters IVs, one at a time. The *Wald* statistic is used to determine variable selection.

Backward: LR—All IVs are entered at once, then variables are removed one at a time. The likelihood-ratio is used to determine variable removal.

Backward: Wald— All IVs are entered at once, then variables are removed one at a time. The *Wald* statistic is used to determine variable removal.

For our example, we selected **Forward: LR**. Next, click **Categorical**.

Logistic Regression: Define Categorical Variable Dialogue Box (see Figure 11.12)

By default in logistic regression, SPSS treats any numerical variable as continuous. Consequently, when an IV is categorical, you need to specify how SPSS should address it. Once in this dialogue box, identify any categorical variables and move them to Categorical Covariates box. Then under Change Contrast, select the method of contrast. SPSS provides several contrast methods. The Indicator method is the default and is the most common. Three of the contrasting methods are described as follows.

Indicator—Indicates the presence or absence of group membership. This is the default.

Simple—Each category of the IV (except the reference category) is compared to the reference category.

Deviation—Each category of the IV (except the reference category) is compared to the overall effect.

If you have selected one of these contrasting methods, you can identify a specific category to be used as the **Reference Category**. Two options are available: **Last** category (the default) or **First** category. If you have selected any options other than defaults for contrasting methods or reference category, you must then click **Change**. Click **Continue**, then **Options**.

Figure 11.11   Logistic Regression Dialogue Box.

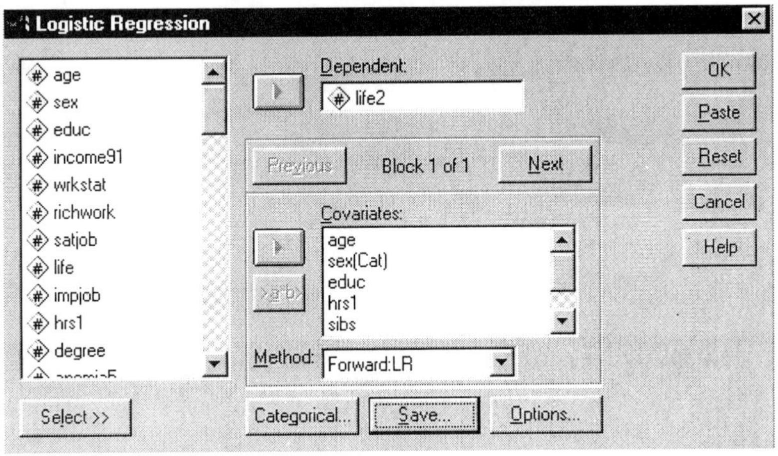

Figure 11.12   Logistic Regression: Define Categorical Variable Diaglogue Box.

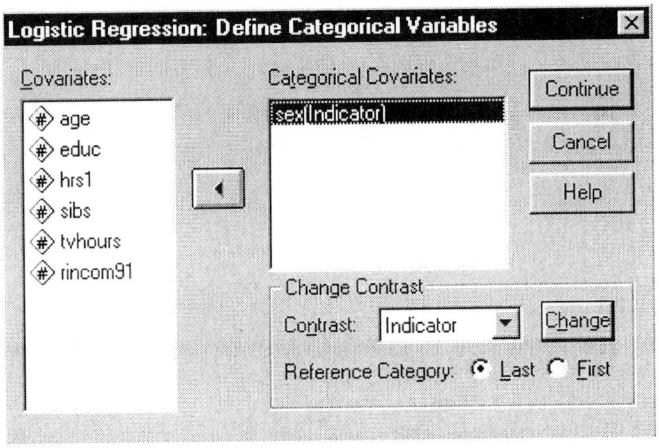

Logistic Regression: Options Dialogue Box (see Figure 11.13)

SPSS provides several options within logistic regression. Commonly used options are described below.

Classification Plots—Graph of actual and predicted values for the DV.

Correlations of Estimates—Correlation matrix of parameter estimates for model variables.

Iteration History—Presents coefficients and log likelihood at each iteration.

CI for exp(B)—Calculates the confidence intervals for the odds ratios of each model variable. You can indicate the level of probability associated with this interval. The default is 95%.

For our example, we did not select any of these options. A display option is also available if a stepping method has been utilized. You may want to select the display **At Last Step**, as we did, to conserve space. Other options available are the probability for stepwise, the maximum number of iterations, and inclusion of constant in the model. We maintained the defaults for these options; however, there may be times when it is necessary to increase the maximum number of iterations in order to generate a complete model. Once you have selected the appropriate options, click **Continue**, then **OK**. The reader should note that SPSS also provides options for saving variables. By clicking **Save**, you can save predicted values, residuals, etc., as new variables.

Figure 11.13 Logistic Regression: Options Dialogue Box.

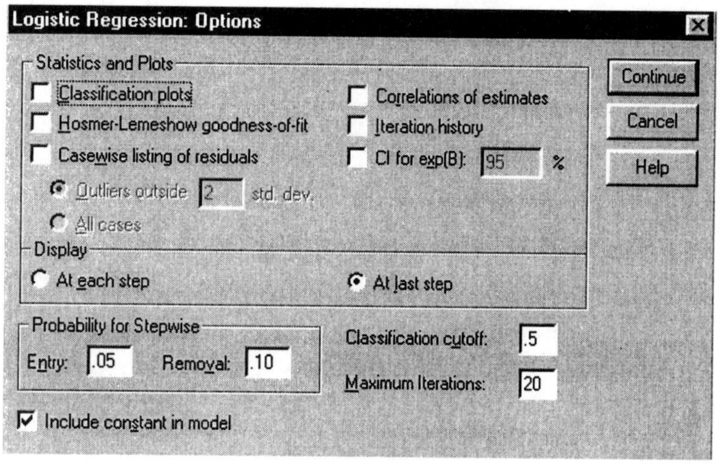

Summary

Logistic regression tests the ability of a model or group of variables to predict group membership as defined by some categorical DV. In binary logistic regression, the DV must be dichotomous, but the IVs may be categorical or continuous. Logistic regression actually predicts the probability of membership occurring, which varies from 0 to 1. A variety of methods can be used to test and develop different models (enter, Forward: LR, Backward: Wald, etc.). Although logistic regression requires fulfillment of few test assumptions, data should be screened for outliers and multicollinearity. Logistic regressions output includes three parts: statistics for overall model fit, classification table, and summary of model variables. Statistics for the overall model provide several indices of model fit: –2 Log Likelihood, Goodness-of-Fit, and Model Chi-Square. The classification table presents the percent of cases correctly classified with the generated model. The summary of model variables provides several variable statistics that indicate variable contribution to the model: *B, Wald, df,* level of significance, and odds ratio. A good fitting model will typically have: fairly low values for –2 Log Likelihood and Goodness-of-Fit, significant Model Chi-Square and variables with odds ratios greater than 1. Figure 11.14 provides a checklist for conducting binary logistic regression.

Figure 11.14 Checklist for Conducting Binary Logistic Regression.

I.   Screen Data
  a. Missing Data?
  b. Multivariate Outliers and Multicollinearity?
    ❑ Run preliminary Linear Regression.
      1. ᧆ **Analyze…Regression…Linear.**
      2. Identify a variable that serves as a case number and move to Dependent Variable box.
      3. Identify all appropriate quantitative variables and move to Independent(s) box.
      4. ᧆ **Statistics.**
      5. Check **Collinearity Diagnostics.**
      6. ᧆ **Continue.**
      7. ᧆ **Save.**
      5. Check **Mahalanobis'.**
      6. ᧆ **Continue.**
      7. ᧆ **OK.**
      8. Determine chi square χ^2 critical value at $p<.001$.
    ❑ Conduct **Explore** to test outliers for Mahalanobis chi square χ^2.
      1. ᧆ **Analyze…Descriptive Statistics…Explore**
      2. Move *mah_1* to Dependent Variable box.
      3. Leave Factor box empty.
      4. ᧆ **Statistics.**
      5. Check **Outliers.**
      6. ᧆ **Continue.**
      7. ᧆ **OK.**
    ❑ Delete outliers for subjects when χ^2 exceeds critical χ^2 at $p<.001$.

II. Conduct Logistic Regression
  a. Run Binary Logistic Regression using **Regression**.
      1. ᧆ **Analyze…** ᧆ **Regression…** ᧆ **Binary Logistic…**
      2. Move the DV to the Dependent Box.
      3. Move IVs to the Covariates Box.
      4. Select Method.
      5. ᧆ **Categorical** (if any IVs are categorical).
      6. Move any categorical IVs to the Categorical Covariates Box.
      7. Select Contrast Method and Reference Category.
      8. ᧆ **Continue.**
      9. ᧆ **Options.**
      10. Check appropriate options.
      11. ᧆ **Continue.**
      12. ᧆ **OK.**

III. Summarize Results
  a. Describe any data elimination or transformation.
  b. Describe the model generated (-2 Log Likelihood, Goodness of Fit, Model chi-square with *df* and *p*-value).
  c. Report the accuracy of classification.
  d. Present the regression coefficients for model variables in table format.
  e. Report odds ratios for model variables.
  f. Draw conclusions.

Exercises for Chapter 11

This exercise utilizes the data set *gss.sav*, which can be downloaded at the SPSS Web site. Open the URL: **www.spss.com/tech/DataSets.html** in your Web browser. Scroll down until you see "Data Used in SPSS Guide to Data Analysis—8.0 and 9.0" and click on the link "dataset.exe." When the "Save As" dialogue appears, select the appropriate folder and save the file. Preferably, this should be a folder created in the SPSS folder of your hard drive for this purpose. Once the file is saved, double-click the "dataset.exe" file to extract the data sets to the folder.

Conduct a Forward: LR logistic regression analysis with the following variables:

IV—*age, educ, hrsl, life, sibs, rincom91*
DV—*satjob2*

1. Develop a research question for the following scenario.

2. Conduct a preliminary **Linear Regression** to identify outliers and evaluate multi-collinearity among the five continuous variables. Complete the following:

   a. Using the Chi-Square table, identify the critical value at p<.001 for identifying outliers. Use **Explore** to determine if there are outliers. Which cases should be eliminated?

   b. Is multicollinearity a problem among the five continuous variables?

3. Conduct **Binary Logistic Regression** using the Forward: LR method. Be sure to identify the variable of *life* as categorical (use the defaults).

   a. Which variables were entered into the model?

   b. To what degree does the model fit the data? Explain.

   c. Is the generated model significantly different from the constant-only model?

   d. How accurate is the model in predicting job satisfaction?

   e. What are the odds ratios for the model variables? Explain.

VARIABLE NAME	VARIABLE LABEL	CAT or QUANT	SCALE
buying.sav			
husbr1	Husband buy videotape unit	CAT	1=Definitely; 2=Very likely 3=Somewhat likely; 4=Indifferent; 5=Somewhat likely; 6=Very unlikely; 7=Definitely not
husbr2	Husband buy pop-top cans	CAT	
husbr3	Husband buy alarm system	CAT	
husbr4	Husband use automatic teller	CAT	
husbr5	Husband buy big TV	CAT	
wifer1	Wife buy videotape unit	CAT	
wifer2	Wife buy pop-top cans	CAT	
wifer3	Wife buy alarm system	CAT	
wifer4	Wife use automatic teller	CAT	
wifer5	Wife buy big screen TV	CAT	
hpredw1	Husband predicts wife buy VTR	CAT	
hpredw2	Husband predicts wife buys pop-top cans	CAT	
hpredw3	Husband predicts wife buy alarm system	CAT	
hpredw4	Husband predicts wife use automatic teller	CAT	
hpredw5	Husband predicts wife buy big screen TV	CAT	
wpredh1	Wife predicts husband buy VTR	CAT	
wpredh2	Wife predicts husband buy pop-top cans	CAT	
wpredh3	Wife predicts husband buy alarm system	CAT	
wpredh4	Wife predicts husband use automatic teller	CAT	
wpredh5	Wife predicts husband buy big screen TV	CAT	
husbinf1	Husband's influence rating	CAT	1 = Husb decide; 2 = Husb more influence; 3 = Husb/wife equal; 4 = Wife more influence; 5 = Wife decide
husbinf2	Husband's influence rating	CAT	
husbinf3	Husband's influence rating	CAT	
husbinf4	Husband's influence rating	CAT	
hsbinf5	Husband's influence rating	CAT	
wifeinf1	Wife's influence rating	CAT	
wifeinf2	Wife's influence rating	CAT	
wifeinf3	Wife's influence rating	CAT	
wifeinf4	Wife's influence rating	CAT	
wifeinf5	Wife's influence rating	CAT	
picture	Picture accompanied question	CAT	0= No pictures; 1= Pictures
hsumbuy	Sum of husband's buying scores	QUANT	
hsuminf	Sum of husband's influence scores	QUANT	
hpredsum	Sum of husband's predictions	QUANT	
wsumbuy	Sum of wife's buying scores	QUANT	
wpredsum	Sum of wife's predictions	QUANT	
wsuminf	Sum of wife's influence scores	QUANT	
famscore	Family buying score	QUANT	
husbsays	Husband's response	CAT	1=Buy; 2 = Indifferent; 3 = Not buy
wifepred	Wife's prediction of husband	CAT	1=Buy; 2 = Indifferent; 3 = Not buy
cntry15.sav			
country	Country	String	
pop92	Population 1992, in millions	QUANT	
urban	Percent urban, 1992	QUANT	
gdp	GDP per capita	QUANT	
lifeexpm	Male life expectancy 1992	QUANT	

lifeexpf	Female life expectancy 1992	QUANT	
birthrat	Birth rate per 1,000 population 1992	QUANT	
deathrat	Death rate per 1,000 population	QUANT	
infmr	Infant mortality rate 1992 per thousand live births	QUANT	
fertrate	Fertility rate per woman 1990	QUANT	
region	Region of the world	CAT	1= Eastern Africa; 2 = Middle Africa; 3 = Northern Africa; 4 = Southern Africa; 5=Western Africa; 6= Caribbean; 7 =Central America; 8 = South America; 9 = North America; 10 = Eastern Asia; 11 = Southeast Asia; 12 = Southern Asia; 13 = Western Asia; 14 = Eastern Europe; 15 = Northern Europe; 16= Southern Europe; 17= Western Europe; 18 = Oceania; 19 = USSR
develop	Status as developing country	CAT	0 =Developed country; 1 = Developing country
radio	Radios per 100 people	QUANT	
phone	Phones per 100 people	QUANT	
hospbed	Hospital beds per 10,000 people	QUANT	
docs	Doctors per 10,000 people	QUANT	
lndocs	Natural log of doctors per 10,000	QUANT	
lnradios	Natural log of radios per 100 people	QUANT	
lnphone	Natural log of phones per 100 people	QUANT	
lngdp	Natural log of GDP	QUANT	
lnbeds	Natural log of hospital beds/10,000	QUANT	
sequence	Arbitrary id	QUANT	
country.sav			
country	Country	String	
pop92	Population 1992 in millions	QUANT	
urban	Percent urban, 1992	QUANT	
gdp	GDP per capita	QUANT	
lifeexpm	Male life expectancy 1992	QUANT	
lifeexpf	Female life expectancy 1992	QUANT	
birthrat	Birth rate per 1,000 population 1992	QUANT	
deathrat	Death rate per 1,000 population	QUANT	
infmr	Infant mortality rate 1992 per thousand live births	QUANT	
fertrate	Fertility rate per woman 1990	QUANT	
region	Region of the world	CAT	1= Eastern Africa; 2 = Middle Africa; 3 = Northern Africa; 4 = Southern Africa; 5=Western Africa; 6= Caribbean; 7 =Central America; 8 = South America; 9 = North America; 10 = Eastern Asia; 11 = Southeast Asia; 12 = Southern Asia; 13 = Western Asia; 14 = Eastern Europe; 15 = Northern Europe; 16= Southern Europe; 17= Western Europe; 18 = Oceania; 19 = USSR
develop	Status as developing country	CAT	0 =Developed country; 1 = Developing country
radio	Radios per 100 people	QUANT	
phone	Phones per 100 people	QUANT	
hospbed	Hospital beds per 10,000 people	QUANT	
docs	Doctors per 10,000 people	QUANT	
lndocs	Natural log of doctors per 10,000	QUANT	
lnradios	Natural log of radios per 100 people	QUANT	
lnphone	Natural log of phones per 100 people	QUANT	
lngdp	Natural log of GDP	QUANT	
sequence	Arbitrary id	QUANT	
lnbeds	Natural log of hospital beds per 10000	QUANT	

Appendix A

electric.sav			
caseid	Case identification number	QUANT	
firstchd	First CHD event	CAT	1= No CHD; 2 = Sudden death; 3 =Nonfatal MI; 5 = Fatal MI; 6 = Other CHD
age	Age at entry	QUANT	
dbp58	Average diast blood pressure 1958	QUANT	
eduyr	Years of education	QUANT	
chol58	Serum cholesterol 1958—MG per DL	QUANT	
cgt58	No. of cigarettes per day in 1958	QUANT	
ht58	Stature 1958 –to nearest 0.1 inch	QUANT	
wt58	Body weight 1958—LBS	QUANT	
dayofwk	Day of death	CAT	1 = Sunday; 2 = Monday; 3 = Tuesday; 4 = Wednesday; 5= Thursday; 6 = Friday; 7 = Saturday; 9 =Missing
vital10	Status at ten years	CAT	0 = Alive; 1 = Dead
famhxcvr	Family history of CHD	CAT	N = No; Y =Yes
chd	Incidence of coronary heart disease	CAT	0 = None; 1 = CHD
hist	Family history of CHD	CAT	0 = No family history
educcat	Highest level of schooling	CAT	1= Grammar school; 2 = High school; 3 = College
endorph.sav			
before	Before	QUANT	
after	After	QUANT	
diff	Diff	QUANT	
gss.sav			
age	Age of respondent	QUANT	
sex	Respondent's sex	CAT	1= Male; 2 = Female
educ	Highest year of school completed	CAT	97 = NAP; 98 =DK; 99 = NA
income91	Total family Income	CAT	0 = Not applicable; 1 = LT$1000; 2 = $1000-$2999; 3 = $3000-$3999; 4 = $4000-$4999; 5 = $5000- $5999; 6= 6000-$6999; 7 = $7000-$7999; 8 = $8000-$9999; 9 = $10000-$12499; 10 = 12500-$14999; 11 = $15000-$17499; 12 = $17500-$19999; 13 = $20000-$22499; 14 = $22500-$24999; 15 = $25000- $29999; 16 = $30000-$34999; 17 = $35000-39999; 18 = $40000-$49999; 19 = $50000-$59999; 20 = $60000-$74999; 21 = $75000+
wrkstat	Labor force status	CAT	0 = Not applicable; 1 = Working full-time; 2 = Working part-time; 3 = Temp not working; 4 = Unemployed/laid off; 5 = Retired ; 6 = School; 7 = Keeping house; 8 = Other; 9 = NA
richwork	If rich, continue or stop working	CAT	0 = Not applicable; 1 = Continue working; 2 = stop working; 8 = Don't know; 9 = No answer
satjob	Job satisfaction	CAT	0 = Not applicable; 1 = Very satisfied ; 2 = Mod satisfied; 3 = A little dissatisfied ; 4 = Very dissatisfied; 8 = Don't know; 9 = No answer
life	Is life exciting or dull	CAT	0= Not applicable; 1 = Dull; 2 = Routine ; 3 = Exciting; 8 = Don't know; 9 = No answer
impjob	Importance to R of having a fulfilling job	CAT	0 = Not applicable; 1 = One of the most important; 2 = Very important; 3 = Somewhat important; 4 = Not too important; 5 = Not at all important; 8 = Don't know; 9 = No answer
hrs1	Number of hours worked last week	CAT	-1= Not applicable; 98 = Don't know; 99 = No answer
degree	RS highest degree	CAT	0 = Less than high school; 1 = High school; 2 = Junior college; 3 = Bachelor; 4 = Graduate ; 7 = Not applicable; 8 = Don't know; 9 =No answer
anomia5	Lot of average man getting worse	CAT	0 = Not applicable; 1 = Agree ; 2 = Disagree; 8 =Don't know; 9 = No answer

Appendix A

degree2	College degree	CAT	0 = No college degree; 1 = College degree; 7 = Not applicable; 8 = Don't know; 9 = No answer
maeduc	Highest year of school completed, Mother	CAT	97 = Not applicable; 98 = Don't know; 99 = No answer
paeduc	Highest year of school completed, Father	CAT	97 = Not applicable; 98 = Don't know; 99 = No answer
macolleg	Mother a college grad	CAT	0 = No; 1 =Yes
pacolleg	Father a college grad	CAT	0 = No; 1 =Yes
satjob2	Job satisfaction	CAT	1= Very satisfied; 2 = Not very satisfied
marital	Marital status	CAT	1 = Married; 2 = Widowed; 3 = Divorced; 4 = Separated; 5 = Never married; 9 = NA
agewed	Age when first married	QUANT	0= Not applicable; 98 = Don't know; 99 = No answer
spwrksta	Spouses labor force status	CAT	0 = Not applicable; 1 = Working full time; 2 =Working part time; 3 = Temp not working; 4 = Unemployed, laid off; 5 = Retired; 6 = School; 7 = Keeping house; 8 = Other; 9 = No answer
sphrs1	No. of hours spouse worked last week	QUANT	-1= Not applicable; 98 = Don't know; 99 = No answer
sibs	Number of brothers and sisters	QUANT	-1= Not applicable; 98 = Don't know; 99 = No answer
zodiac	Respondents' astrological sign	CAT	0 = Not applicable; 1 = Aries; 2 = Taurus; 3 = Gemini; 4 = Cancer; 5 = Leo; 6 = Virgo ; 7 = Libra; 8 = Scorpio; 9 = Sagittarius; 10 = Capricorn; 11= Aquarius; 12 = Pisces; 98 = Don't know; 99 = No answer
speduc	Highest year school completed, spouse	QUANT	97 = Not applicable; 98 = Don't know; 99 = No answer
spdeg	Spouse highest degree	CAT	0 = Less than high school; 1 = High school; 2 = Junior college; 3 = Bachelor; 4 = Graduate; 7 = Not applicable; 8 = Don't know; 9 =No answer
partyid	Political party affiliation	CAT	0 = Strong democrat; 1 = Not str democrat; 2 = Ind, near dem; 3 = Independent; 4 = Ind, near rep; 5 = Not str republican; 6 = Strong republican; 7 = Other party; 8 = Don't know 9 = No Answer
vote92	Did R vote in 1992 election	CAT	0 = Not applicable; 1 = Voted; 2 = Did not vote; 3 = Not eligible; 4 = Refused ; 8 = Don't know; 9 = No answer
pres92	Vote for Clinton, Bush, Perot	CAT	0 = Not applicable; 1 = Clinton; 2 = Bush; 3 = Perot; 4 = Other; 6 = No pres vote; 8 = Don't know; 9 = No Answer
postlife	Belief in life after death	CAT	0 = Not applicable; 1 = Yes; 2 = No; 8 = Don't know; 9 = No answer;
happy	General happiness	CAT	0 = Not applicable; 1 = Very happy; 2 = Pretty happy; 3 = Not too happy; 8 = Don't know; 9 = No answer
hapmar	Happiness of marriage	CAT	0 = Not applicable; 1 = Very happy; 2 = Pretty happy; 3 = Not too happy; 8 = Don't know; 9 = No answer
jobinc	How important is a high income	CAT	0 = Not applicable; = Most impt; 2 = Second; 3 = Third; 4 = Fourth; 5 = Fifth; 8 = DK; 9 = NA
classicl	Like or dislike classical music	CAT	0 = Not applicable; 1 = Like very much; 2 = Like it; 3 = Mixed feelings; 4 = Dislike it; 5 = Dislike it very much; 8= DK much about it; 9 = No answer
opera	Like or dislike opera	CAT	0 = Not applicable; 1 = Like very much; 2 = Like it; 3 = Mixed feelings; 4 = Dislike it; 5 = Dislike it very much; 8= DK much about it; 9 = No answer
country	Like or dislike country western music	CAT	0 = Not applicable; 1 = Like very much; 2 = Like it; 3 = Mixed feelings; 4 = Dislike it; 5 = Dislike it very much; 8= DK much about it; 9 = No answer
rincom91	Respondents' income	CAT	0 = Not applicable; 1 = LT$1000; 2 = $1000-$2999; 3 = $3000-$3999; 4 = $4000-$4999; 5 = $5000- $5999; 6= 6000-$6999; 7 = $7000-$7999; 8 = $8000-$9999; 9 = $10000-$12499; 10 = 12500-$14999; 11 = $15000-$17499; 12 = $17500-$19999; 13 = $20000-

			$22499; 14 = $22500-$24999; 15 = $25000- $29999; 16 = $30000-$34999; 17 = $35000-39999; 18 = $40000-$49999; 19 = $50000-$59999; 20 = $60000-$74999; 21 = $75000+
tvhours	Hours per day watching TV	QUANT	
finrela	Opinion of family income	CAT	1= Far below average; 2 = Below average; 3 = Average; 4 = Above average; 5 = Far above average; 8= DK; 9 = NA
wifeduc	Wife's education	QUANT	
husbeduc	Husband's education	QUANT	
incomdol	Family income recoded to dollars	QUANT	
rincomdol	Respondent's income recoded to dollars	QUANT	
id	Arbitrary id numbers	QUANT	
husbhr	Hrs worked last week by husband	QUANT	
wifehr	Hrs worked last week by wife	QUANT	
husbft	Husband employed full-time	CAT	0 = No; 1 = Yes
wifeft	Wife employed full-time	CAT	0 = No; 1 = Yes
educdiff	Educdiff	QUANT	
income4	Total family income in quartiles	QUANT	
gssft.sav			
age	Age of respondent	QUANT	
educ	Highest year of school completed	CAT	97 = NAP; 98 = DK; 99 = NA
sex	Respondent's sex	CAT	0 = Males; 1 = Females
degree	RS highest degree	CAT	0 = Less than high school; 1 = High school; 2 = Junior college; 3 = Bachelor; 4 = Graduate; 7 = Not applicable; 8 = Don't know; 9 =No answer
satjob	Job satisfaction	CAT	0 = Not applicable; 1 = Very satisfied; 2 = Mod satisfied; 3 = A little dissatisfied; 4 = Very dissatisfied; 8 = Don't know; 9 = No answer
satjob2	Job satisfaction	CAT	1= Very satisfied; 2 = Not very satisfied
income4	Total family income in quartiles	CAT	1 = 24999 or less; 2 = 25000 to 39,999; 3 = 40,000 to 59,999; 4 = 60,000 or more
rincom91	Respondent's income	CAT	0 = Not applicable; 1 = LT$1000; 2 = $1000-$2999; 3 = $3000-$3999; 4 = $4000-$4999; 5 = $5000- $5999; 6= $6000-$6999; 7 = $7000-$7999; 8 = $8000-$9999; 9 = $10000-$12499; 10 = $12500-$14999; 11 = $15000-$17499; 12 = $17500-$19999; 13 = $20000-$22499; 14 = $22500-$24999; 15 = $25000- $29999; 16 = $30000-$34999; 17 = $35000-39999; 18 = $40000-$49999; 19 = $50000-$59999; 20 = $60000-$74999; 21 = $75000+
wrkstat	Labor force status	CAT	0 = Not applicable; 1 = Working full time; 2 =Working part time; 3 = Temp not working; 4 = Unemployed, laid off; 5 = Retired; 6 = School; 7 = Keeping house; 8 = Other; 9 = No answer
jobinc	Importance of high income	CAT	0 = Not applicable; 1 = Most impt; 2 = Second; 3 = Third; 4 = Fourth; 5 = Fifth; 8 = DK; 9 = NA
impjob	Importance to respondent of having a fulfilling Job	CAT	0 = Not applicable; 1 = One of the most important; 2 = Very important; 3 =Somewhat important; 4 = Not too important; 5 = Not at all important; 8 = Don't know; 9 = No answer
bothft	Both employed full-time	CAT	0 = No; 1 = Yes
husbhr	Hrs worked last week by husband	QUANT	
wifehr	Hrs worked last week by wife	QUANT	
husbft	Husband employed full-time	CAT	0 = No; 1 = Yes
wifeft	Wife employed full-time	CAT	0 = No; 1 = Yes
id	Id	QUANT	
agecat94	4 Categories of Age	CAT	1 = 18-29; 2 = 30 – 39; 3 = 40- 49; 4 =50+
rincom2	Rincom2	QUANT	

iq.sav			
id	Student ID number	QUANT	
score		QUANT	
lambda.sav			
life	Life	CAT	1.00= Exciting; 2.00 = Pretty routine; 3.00 = Dull
city	City	CAT	1.00= Chicago; 2.00= New England; 3.00 = Miami, New York; 4.00 = Other
weight	Weight	QUANT	
renal .sav			
id	Id	QUANT	
type	Case/Control	CAT	0 = Control; 1= Case
age	Age	QUANT	
alive	Alive discharge	CAT	0 = No; 1 = Yes
los	Length of stay in hospital	QUANT	
iculos	ICU length of stay	QUANT	
bypass	Bypass surgery only	CAT	0 = No; 1 = Yes
gender	Gender	CAT	0 = Males; 1 = Females
hypertns	Preexisting hypertension	CAT	0 = No; 1 = Yes
diabetes	Adult onset diabetes mellitus	CAT	0 = No; 1 = Yes
chf	Preexisting congestive heart failure	CAT	0 = No; 1 = Yes
ami	Prior AMI	CAT	0 = No; 1 = Yes
priorsur	Prior cardiac surgery	CAT	0 = No; 1 = Yes
admbun	Admission BUN	QUANT	
admsrea	Admission creatine	QUANT	
preuric	Uric acid preop	QUANT	
prebun	Preoperative BUN	QUANT	
precreat	Preoperative creatine	QUANT	
maxbun	Highest BUN	QUANT	
maxcreat	Maximum creatine	QUANT	
daystomx	Days to maximum creatine	QUANT	
contdye	Preop contrast dye before surgery	CAT	0 = No; 1 = Yes
balnpre	Balloon pump preop	CAT	0 = No; 1 = Yes
orhours	Hours in operating room	QUANT	
pumphrs	Hours on bypass pump	QUANT	
hypotnsn	# of hypotensive episodes	QUANT	
balnpost	Balloon pump post operatively	QUANT	
reop	Reoperation	CAT	0 = No; 1 = Yes
sepsis	Sepsis	CAT	0 = No; 1 = Yes
respfail	Respiratory failure	CAT	0 = No; 1 = Yes
pneum	Pneumonia	CAT	0 = No; 1 = Yes
extub24	Extubation with 24 hours	CAT	0 = No; 1 = Yes
extub48	Extubation between 48 hours	CAT	0 = No; 1 = Yes
compnumb	Number of complications	QUANT	
recovery	Recovery of renal function	CAT	0 = No; 1 = Yes
oliguria	Urine output 400ml/day	CAT	0 = No; 1 = Yes
recovday	Days to recovery of renal function	QUANT	
finbun	Final BUN	QUANT	
fincreat	Final creatine value	QUANT	
valnbyps	Valve and bypass surgery	CAT	0 = No; 1 = Yes
laterf	Late acute renal failure	CAT	0 = No; 1 = Yes
riskfact	No of preop risk factors	QUANT	
cardrsk	No of cardiac risk factors	QUANT	
totpulm	Total pulmonary complication	QUANT	
extub2da	Extubation within 48 hours	CAT	0 = No; 1 = Yes

Appendix A

salary.sav			
id	Employee code	QUANT	
salbeg	Beginning salary	QUANT	
sex	Sex of employee	CAT	0 = Male; 1 = Female
time	Job seniority	QUANT	
age	Age of employee	QUANT	
salnow	Current salary	QUANT	
edlevel	Educational level	QUANT	
work	Work experience	QUANT	
jobcat	Employment category	CAT	1 = Clerical; 2 = Office trainee; 3 = Security officer; 4 = College trainee; 5 = Exempt employee; 6 = MBA trainee; 7 = Technical
minority	Minority classification	CAT	0 = White; 1 = Nonwhite
sexrace	Sex-race classification	CAT	1 = White males; 2 = Minority males; 3 = White females; 4 = Minority females
schools.sav			
school	School name		
loinc93	Percent low income	QUANT	
lep93	Percent limited English proficiency	QUANT	
lep94	Percent limited English proficiency 1994	QUANT	
grad93	Percent Graduating 1993	QUANT	
grad94	Percent Graduating 1994	QUANT	
act94	Average ACT score 1994	QUANT	
act93	Average ACT score 1993	QUANT	
pctact93	Percent taking ACT 1993	QUANT	
pctact94	Percent taking ACT 1994	QUANT	
math93	10th grade average math score	QUANT	
math94me	Percent meet or exceed state standards	QUANT	
mathch94	Change in percent meet/exceed 1994-1993	QUANT	
read93	10th average reading score	QUANT	
read94me	Percent meet or exceed state standards	QUANT	
readch94	Change in percent meet/exceed 1994-1993	QUANT	
scienc93	11th grade average science score	QUANT	
sci94me	Percent meet/exceed 1994-1993	QUANT	
scich94	Change in percent meet/exceed 1994-1993	QUANT	
id	Id	QUANT	
medloinc	Above or below median low income	CAT	0 = Below the median low income in 1993; 1 = Above the median low income in 1993

Appendix B: The Chi-Square Distribution

df	Proportion in critical region					
	0.10	0.05	0.025	0.01	0.005	0.001
1	2.71	3.84	5.02	6.63	7.88	10.828
2	4.61	5.99	7.38	9.21	10.60	13.816
3	6.25	7.81	9.35	11.34	12.84	16.266
4	7.78	9.49	11.14	13.28	14.86	18.467
5	9.24	11.07	12.83	15.09	16.75	20.515
6	10.64	12.59	14.45	16.81	18.55	22.458
7	12.02	14.07	16.01	18.48	20.28	24.322
8	13.36	15.51	17.53	20.09	21.96	26.125
9	14.68	16.92	19.02	21.67	23.59	27.877
10	15.99	18.31	20.48	23.21	25.19	29.588
11	17.28	19.68	21.92	24.72	26.76	31.264
12	18.55	21.03	23.34	26.22	28.30	32.909
13	19.81	22.36	24.74	27.69	29.82	34.528
14	21.06	23.68	26.12	29.14	31.32	36.123
15	22.31	25.00	27.49	30.58	32.80	37.697
16	23.54	26.30	28.85	32.00	34.27	39.252
17	24.77	27.59	30.19	33.41	35.72	40.790
18	25.99	28.87	31.53	34.81	37.16	42.312
19	27.20	30.14	32.85	36.19	38.58	43.820
20	28.41	31.41	34.17	37.57	40.00	45.315
21	29.62	32.67	35.48	38.93	41.40	46.797
22	30.81	33.92	36.78	40.29	42.80	48.268
23	32.01	35.17	38.08	41.64	44.18	49.728
24	33.20	36.42	39.36	42.98	45.56	51.179
25	34.38	37.65	40.65	44.31	46.93	52.620
26	35.56	38.89	41.92	45.64	48.29	54.052
27	36.74	40.11	43.19	46.96	49.64	55.476
28	37.92	41.34	44.46	48.28	50.99	56.892
29	39.09	42.56	45.72	49.59	52.34	58.302
30	40.26	43.77	46.98	50.89	53.67	59.703
40	51.81	55.76	59.34	63.69	66.77	73.402
50	63.17	67.50	71.42	76.15	79.49	86.661
60	74.40	79.08	83.30	88.38	91.95	99.607
70	85.53	90.53	95.02	100.42	104.22	112.317
80	96.58	101.88	106.63	112.33	116.32	124.839
90	107.56	113.14	118.14	124.12	128.30	137.208
100	118.50	124.34	129.56	135.81	140.17	149.449

Glossary

adjusted squared multiple correlation (R^2) - the unbiased estimate of R^2; usual estimate of R^2 is positively biased

backward deletion (in multiple regression) - form of statistical regression where order of entry of variables into solution is based entirely on statistical criteria; equation starts out with all IVs in the solution and variables are deleted one at a time (if they do not contribute significantly to the regression solution)

beta coefficients - see *regression coefficients*

beta weights - see *regression coefficients*

causal modeling - statistical technique, using regression analysis, which examines patterns of intercorrelations among variables in order to determine if they "fit" the researcher's underlying theory of which variables are causing which other variables

coefficient of determination - see *squared multiple correlation*

communality (h_j) - amount of variance in each variable accounted for by the factors; equal to the squared multiple correlation of the variable as predicted from the factors; also equal to the sum of squared loadings for a variable across all factors; provided for each variable

concomitant variable - an accompanying variable not central to an analysis; also referred to as an extraneous variable

confirmatory factor analysis - more advanced than exploratory factor analysis; used to test a theory about latent (i.e., underlying, unobservable) processes

continuous variable - measured on a scale that changes smoothly over possible values rather than in steps; also referred to as "interval" or "quantitative"

control variable - a variable whose effect on a DV is removed; also referred to as a variable whose effect has been "partialed out"

correlation matrix (observed) - a square, symmetrical matrix; each row and each column represent different variables; located at each intersection of a row and column is the bivariate correlation between the two variables

covariate - IV used as the basis for the adjustment of DV scores (as a statistical control mechanism) prior to examining main effects and interactions between IVs with respect to some DV

cross-validation - a procedure in which the predictive power of a regression equation is typically assessed by using the equation derived from one sample to predict *Y* values in a second sample (called the *cross-validation sample*)

data matrix - organization of raw scores, or data, where rows represent cases (subjects) and columns represent variables

deflated correlations - occur when variables have a restricted range in the sampling of cases or when there exist very uneven splits in the cases in categories of discrete or dichotomous variables

dichotomous variable - a discrete variable with only two possible values

discrepancy - measure of the impact of cases on a solution; measures the extent to which a case is in line with the others

discrete variable - measured in finite, separate categories with no smooth transition; also known as "nominal," "categorical," or "qualitative"

discriminant variate (or function) - maximization of the linear combination of IVs used to discriminate among groups

effect size - the size of the treatment effect the researcher wishes to detect with respect to a given level of power

eigenvalue (in factor analysis) - amount of total variance explained by each factor

endogenous variable - the variable being explained by a causal model; in other words, having its variance explained by other variables included in the model; also referred to as the dependent variable

eta-squared (η^2) - a measure of strength of association (also known as "effect size"); statistical significance measures whether or not there is an association between IVs and DVs, but eta-squared measures *how much* association there is; shows the proportion of variance in the DV that is attributable to the effect (IV)

exogenous variable - a variable not being explained by a causal model, whose variance is accounted for by other variables outside of the model; also referred to as an independent variable

exploratory factor analysis - goal is to describe and summarize data by grouping together variables that are correlated; variables may or may not have been chosen with these underlying structures in mind

factor analysis - a mathematical model is created, resulting in the estimation of *factors*; contrast with principal components analysis

factor extraction - the specific mathematical procedure by which factors are determined

factor loading - the Pearson correlation of an original variable with a factor

factor matrix - provides correlation coefficients between each IV and each factor in the solution; values can also be interpreted as that amount that each IV contributes to each factor

factor scores - estimates of the scores subjects would have received on each of the factors had they been measured directly

factorial analysis - statistical analysis that involves more than one IV

first principal component - the initial linear combination of IVs, accounts for the largest amount of total variance; equal to largest eigenvalue for the solution

fixed effects - in ANOVA, levels of each IV are selected based on interest in testing for significance of the particular IV

forward selection (in multiple regression) - form of statistical regression where order of entry of variables into solution is based entirely on statistical criteria; equation starts out empty and variables are entered one at a time (if they are statistically significant); once entered into the equation, the variable stays

group similarity (in classification) - represented by the classical probability of the data given group membership; i.e., $P(X_i/G_k)$

hierarchical multiple regression - see *sequential analysis*

hit rate - the number of correct classifications in a discriminant function analysis

homoscedasticity - assumption that the variability in scores for one continuous variable is roughly the same at all values of another continuous variable

honest correlations - accurate bivariate correlations between pairs of variables

inflated correlations - correlations between composite variables can be inflated when variables are reused in making up the composite variables

influence - product of leverage and discrepancy; assesses change in regression coefficients when a case is deleted; cases with influence scores greater than 1.00 are possible outliers with great influence on the solution

kurtosis - degree of peakedness of a distribution; equal to zero when a distribution is normal

least-squares solution - the regression model with coefficient values that minimize the sum of squared residuals

leverage - measure of the impact of cases on a solution; closely related to Mahalanobis distance; cases with high leverage are far away from other cases but essentially on the same line

likelihood ratio test - test of overall logistic relationship; analogous to F-test of ΔR^2 in multiple regression

linearity - assumption that there is a straight line relationship between two variables

logit - the natural logarithm of the odds

Mahalanobis distance - statistical measure of an outlier; distance of a case from the centroid of the remaining cases where the centroid is the point created by the means of all the variables

multicollinearity - problem created when IVs are very highly correlated (r .90) with each other

multiple correlation (R) - correlation between the observed and predicted values of the DV

multivariate analysis - statistical analysis that involves more than one DV

multivariate outlier - cases with an unusual pattern of scores; values on individual variables may look reasonable, but the combinations of two variables produce values that look unusual or discrepant

normality - assumption that each variable and linear combinations of variables are normally distributed

oblique rotation - rotation of factors resulting in factors being correlated with each other; results in the production of several matrices; one is a *factor correlation* matrix (i.e., a matrix of correlations between all factors); loading matrix is separated into *structure* matrix (i.e., correlations between factors and variables) and *pattern* matrix (i.e., unique relationships with no overlapping among factors between each factor and each observed variable); interpretation of factors is obtained from the pattern matrix

odds (in logistic regression) - ratio of the probability of outcome Y = 1 (e.g., program completion) to the probability of outcome Y = 2 (e.g., program noncompletion)

odds ratio (in logistic regression) - the ratio of odds of Y = 1, for example, for two different values of the IV; symbolized by Ψ

orthogonal rotation - rotation of factors resulting in factors being uncorrelated with each other; result is a *loading* matrix (i.e., a matrix of correlations between all observed variables and factors) where the size of the loading reflects the extent of the relationship between each observed variable and each factor; interpretation of factors is obtained from the loading matrix

orthogonality - perfect nonassociation between variables

outlier - cases with extreme values on one variable or on a combination of variables so that they distort resulting statistics or unduly influence solutions or models; would result in an excessively large residual

partial correlation - a measure of the relationship between an IV and the DV, holding constant (or partialing out the effect of) all other IVs

path analysis – a method of analyzing correlations among a set of variables in order to examine the pattern of causal relationships; usually depicted with a path diagram with arrows showing the direction of causation among variables

path coefficient - the standardized regression coefficients associated with causal paths in a causal model; see also *structural coefficient*

path decomposition - process that results in the reproduced correlations in a path analysis; the reproduced correlations are equal to the product of all coefficients in a given path; see also *path tracing*

path diagram - a pictorial representation of a causal model, showing the direction of causation and associated path coefficients

path tracing - process that results in the reproduced correlations in a path analysis; the reproduced correlations are equal to the product of all coefficients in a given path; see also *path decomposition*

power - the probability of rejecting H_o when H_o is, in fact, false; equal to $1-\beta$

principal components analysis - most common for extracting factors in factor analysis; original variables are transformed into a new set of linear combinations by extracting the maximum variance from the data set with each component; results in *components*; contrast with factor analysis

prior probability (in classification) - represented by the probability of membership in group k prior to collection of the data; i.e., P_k

random effects - in ANOVA, it may be desirable to generalize to a population of only certain *levels* of the IV (as opposed to the entire IV), so levels of the IV are randomly selected from the population

regression coefficients - the values for the constants in a regression equation; also function as the weights attached to each IV; also known as *beta coefficients* or *beta weights*

regression line - the prediction line, as defined by the best-fitting line, through a series of points in a scatterplot

reproduced correlation matrix - the correlation matrix produced from the factor solution or from the theoretical model in path analysis

residual correlation matrix - the difference between the observed and reproduced correlation matrices

residuals - portions of scores not accounted for by the analysis; also a measure of the difference between the obtained and predicted values on the DV, therefore referred to as "prediction error"

robustness - the degree to which a statistical test is still appropriate to apply when some of its assumptions are not met

rotation of factors - process by which the solution of a factor analysis is made more interpretable without altering the underlying mathematical structure

scree plot - a graph of the magnitude of each eigenvalue (vertical axis) plotted against their ordinal numbers (horizontal axis)

sequential analysis - researcher assigns priority for the entry of variables into the equation (solution); first variable to be entered is assigned both its unique variance and any overlapping variance it has with other IVs; upon entry, lower-priority variables are then assigned their unique variance and any remaining overlapping variance

singularity - problem created when variables are redundant (i.e., one of the variables is a combination of two or more of the other variables)

skewness - degree of symmetry of a distribution; equal to zero when distribution is normal

spurious effect - causal effect between two variables that is due largely to a common third variable

squared multiple correlation (R^2) - the proportion of variability in the DV explained by the model (i.e., the combination of IVs); also referred to as the *coefficient of determination*

standard analysis - all variables are entered into the solution simultaneously; overlapping variance (i.e., variance in the DV explained by more than one IV) is ignored when assessing the contributions of individual IVs to the variability in the DV

standardized discriminant function coefficient - represents the weight applied to each IV for a given discriminant function; in z-score form

standardized regression coefficient (β) - regression coefficient expressed in z-score form; interpreted as the amount of change in the DV associated with a one standard deviation unit change in that IV, with all other IVs held constant

statistical regression - order of entry of variables into solution is based entirely on statistical criteria; the meaning or interpretation of the inclusion of specific variables is not relevant

stepwise selection (in multiple regression) - form of statistical regression where order of entry of variables into solution is based entirely on statistical criteria; combination of forward and backward regression; equation starts out empty and variables are entered one at a time (if they are statistically significant), but may also be removed (if they no longer contribute significantly to the solution); results in a better solution when compared to forward and backward procedures

structural coefficient - the standardized regression coefficients associated with causal paths in a causal model; see also *path coefficient*

structural equation - another name for a specific causal model when stated as an equation

structural equation modeling - sophisticated version of path analysis incorporating unobservable, unmeasurable (latent) variables into the path model

sum-of-squares and cross-products matrix - precursor to the variance-covariance matrix where the deviations have not yet been averaged

systematic bias - predictable variability that results from intact groups that differ systematically on several variables; is best addressed through means of random assignment of subjects to groups, but can be addressed by using a covariance analysis

tolerance - a measure of collinearity among IVs, ranging from 0 (indicating multicollinearity) to 1 (indicating independence among IVs)

univariate outlier - cases with very large standardized scores on a single variable

unstandardized discriminant function coefficient - represents the weight applied to each IV for a given discriminant function; in raw score form

unstandardized regression coefficient (B) - the raw score weight associated with a specific IV in a regression equation; interpreted as the amount of change in the DV associated with a one-unit change in that IV, with all other IVs held constant

variance-covariance matrix - used when scores are measured along a continuous scale; a square, symmetrical matrix where the elements on the main diagonal are the variances for each variable and the elements on the off-diagonals are the covariances between pairs of variables

variance inflation factor (VIF) - a measure of the extent to which there exists multicollinear relationships for given predictor IVs

References

Agresti, A. & Finlay, B. (1997). *Statistical methods for the social sciences (3rd ed.).* Upper Saddle River, NJ: Prentice Hall.

Aron, A. & Aron, E. N. (1999). *Statistics for psychology (2nd ed.).* Upper Saddle River, NJ: Prentice Hall.

Aron, A. & Aron, E. N. (1997). *Statistics for the behavioral and social sciences: A brief course.* Upper Saddle River, NJ: Prentice Hall.

Asher, H. B. (1983). *Causal modeling.* Sage University Paper series on Quantitative Application in the Social Sciences, series no. 07-003. Beverly Hills and London: Sage.

Brewer, J. K. (1978). *Everything you always wanted to know about statistics, but didn't know how to ask.* Dubuque, IA: Kendall/Hunt.

Bruning, J. L. & Kintz, B. L. (1997). *Computational handbook of statistics (4th ed.).* New York: Longman.

Cohen, J. (1969). *Statistical power analysis for the behavioral sciences.* New York: Academic Press.

Cohen, J. (1988). *Statistical power analysis for the behavioral sciences (2nd ed.).* Hillsdale, NJ: Lawrence Erlbaum Associates.

George, D. & Mallery, P. (2000). *SPSS for Windows step-by-step: A simple guide and reference (2nd ed.).* Boston: Allyn and Bacon.

Gravetter, F. J. & Wallnau, L. B. (1999). *Essentials of statistics for the behavioral sciences (3rd ed.).* Pacific Grove, CA: Brooks/Cole.

Harris, M. B. (1998). *Basic statistics for behavioral science research (2nd ed.).* Boston: Allyn and Bacon.

Huitema, B. (1980). *The analysis of covariance and alternatives.* New York: Wiley.

Johnson, R. A. & Wichern, D. W. (1998). *Applied multivariate statistical analysis (4th ed.).* Upper Saddle River, NJ: Prentice Hall.

Kennedy, J. J. & Bush, A. J. (1985). *An introduction to the design and analysis of experiments in behavioral research.* Lanham, MD: University Press of America.

Levin, J. & Fox, J. A. (2000). *Elementary statistics in social research (8th ed.).* Boston: Allyn and Bacon.

Long, J. S. (1983). *Covariance structure models: An introduction to LISREL.* Sage University Paper series on Quantitative Application in the Social Sciences, series no. 07-034. Beverly Hills and London: Sage.

Newman, I. & Newman, C. (1994). *Conceptual statistics for beginners (2nd ed.).* Lanham, MD: University Press of America.

Norusis, M. J. (1998). *SPSS 8.0: Guide to data analysis.* Upper Saddle River, NJ: Prentice Hall.

Pedhazur, E. J. (1982). *Multiple regression in behavioral research: Explanation and prediction.* Fort Worth, TX: Holt, Rinehart, and Winston.

Sprinthall, R. C. (2000). *Basic statistical analysis (6th ed.).* Boston: Allyn and Bacon.

Stevens, J. (1992). *Applied multivariate statistics for the social sciences (2nd ed.).* Hillsdale, NJ: Lawrence Erlbaum Associates.

Tabachnick, B. G. & Fidell, L. S. (1996). *Using multivariate statistics (3rd ed.).* New York: HarperCollins.

Tate, R. (1992). *General linear model applications.* Unpublished manuscript, Florida State University.

Tatsuoka, M. M. (1988). *Multivariate analysis: Techniques for educational and psychological research (2nd ed.).* New York: Macmillan.

Thompson, B. (1998, April). *Five methodology errors in educational research: The pantheon of statistical significance and other faux pas.* Paper presented at the annual meeting of the American Educational Research Association, San Diego, CA.

Vogt, W. P. (1993). *Dictionary of statistics and methodology: A nontechnical guide for the social sciences.* Newbury Park, CA: Sage.

Williams, F. (1992). *Reasoning with statistics: How to read quantitative research (4th ed.).* Fort Worth, TX: Harcourt Brace Jovanovich.